DODGE | NEON
2000-05 REPAIR MANUAL

**Covers U.S. and Ca... ...tios of
Dodge and Plymouth Neon**

Does not include information specific to SRT-4 models

by Larry Warren

CHILTON *Automotive Books*

PUBLISHED BY **HAYNES NORTH AMERICA, Inc.**

Manufactured in USA
© 2003, 2007 Haynes North America, Inc.
ISBN-13: 978-1-56392-670-9
ISBN-10: 1-56392-670-9
Library of Congress Control No. 2007922880

Haynes Publishing Group
Sparkford Nr Yeovil
Somerset BA22 7JJ England

Haynes North America, Inc
861 Lawrence Drive
Newbury Park
California 91320 USA

ABCDE
FGHIJ
KLMNO
PQ

1K2

Contents

1 TUNE-UP AND ROUTINE MAINTENANCE – 1-1

2 ENGINE – 2A-1
GENERAL ENGINE OVERHAUL PROCEDURES – 2B-1

3 COOLING, HEATING AND AIR-CONDITIONING SYSTEMS – 3-1

4 FUEL AND EXHAUST SYSTEMS – 4-1

5 ENGINE ELECTRICAL SYSTEMS – 5-1

6 EMISSIONS AND ENGINE CONTROL SYSTEMS – 6-1

Mechanic, author and photographer with a 2000 Dodge Neon

ACKNOWLEDGMENTS

Wiring diagrams provided exclusively for Haynes North America, Inc. by Valley Forge Technical Information Services. Technical writers who contributed to this project include Mike Stubblefield and Robert Maddox. Technical consultants include Jamie Sarté, Jr. and Brad Conn.

About this manual

ITS PURPOSE

The purpose of this manual is to help you get the best value from your vehicle. It can do so in several ways. It can help you decide what work must be done, even if you choose to have it done by a dealer service department or a repair shop; it provides information and procedures for routine maintenance and servicing; and it offers diagnostic and repair procedures to follow when trouble occurs.

We hope you use the manual to tackle the work yourself. For many simpler jobs, doing it yourself may be quicker than arranging an appointment to get the vehicle into a shop and making the trips to leave it and pick it up. More importantly, a lot of money can be saved by avoiding the expense the shop must pass on to you to cover its labor and overhead costs. An added benefit is the sense of satisfaction and accomplishment that you feel after doing the job yourself.

USING THE MANUAL

The manual is divided into Chapters. Each Chapter is divided into numbered Sections. Each Section consists of consecutively numbered paragraphs.

At the beginning of each numbered Section you will be referred to any illustrations which apply to the procedures in that Section. The reference numbers used in illustration captions pinpoint the pertinent Section and the Step within that Section. That is, illustration 3.2 means the illustration refers to Section 3 and Step (or paragraph) 2 within that Section.

Procedures, once described in the text, are not normally repeated. When it's necessary to refer to another Chapter, the reference will be given as Chapter and Section number. Cross references given without use of the word "Chapter" apply to Sections and/or paragraphs in the same Chapter. For example, "see Section 8" means in the same Chapter.

References to the left or right side of the vehicle assume you are sitting in the driver's seat, facing forward.

Even though we have prepared this manual with extreme care, neither the publisher nor the author can accept responsibility for any errors in, or omissions from, the information given.

➡ **NOTE**

A *Note* provides information necessary to properly complete a procedure or information which will make the procedure easier to understand.

✳✳ **CAUTION**

A *Caution* provides a special procedure or special steps which must be taken while completing the procedure where the Caution is found. Not heeding a Caution can result in damage to the assembly being worked on.

✳✳ **WARNING**

A *Warning* provides a special procedure or special steps which must be taken while completing the procedure where the Warning is found. Not heeding a Warning can result in personal injury.

Introduction to the Dodge Neon

The Dodge Neon models covered by this manual are available in a four-door sedan body style only.

The 2.0L four-cylinder engine is a Single Overhead Camshaft (SOHC) model and is transversely mounted. There are two versions available: The 2.0L 132 horsepower version and the 2.0L High Output (HO) 150 horsepower version. Both models are equipped with sequential multi-port fuel injection.

The engine transmits power to the front wheels through either a five-speed manual transaxle or a three- or four-speed automatic transaxle via independent driveaxles.

The Dodge Neon features MacPherson strut-type suspension front and rear. The rack-and-pinion steering unit is mounted behind the engine with power-assist available as standard equipment.

All models have a power assisted brake system with disc brakes at the front and either disc or drum brakes at the rear. An Anti-lock Brake System is available as an option.

Vehicle identification numbers

Modifications are a continuing and unpublicized process in vehicle manufacturing. Since spare parts manuals and lists are compiled on a numerical basis, the individual vehicle numbers are essential to correctly identify the component required.

VEHICLE IDENTIFICATION NUMBER (VIN)

This very important identification number is located on a plate attached to the dashboard inside the windshield on the driver's side of the vehicle (see illustration). The VIN also appears on the Vehicle Certificate of Title and Registration. It contains information such as where and when the vehicle was manufactured, the model year and the body style.

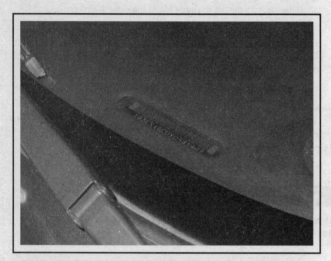

The Vehicle Identification Number (VIN) is stamped into a metal plate fastened to the dashboard on the driver's side - it's visible through the windshield

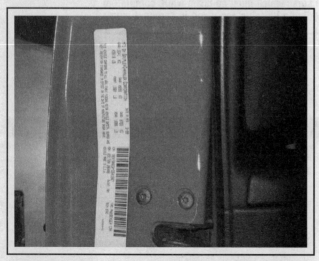

The Vehicle Safety Certification label is on the end of the driver's door

VIN engine and model year codes

Two particularly important pieces of information found in the VIN are the engine code and the model year code. Counting from the left, the engine code letter designation is the 8th digit and the model year code designation is the 10th digit.

On the models covered by this manual the engine codes are:
C . . . . .2.0L SOHC (132 hp)
F .2.0L High Output (HO) SOHC (150 hp)

On the models covered by this manual the model year codes are:
Y 2000
1 2001
2 2002
3 2003
4 2004
5 2005

VEHICLE SAFETY CERTIFICATION LABEL

The Vehicle Safety Certification label is attached to the end of the driver's door. The label contains the name of the manufacturer, the year, month, day and hour of production, the Gross Vehicle Weight Rating (GVWR), the Gross Axle Weight Rating (GAWR) and the certification statement. It also contains the paint code, trim code and vehicle order number, as well as the VIN.

ENGINE IDENTIFICATION NUMBERS

The engine identification numbers can be found stamped on a pad on the left rear of the engine block (the front side of the block as it sits in the car), near the starter (see illustration).

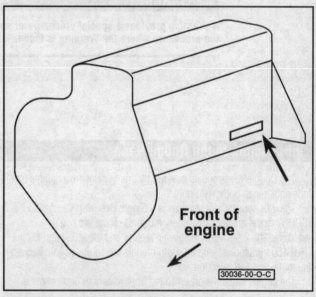

Front of engine

30036-00-O-C

Location of the engine identification number

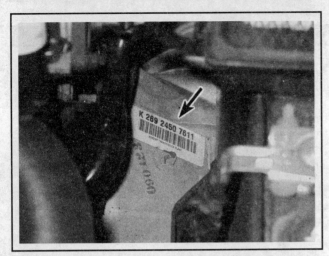

The identification number on the automatic transaxle is located on a label affixed to the transaxle torque converter housing

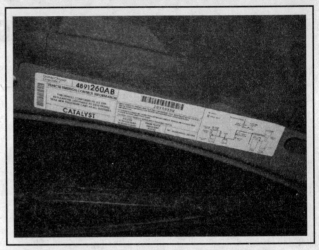

The emissions control information label is found on the underside of the hood

TRANSAXLE IDENTIFICATION NUMBERS

The manual transaxle identification information can be found on a bar code label located on the front of the transaxle or stamped onto a plate bolted to the end cover of the transaxle. On automatic transaxles, the identification number is located on a label affixed to the torque converter housing or stamped into the case near the transfer gear cover (see illustration).

VEHICLE EMISSIONS CONTROL INFORMATION (VECI) LABEL

The emissions control information label is found under the hood, normally on the radiator support or the bottom side of the hood. This label contains information on the emissions control equipment installed on the vehicle, as well as tune-up specifications (see Chapter 6).

Buying parts

Replacement parts are available from many sources, which generally fall into one of two categories - authorized dealer parts departments and independent retail auto parts stores. Our advice concerning these parts is as follows:

Retail auto parts stores: Good auto parts stores will stock frequently needed components which wear out relatively fast, such as clutch components, exhaust systems, brake parts, tune-up parts, etc. These stores often supply new or reconditioned parts on an exchange basis, which can save a considerable amount of money. Discount auto parts stores are often very good places to buy materials and parts needed for general vehicle maintenance such as oil, grease, filters, spark plugs, belts, touch-up paint, bulbs, etc. They also usually sell

tools and general accessories, have convenient hours, charge lower prices and can often be found not far from home.

Authorized dealer parts department: This is the best source for parts which are unique to the vehicle and not generally available elsewhere (such as major engine parts, transmission parts, trim pieces, etc.).

Warranty information: If the vehicle is still covered under warranty, be sure that any replacement parts purchased - regardless of the source - do not invalidate the warranty!

To be sure of obtaining the correct parts, have engine and chassis numbers available and, if possible, take the old parts along for positive identification.

Maintenance techniques, tools and working facilities

MAINTENANCE TECHNIQUES

There are a number of techniques involved in maintenance and repair that will be referred to throughout this manual. Application of these techniques will enable the home mechanic to be more efficient, better organized and capable of performing the various tasks properly, which will ensure that the repair job is thorough and complete.

Fasteners

Fasteners are nuts, bolts, studs and screws used to hold two or more parts together. There are a few things to keep in mind when working with fasteners. Almost all of them use a locking device of some type, either a lockwasher, locknut, locking tab or thread adhesive. All threaded fasteners should be clean and straight, with undamaged threads and undamaged corners on the hex head where the wrench fits. Develop the habit of replacing all damaged nuts and bolts with new ones. Special locknuts with nylon or fiber inserts can only be used once. If they are removed, they lose their locking ability and must be replaced with new ones.

Rusted nuts and bolts should be treated with a penetrating fluid to ease removal and prevent breakage. Some mechanics use turpentine in a spout-type oil can, which works quite well. After applying the rust penetrant, let it work for a few minutes before trying to loosen the nut or bolt. Badly rusted fasteners may have to be chiseled or sawed off or removed with a special nut breaker, available at tool stores.

If a bolt or stud breaks off in an assembly, it can be drilled and removed with a special tool commonly available for this purpose. Most automotive machine shops can perform this task, as well as other repair procedures, such as the repair of threaded holes that have been stripped out.

Flat washers and lockwashers, when removed from an assembly, should always be replaced exactly as removed. Replace any damaged washers with new ones. Never use a lockwasher on any soft metal surface (such as aluminum), thin sheet metal or plastic.

Fastener sizes

For a number of reasons, automobile manufacturers are making wider and wider use of metric fasteners. Therefore, it is important to be able to tell the difference between standard (sometimes called U.S. or SAE) and metric hardware, since they cannot be interchanged.

All bolts, whether standard or metric, are sized according to diameter, thread pitch and length. For example, a standard 1/2 - 13 x 1 bolt is 1/2 inch in diameter, has 13 threads per inch and is 1 inch long. An M12 - 1.75 x 25 metric bolt is 12 mm in diameter, has a thread pitch of 1.75 mm (the distance between threads) and is 25 mm long. The two bolts are nearly identical, and easily confused, but they are not interchangeable.

In addition to the differences in diameter, thread pitch and length, metric and standard bolts can also be distinguished by examining the bolt heads. To begin with, the distance across the flats on a standard bolt head is measured in inches, while the same dimension on a metric bolt is sized in millimeters (the same is true for nuts). As a result, a standard wrench should not be used on a metric bolt and a metric wrench should not be used on a standard bolt. Also, most standard bolts have slashes radiating out from the center of the head to denote the grade or strength of the bolt, which is an indication of the amount of torque that can be applied to it. The greater the number of slashes, the greater the strength of the bolt. Grades 0 through 5 are commonly used on automobiles. Metric bolts have a property class (grade) number, rather than a slash, molded into their heads to indicate bolt strength. In this case, the higher the number, the stronger the bolt. Property class numbers 8.8, 9.8 and 10.9 are commonly used on automobiles.

Strength markings can also be used to distinguish standard hex nuts from metric hex nuts. Many standard nuts have dots stamped into one side, while metric nuts are marked with a number. The greater the number of dots, or the higher the number, the greater the strength of the nut.

Metric studs are also marked on their ends according to property class (grade). Larger studs are numbered (the same as metric bolts), while smaller studs carry a geometric code to denote grade.

It should be noted that many fasteners, especially Grades 0 through 2, have no distinguishing marks on them. When such is the case, the only way to determine whether it is standard or metric is to measure the thread pitch or compare it to a known fastener of the same size.

Standard fasteners are often referred to as SAE, as opposed to metric. However, it should be noted that SAE technically refers to a non-metric fine thread fastener only. Coarse thread non-metric fasteners are referred to as USS sizes.

Since fasteners of the same size (both standard and metric) may have different strength ratings, be sure to reinstall any bolts, studs or nuts removed from your vehicle in their original locations. Also, when replacing a fastener with a new one, make sure that the new one has a strength rating equal to or greater than the original.

Tightening sequences and procedures

Most threaded fasteners should be tightened to a specific torque value (torque is the twisting force applied to a threaded component such as a nut or bolt). Overtightening the fastener can weaken it and cause it to break, while undertightening can cause it to eventually come loose. Bolts, screws and studs, depending on the material they are made of and their thread diameters, have specific torque values, many of which are noted in the Specifications at the end of each Chapter. Be sure to follow the torque recommendations closely. For fasteners not assigned a specific torque, a general torque value chart is presented here as a guide. These torque values are for dry (unlubricated) fasteners threaded into steel or cast iron (not aluminum). As was previously mentioned, the size and grade of a fastener determine the amount of torque that can safely be applied to it. The figures listed here are approximate for Grade 2 and Grade 3 fasteners. Higher grades can tolerate higher torque values.

Fasteners laid out in a pattern, such as cylinder head bolts, oil pan bolts, differential cover bolts, etc., must be loosened or tightened in sequence to avoid warping the component. This sequence will normally be shown in the appropriate Chapter. If a specific pattern is not given, the following procedures can be used to prevent warping.

Initially, the bolts or nuts should be assembled finger-tight only. Next, they should be tightened one full turn each, in a criss-cross or diagonal pattern. After each one has been tightened one full turn, return to the first one and tighten them all one-half turn, following the same

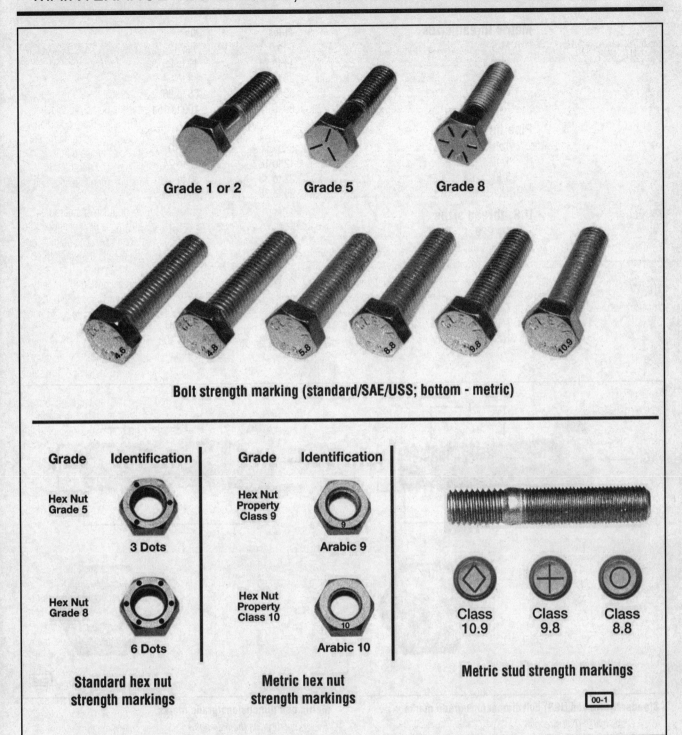

Grade 1 or 2   Grade 5   Grade 8

Bolt strength marking (standard/SAE/USS; bottom - metric)

Grade	Identification
Hex Nut Grade 5	3 Dots
Hex Nut Grade 8	6 Dots

Standard hex nut strength markings

Grade	Identification
Hex Nut Property Class 9	Arabic 9
Hex Nut Property Class 10	Arabic 10

Metric hex nut strength markings

Class 10.9   Class 9.8   Class 8.8

Metric stud strength markings

00-1

pattern. Finally, tighten each of them one-quarter turn at a time until each fastener has been tightened to the proper torque. To loosen and remove the fasteners, the procedure would be reversed.

Component disassembly

Component disassembly should be done with care and purpose to help ensure that the parts go back together properly. Always keep track of the sequence in which parts are removed. Make note of special characteristics or marks on parts that can be installed more than one way, such as a grooved thrust washer on a shaft. It is a good idea to lay the disassembled parts out on a clean surface in the order that they were removed. It may also be helpful to make sketches or take instant photos of components before removal.

When removing fasteners from a component, keep track of their locations. Sometimes threading a bolt back in a part, or putting the washers and nut back on a stud, can prevent mix-ups later. If nuts and bolts cannot be returned to their original locations, they should be kept in a compartmented box or a series of small boxes. A cupcake or muffin tin is ideal for this purpose, since each cavity can hold the bolts and nuts from a particular area (i.e. oil pan bolts, valve cover bolts, engine

Metric thread sizes

	Ft-lbs	Nm
M-6	6 to 9	9 to 12
M-8	14 to 21	19 to 28
M-10	28 to 40	38 to 54
M-12	50 to 71	68 to 96
M-14	80 to 140	109 to 154

Pipe thread sizes

1/8	5 to 8	7 to 10
1/4	12 to 18	17 to 24
3/8	22 to 33	30 to 44
1/2	25 to 35	34 to 47

U.S. thread sizes

1/4 - 20	6 to 9	9 to 12
5/16 - 18	12 to 18	17 to 24
5/16 - 24	14 to 20	19 to 27
3/8 - 16	22 to 32	30 to 43
3/8 - 24	27 to 38	37 to 51
7/16 - 14	40 to 55	55 to 74
7/16 - 20	40 to 60	55 to 81
1/2 - 13	55 to 80	75 to 108

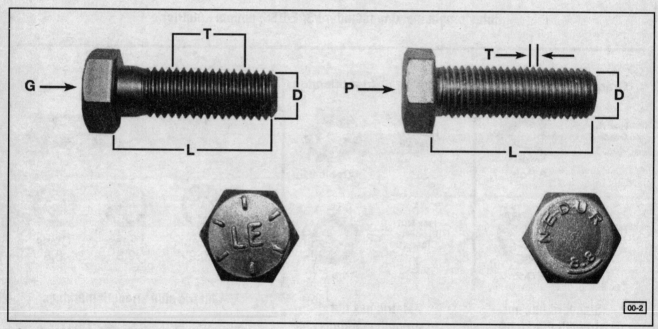

00-2

Standard (SAE and USS) bolt dimensions/grade marks

G   *Grade marks (bolt strength)*
L   *Length (in inches)*
T   *Thread pitch (number of threads per inch)*
D   *Nominal diameter (in inches)*

Metric bolt dimensions/grade marks

P   *Property class (bolt strength)*
L   *Length (in millimeters)*
T   *Thread pitch (distance between threads in millimeters)*
D   *Diameter*

mount bolts, etc.). A pan of this type is especially helpful when working on assemblies with very small parts, such as the carburetor, alternator, valve train or interior dash and trim pieces. The cavities can be marked with paint or tape to identify the contents.

Whenever wiring looms, harnesses or connectors are separated, it is a good idea to identify the two halves with numbered pieces of masking tape so they can be easily reconnected.

Gasket sealing surfaces

Throughout any vehicle, gaskets are used to seal the mating surfaces between two parts and keep lubricants, fluids, vacuum or pressure contained in an assembly.

Many times these gaskets are coated with a liquid or paste-type gasket sealing compound before assembly. Age, heat and pressure can sometimes cause the two parts to stick together so tightly that they are

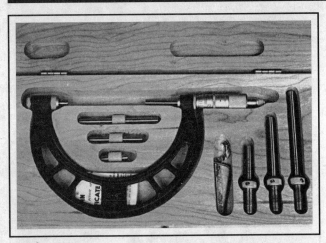

Micrometer set

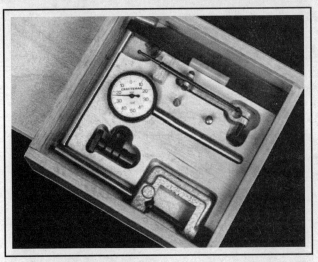

Dial indicator set

very difficult to separate. Often, the assembly can be loosened by striking it with a soft-face hammer near the mating surfaces. A regular hammer can be used if a block of wood is placed between the hammer and the part. Do not hammer on cast parts or parts that could be easily damaged. With any particularly stubborn part, always recheck to make sure that every fastener has been removed.

Avoid using a screwdriver or bar to pry apart an assembly, as they can easily mar the gasket sealing surfaces of the parts, which must remain smooth. If prying is absolutely necessary, use an old broom handle, but keep in mind that extra clean up will be necessary if the wood splinters.

After the parts are separated, the old gasket must be carefully scraped off and the gasket surfaces cleaned. Stubborn gasket material can be soaked with rust penetrant or treated with a special chemical to soften it so it can be easily scraped off.

✶✶ CAUTION:

Never use gasket removal solutions or caustic chemicals on plastic or other composite components.

A scraper can be fashioned from a piece of copper tubing by flattening and sharpening one end. Copper is recommended because it is usually softer than the surfaces to be scraped, which reduces the chance of gouging the part. Some gaskets can be removed with a wire brush, but regardless of the method used, the mating surfaces must be left clean and smooth. If for some reason the gasket surface is gouged, then a gasket sealer thick enough to fill scratches will have to be used during reassembly of the components. For most applications, a non-drying (or semi-drying) gasket sealer should be used.

Hose removal tips

✶✶ WARNING:

If the vehicle is equipped with air conditioning, do not disconnect any of the A/C hoses without first having the system depressurized by a dealer service department or a service station.

Hose removal precautions closely parallel gasket removal precautions. Avoid scratching or gouging the surface that the hose mates against or the connection may leak. This is especially true for radiator hoses. Because of various chemical reactions, the rubber in hoses can bond itself to the metal spigot that the hose fits over. To remove a hose, first loosen the hose clamps that secure it to the spigot. Then, with slip-joint pliers, grab the hose at the clamp and rotate it around the spigot. Work it back and forth until it is completely free, then pull it off. Silicone or other lubricants will ease removal if they can be applied between the hose and the outside of the spigot. Apply the same lubricant to the inside of the hose and the outside of the spigot to simplify installation.

As a last resort (and if the hose is to be replaced with a new one anyway), the rubber can be slit with a knife and the hose peeled from the spigot. If this must be done, be careful that the metal connection is not damaged.

If a hose clamp is broken or damaged, do not reuse it. Wire-type clamps usually weaken with age, so it is a good idea to replace them with screw-type clamps whenever a hose is removed.

TOOLS

A selection of good tools is a basic requirement for anyone who plans to maintain and repair his or her own vehicle. For the owner who has few tools, the initial investment might seem high, but when compared to the spiraling costs of professional auto maintenance and repair, it is a wise one.

To help the owner decide which tools are needed to perform the tasks detailed in this manual, the following tool lists are offered: *Maintenance and minor repair, Repair/overhaul and Special.*

The newcomer to practical mechanics should start off with the *maintenance and minor repair* tool kit, which is adequate for the simpler jobs performed on a vehicle. Then, as confidence and experience grow, the owner can tackle more difficult tasks, buying additional tools as they are needed. Eventually the basic kit will be expanded into the *repair and overhaul* tool set. Over a period of time, the experienced do-it-yourselfer will assemble a tool set complete enough for most repair and overhaul procedures and will add tools from the special category when it is felt that the expense is justified by the frequency of use.

Maintenance and minor repair tool kit

The tools in this list should be considered the minimum required for performance of routine maintenance, servicing and minor repair work. We recommend the purchase of combination wrenches (box-end and open-end combined in one wrench). While more expensive than open end wrenches, they offer the advantages of both types of wrench.

Combination wrench set (1/4-inch to 1 inch or 6 mm to 19 mm)

Adjustable wrench, 8 inch
Spark plug wrench with rubber insert
Spark plug gap adjusting tool
Feeler gauge set
Brake bleeder wrench
Standard screwdriver (5/16-inch x 6 inch)
Phillips screwdriver (No. 2 x 6 inch)
Combination pliers - 6 inch

Hacksaw and assortment of blades
Tire pressure gauge
Grease gun
Oil can
Fine emery cloth
Wire brush
Battery post and cable cleaning tool
Oil filter wrench

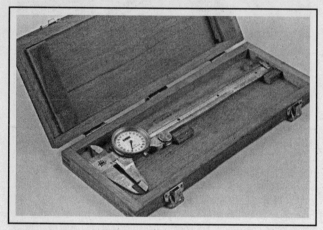

Dial caliper

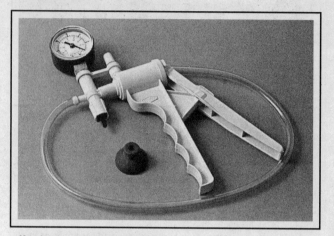

Hand-operated vacuum pump

Timing light

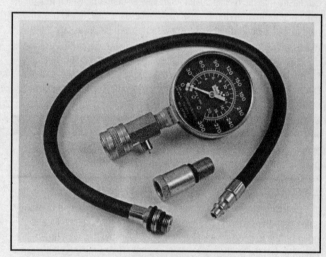

Compression gauge with spark plug hole adapter

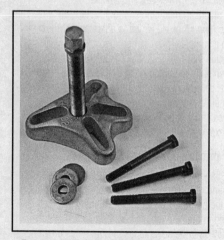

Damper/steering wheel puller

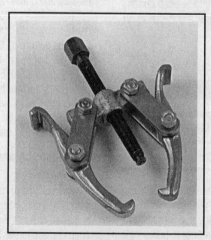

General purpose puller

Hydraulic lifter removal tool

Funnel (medium size)
Safety goggles
Jackstands (2)
Drain pan

➡**Note: If basic tune-ups are going to be part of routine maintenance, it will be necessary to purchase a good quality stroboscopic timing light and combination tachometer/dwell meter. Although they are included in the list of special tools, it is mentioned here because they are absolutely necessary for tuning most vehicles properly.**

Repair and overhaul tool set

These tools are essential for anyone who plans to perform major repairs and are in addition to those in the maintenance and minor repair tool kit. Included is a comprehensive set of sockets which, though expensive, are invaluable because of their versatility, especially when various extensions and drives are available. We recommend the 1/2-inch drive over the 3/8-inch drive. Although the larger drive is bulky and more expensive, it has the capacity of accepting a very wide range of large sockets. Ideally, however, the mechanic should have a 3/8-inch drive set and a 1/2-inch drive set.

Valve spring compressor

Valve spring compressor

Ridge reamer

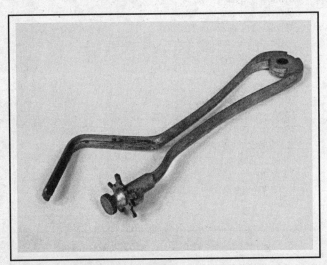

Piston ring groove cleaning tool

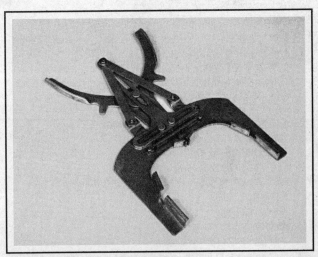

Ring removal/installation tool

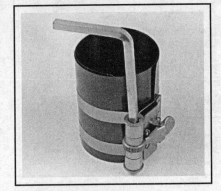

Ring compressor

Cylinder hone

Brake hold-down spring tool

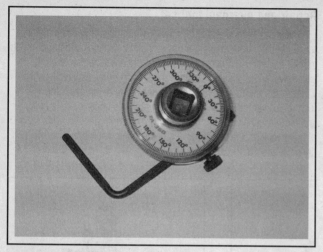

Torque angle gauge

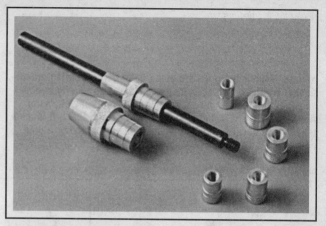

Clutch plate alignment tool

Socket set(s)
Reversible ratchet
Extension - 10 inch
Universal joint
Torque wrench (same size drive as sockets)
Ball peen hammer - 8 ounce
Soft-face hammer (plastic/rubber)
Standard screwdriver (1/4-inch x 6 inch)
Standard screwdriver (stubby - 5/16-inch)
Phillips screwdriver (No. 3 x 8 inch)
Phillips screwdriver (stubby - No. 2)
Pliers - vise grip
Pliers - lineman's
Pliers - needle nose
Pliers - snap-ring (internal and external)
Cold chisel - 1/2-inch
Scribe
Scraper (made from flattened copper tubing)
Centerpunch
Pin punches (1/16, 1/8, 3/16-inch)
Steel rule/straightedge - 12 inch
Allen wrench set (1/8 to 3/8-inch or 4 mm to 10 mm)
A selection of files
Wire brush (large)
Jackstands (second set)
Jack (scissor or hydraulic type)

➡**Note: Another tool which is often useful is an electric drill with a chuck capacity of 3/8-inch and a set of good quality drill bits.**

Special tools

The tools in this list include those which are not used regularly, are expensive to buy, or which need to be used in accordance with their manufacturer's instructions. Unless these tools will be used frequently, it is not very economical to purchase many of them. A consideration would be to split the cost and use between yourself and a friend or friends. In addition, most of these tools can be obtained from a tool rental shop on a temporary basis.

This list primarily contains only those tools and instruments widely available to the public, and not those special tools produced by the vehicle manufacturer for distribution to dealer service departments. Occasionally, references to the manufacturer's special tools are included in the text of this manual. Generally, an alternative method of doing the job without the special tool is offered. However, sometimes there is no alternative to their use. Where this is the case, and the tool cannot be purchased or borrowed, the work should be turned over to the dealer service department or an automotive repair shop.

Valve spring compressor
Piston ring groove cleaning tool
Piston ring compressor
Piston ring installation tool
Cylinder compression gauge
Cylinder ridge reamer
Cylinder surfacing hone
Cylinder bore gauge
Micrometers and/or dial calipers
Hydraulic lifter removal tool
Balljoint separator
Universal-type puller
Impact screwdriver
Dial indicator set
Stroboscopic timing light (inductive pick-up)
Hand operated vacuum/pressure pump
Tachometer/dwell meter
Universal electrical multimeter
Cable hoist
Brake spring removal and installation tools
Floor jack

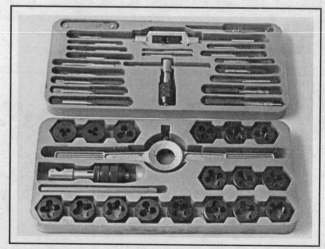

Tap and die set

Buying tools

For the do-it-yourselfer who is just starting to get involved in vehicle maintenance and repair, there are a number of options available when purchasing tools. If maintenance and minor repair is the extent of the work to be done, the purchase of individual tools is satisfactory. If, on the other hand, extensive work is planned, it would be a good idea to purchase a modest tool set from one of the large retail chain stores. A set can usually be bought at a substantial savings over the individual tool prices, and they often come with a tool box. As additional tools are needed, add-on sets, individual tools and a larger tool box can be purchased to expand the tool selection. Building a tool set gradually allows the cost of the tools to be spread over a longer period of time and gives the mechanic the freedom to choose only those tools that will actually be used.

Tool stores will often be the only source of some of the special tools that are needed, but regardless of where tools are bought, try to avoid cheap ones, especially when buying screwdrivers and sockets, because they won't last very long. The expense involved in replacing cheap tools will eventually be greater than the initial cost of quality tools.

Care and maintenance of tools

Good tools are expensive, so it makes sense to treat them with respect. Keep them clean and in usable condition and store them properly when not in use. Always wipe off any dirt, grease or metal chips before putting them away. Never leave tools lying around in the work area. Upon completion of a job, always check closely under the hood for tools that may have been left there so they won't get lost during a test drive.

Some tools, such as screwdrivers, pliers, wrenches and sockets, can be hung on a panel mounted on the garage or workshop wall, while others should be kept in a tool box or tray. Measuring instruments, gauges, meters, etc. must be carefully stored where they cannot be damaged by weather or impact from other tools.

When tools are used with care and stored properly, they will last a very long time. Even with the best of care, though, tools will wear out if used frequently. When a tool is damaged or worn out, replace it. Subsequent jobs will be safer and more enjoyable if you do.

HOW TO REPAIR DAMAGED THREADS

Sometimes, the internal threads of a nut or bolt hole can become stripped, usually from overtightening. Stripping threads is an all-too-common occurrence, especially when working with aluminum parts, because aluminum is so soft that it easily strips out.

Usually, external or internal threads are only partially stripped. After they've been cleaned up with a tap or die, they'll still work. Sometimes, however, threads are badly damaged. When this happens, you've got three choices:

1) *Drill and tap the hole to the next suitable oversize and install a larger diameter bolt, screw or stud.*

2) *Drill and tap the hole to accept a threaded plug, then drill and tap the plug to the original screw size. You can also buy a plug already threaded to the original size. Then you simply drill a hole to the specified size, then run the threaded plug into the hole with a bolt and jam nut. Once the plug is fully seated, remove the jam nut and bolt.*

3) *The third method uses a patented thread repair kit like Heli-Coil or Slimsert. These easy-to-use kits are designed to repair damaged threads in straight-through holes and blind holes. Both are available as kits which can handle a variety of sizes and thread patterns. Drill the hole, then tap it with the special included tap. Install the Heli-Coil and the hole is back to its original diameter and thread pitch.*

Regardless of which method you use, be sure to proceed calmly and carefully. A little impatience or carelessness during one of these relatively simple procedures can ruin your whole day's work and cost you a bundle if you wreck an expensive part.

WORKING FACILITIES

Not to be overlooked when discussing tools is the workshop. If anything more than routine maintenance is to be carried out, some sort of suitable work area is essential.

It is understood, and appreciated, that many home mechanics do not have a good workshop or garage available, and end up removing an engine or doing major repairs outside. It is recommended, however, that the overhaul or repair be completed under the cover of a roof.

A clean, flat workbench or table of comfortable working height is an absolute necessity. The workbench should be equipped with a vise that has a jaw opening of at least four inches.

As mentioned previously, some clean, dry storage space is also required for tools, as well as the lubricants, fluids, cleaning solvents, etc. which soon become necessary.

Sometimes waste oil and fluids, drained from the engine or cooling system during normal maintenance or repairs, present a disposal problem. To avoid pouring them on the ground or into a sewage system, pour the used fluids into large containers, seal them with caps and take them to an authorized disposal site or recycling center. Plastic jugs, such as old antifreeze containers, are ideal for this purpose.

Always keep a supply of old newspapers and clean rags available. Old towels are excellent for mopping up spills. Many mechanics use rolls of paper towels for most work because they are readily available and disposable. To help keep the area under the vehicle clean, a large cardboard box can be cut open and flattened to protect the garage or shop floor.

Whenever working over a painted surface, such as when leaning over a fender to service something under the hood, always cover it with an old blanket or bedspread to protect the finish. Vinyl covered pads, made especially for this purpose, are available at auto parts stores.

Jacking and towing

JACKING

⁂ WARNING:

The jack supplied with the vehicle should only be used for changing a tire or placing jackstands under the frame. Never work under the vehicle or start the engine while this jack is being used as the only means of support.

The vehicle should be on level ground. Place the shift lever in Park, if you have an automatic, or Reverse if you have a manual transaxle. Block the wheel diagonally opposite the wheel being changed. Set the parking brake.

Remove the jack, wrench and spare tire from their stowage area in the luggage compartment.

Remove the wheel cover and trim ring (if so equipped) with the tapered end of the lug nut wrench by inserting and twisting the handle and then prying against the back of the wheel cover. Loosen the wheel lug nuts about 1/4-to-1/2 turn each.

Place the scissors-type jack under the side of the vehicle and adjust the jack height until it engages the jacking point. There is a front and rear jacking point on each side of the vehicle (see illustrations); at the front, the jack head fits over the rocker panel flange, in the area between the two raised darts. At the rear, the jack head fits over the rocker panel flange, just ahead of the single raised dart.

Turn the jack handle clockwise until the tire clears the ground. Remove the lug nuts and pull the wheel off. Install the spare.

Install the lug nuts with the beveled edges facing in. Tighten them snugly. Don't attempt to tighten them completely until the vehicle is lowered or it could slip off the jack. Turn the jack handle counterclockwise to lower the vehicle. Remove the jack and tighten the lug nuts in a diagonal pattern.

Install the cover (and trim ring, if used) and be sure it's snapped into place all the way around.

Place the tire in the luggage compartment and stow the jack and wrench. Unblock the wheel.

TOWING

As a general rule, the vehicle should be towed with the front (drive) wheels off the ground. If they can't be raised, place them on a dolly. The ignition key must be in the ACC position, since the steering lock mechanism isn't strong enough to hold the front wheels straight while towing.

Vehicles equipped with an automatic transaxle can be towed from the front only with all four wheels on the ground, provided that speeds don't exceed 25 mph and the distance is not over 15 miles. Before towing, check the transaxle fluid level (see Chapter 1). If the level is below the HOT line on the dipstick, add fluid or use a towing dolly.

⁂ CAUTION:

Never tow a vehicle with an automatic transaxle from the rear with the front wheels on the ground.

When towing a vehicle equipped with a manual transaxle with all four wheels on the ground, be sure to place the shift lever in neutral and release the parking brake.

Equipment specifically designed for towing should be used. It should be attached to the main structural members of the vehicle, not the bumpers or brackets. The preferred way to tow a vehicle is with a flat bed type tow truck. Never have this vehicle towed with a sling-type tow truck.

Safety is a major consideration when towing and all applicable state and local laws must be obeyed. A safety chain system must be used at all times.

Front jacking location

Rear jacking location

Booster battery (jump) starting

Observe the following precautions when using a booster battery to start a vehicle:

a) *Before connecting the booster battery, make sure the ignition switch is in the Off position.*

b) *Turn off the lights, heater and other electrical loads.*

c) *Your eyes should be shielded. Safety goggles are a good idea.*

d) *Make sure the booster battery is the same voltage as the dead one in the vehicle.*

e) *The two vehicles MUST NOT TOUCH each other.*

f) *Make sure the transmission is in Neutral (manual transaxle) or Park (automatic transaxle).*

g) *If the booster battery is not a maintenance-free type, remove the vent caps and lay a cloth over the vent holes.*

Connect the red jumper cable to the positive (+) terminals of each battery.

Connect one end of the black cable to the negative (-) terminal of the booster battery. The other end of this cable should be connected to a good ground on the engine block (see illustration). Make sure the cable will not come into contact with the fan, drivebelts or other moving parts of the engine.

Start the engine using the booster battery, then, with the engine running at idle speed, disconnect the jumper cables in the reverse order of connection.

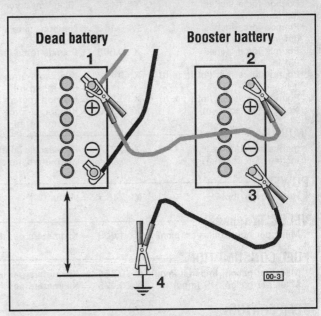

Make the booster battery cable connections in the numerical order shown (note that the negative cable of the booster battery is NOT attached to the negative terminal of the dead battery)

CONVERSION FACTORS

LENGTH (distance)

Inches (in)	X	25.4	= Millimeters (mm)	X 0.0394	= Inches (in)
Feet (ft)	X	0.305	= Meters (m)	X 3.281	= Feet (ft)
Miles	X	1.609	= Kilometers (km)	X 0.621	= Miles

VOLUME (capacity)

Cubic inches (cu in; in^3)	X	16.387	= Cubic centimeters (cc; cm^3)	X 0.061	= Cubic inches (cu in; in^3)
Imperial pints (Imp pt)	X	0.568	= Liters (l)	X 1.76	= Imperial pints (Imp pt)
Imperial quarts (Imp qt)	X	1.137	= Liters (l)	X 0.88	= Imperial quarts (Imp qt)
Imperial quarts (Imp qt)	X	1.201	= US quarts (US qt)	X 0.833	= Imperial quarts (Imp qt)
US quarts (US qt)	X	0.946	= Liters (l)	X 1.057	= US quarts (US qt)
Imperial gallons (Imp gal)	X	4.546	= Liters (l)	X 0.22	= Imperial gallons (Imp gal)
Imperial gallons (Imp gal)	X	1.201	= US gallons (US gal)	X 0.833	= Imperial gallons (Imp gal)
US gallons (US gal)	X	3.785	= Liters (l)	X 0.264	= US gallons (US gal)

MASS (weight)

Ounces (oz)	X	28.35	= Grams (g)	X 0.035	= Ounces (oz)
Pounds (lb)	X	0.454	= Kilograms (kg)	X 2.205	= Pounds (lb)

FORCE

Ounces-force (ozf; oz)	X	0.278	= Newtons (N)	X 3.6	= Ounces-force (ozf; oz)
Pounds-force (lbf; lb)	X	4.448	= Newtons (N)	X 0.225	= Pounds-force (lbf; lb)
Newtons (N)	X	0.1	= Kilograms-force (kgf; kg)	X 9.81	= Newtons (N)

PRESSURE

Pounds-force per square inch (psi; lbf/in^2; lb/in^2)	X	0.070	= Kilograms-force per square centimeter (kgf/cm^2; kg/cm^2)	X 14.223	= Pounds-force per square inch (psi; lbf/in^2; lb/in^2)
Pounds-force per square inch (psi; lbf/in^2; lb/in^2)	X	0.068	= Atmospheres (atm)	X 14.696	= Pounds-force per square inch (psi; lbf/in^2; lb/in^2)
Pounds-force per square inch (psi; lbf/in^2; lb/in^2)	X	0.069	= Bars	X 14.5	= Pounds-force per square inch (psi; lbf/in^2; lb/in^2)
Pounds-force per square inch (psi; lbf/in^2; lb/in^2)	X	6.895	= Kilopascals (kPa)	X 0.145	= Pounds-force per square inch (psi; lbf/in^2; lb/in^2)
Kilopascals (kPa)	X	0.01	= Kilograms-force per square centimeter (kgf/cm^2; kg/cm^2)	X 98.1	= Kilopascals (kPa)

TORQUE (moment of force)

Pounds-force inches (lbf in; lb in)	X	1.152	= Kilograms-force centimeter (kgf cm; kg cm)	X 0.868	= Pounds-force inches (lbf in; lb in)
Pounds-force inches (lbf in; lb in)	X	0.113	= Newton meters (Nm)	X 8.85	= Pounds-force inches (lbf in; lb in)
Pounds-force inches (lbf in; lb in)	X	0.083	= Pounds-force feet (lbf ft; lb ft)	X 12	= Pounds-force inches (lbf in; lb in)
Pounds-force feet (lbf ft; lb ft)	X	0.138	= Kilograms-force meters (kgf m; kg m)	X 7.233	= Pounds-force feet (lbf ft; lb ft)
Pounds-force feet (lbf ft; lb ft)	X	1.356	= Newton meters (Nm)	X 0.738	= Pounds-force feet (lbf ft; lb ft)
Newton meters (Nm)	X	0.102	= Kilograms-force meters (kgf m; kg m)	X 9.804	= Newton meters (Nm)

VACUUM

Inches mercury (in. Hg)	X	3.377	= Kilopascals (kPa)	X 0.2961	= Inches mercury
Inches mercury (in. Hg)	X	25.4	= Millimeters mercury (mm Hg)	X 0.0394	= Inches mercury

POWER

Horsepower (hp)	X	745.7	= Watts (W)	X 0.0013	= Horsepower (hp)

VELOCITY (speed)

Miles per hour (miles/hr; mph)	X	1.609	= Kilometers per hour (km/hr; kph)	X 0.621	= Miles per hour (miles/hr; mph)

FUEL CONSUMPTION *

Miles per gallon, Imperial (mpg)	X	0.354	= Kilometers per liter (km/l)	X 2.825	= Miles per gallon, Imperial (mpg)
Miles per gallon, US (mpg)	X	0.425	= Kilometers per liter (km/l)	X 2.352	= Miles per gallon, US (mpg)

TEMPERATURE

Degrees Fahrenheit = (°C x 1.8) + 32   Degrees Celsius (Degrees Centigrade; °C) = (°F - 32) x 0.56

*It is common practice to convert from miles per gallon (mpg) to liters/100 kilometers (l/100km), where mpg (Imperial) x l/100 km = 282 and mpg (US) x l/100 km = 235

FRACTION/DECIMAL/MILLIMETER EQUIVALENTS

DECIMALS TO MILLIMETERS

Decimal	mm	Decimal	mm
0.001	0.0254	0.500	12.7000
0.002	0.0508	0.510	12.9540
0.003	0.0762	0.520	13.2080
0.004	0.1016	0.530	13.4620
0.005	0.1270	0.540	13.7160
0.006	0.1524	0.550	13.9700
0.007	0.1778	0.560	14.2240
0.008	0.2032	0.570	14.4780
0.009	0.2286	0.580	14.7320
		0.590	14.9860
0.010	0.2540		
0.020	0.5080		
0.030	0.7620		
0.040	1.0160	0.600	15.2400
0.050	1.2700	0.610	15.4940
0.060	1.5240	0.620	15.7480
0.070	1.7780	0.630	16.0020
0.080	2.0320	0.640	16.2560
0.090	2.2860	0.650	16.5100
		0.660	16.7640
0.100	2.5400	0.670	17.0180
0.110	2.7940	0.680	17.2720
0.120	3.0480	0.690	17.5260
0.130	3.3020		
0.140	3.5560		
0.150	3.8100		
0.160	4.0640	0.700	17.7800
0.170	4.3180	0.710	18.0340
0.180	4.5720	0.720	18.2880
0.190	4.8260	0.730	18.5420
		0.740	18.7960
0.200	5.0800	0.750	19.0500
0.210	5.3340	0.760	19.3040
0.220	5.5880	0.770	19.5580
0.230	5.8420	0.780	19.8120
0.240	6.0960	0.790	20.0660
0.250	6.3500		
0.260	6.6040		
0.270	6.8580	0.800	20.3200
0.280	7.1120	0.810	20.5740
0.290	7.3660	0.820	21.8280
		0.830	21.0820
0.300	7.6200	0.840	21.3360
0.310	7.8740	0.850	21.5900
0.320	8.1280	0.860	21.8440
0.330	8.3820	0.870	22.0980
0.340	8.6360	0.880	22.3520
0.350	8.8900	0.890	22.6060
0.360	9.1440		
0.370	9.3980		
0.380	9.6520		
0.390	9.9060	0.900	22.8600
0.400	10.1600	0.910	23.1140
0.410	10.4140	0.920	23.3680
0.420	10.6680	0.930	23.6220
0.430	10.9220	0.940	23.8760
0.440	11.1760	0.950	24.1300
0.450	11.4300	0.960	24.3840
0.460	11.6840	0.970	24.6380
0.470	11.9380	0.980	24.8920
0.480	12.1920	0.990	25.1460
0.490	12.4460	1.000	25.4000

FRACTIONS TO DECIMALS TO MILLIMETERS

Fraction	Decimal	mm	Fraction	Decimal	mm
1/64	0.0156	0.3969	33/64	0.5156	13.0969
1/32	0.0312	0.7938	17/32	0.5312	13.4938
3/64	0.0469	1.1906	35/64	0.5469	13.8906
1/16	0.0625	1.5875	9/16	0.5625	14.2875
5/64	0.0781	1.9844	37/64	0.5781	14.6844
3/32	0.0938	2.3812	19/32	0.5938	15.0812
7/64	0.1094	2.7781	39/64	0.6094	15.4781
1/8	0.1250	3.1750	5/8	0.6250	15.8750
9/64	0.1406	3.5719	41/64	0.6406	16.2719
5/32	0.1562	3.9688	21/32	0.6562	16.6688
11/64	0.1719	4.3656	43/64	0.6719	17.0656
3/16	0.1875	4.7625	11/16	0.6875	17.4625
13/64	0.2031	5.1594	45/64	0.7031	17.8594
7/32	0.2188	5.5562	23/32	0.7188	18.2562
15/64	0.2344	5.9531	47/64	0.7344	18.6531
1/4	0.2500	6.3500	3/4	0.7500	19.0500
17/64	0.2656	6.7469	49/64	0.7656	19.4469
9/32	0.2812	7.1438	25/32	0.7812	19.8438
19/64	0.2969	7.5406	51/64	0.7969	20.2406
5/16	0.3125	7.9375	13/16	0.8125	20.6375
21/64	0.3281	8.3344	53/64	0.8281	21.0344
11/32	0.3438	8.7312	27/32	0.8438	21.4312
23/64	0.3594	9.1281	55/64	0.8594	21.8281
3/8	0.3750	9.5250	7/8	0.8750	22.2250
25/64	0.3906	9.9219	57/64	0.8906	22.6219
13/32	0.4062	10.3188	29/32	0.9062	23.0188
27/64	0.4219	10.7156	59/64	0.9219	23.4156
7/16	0.4375	11.1125	15/16	0.9375	23.8125
29/64	0.4531	11.5094	61/64	0.9531	24.2094
15/32	0.4688	11.9062	31/32	0.9688	24.6062
31/64	0.4844	12.3031	63/64	0.9844	25.0031
1/2	0.5000	12.7000	1	1.0000	25.4000

Automotive chemicals and lubricants

A number of automotive chemicals and lubricants are available for use during vehicle maintenance and repair. They include a wide variety of products ranging from cleaning solvents and degreasers to lubricants and protective sprays for rubber, plastic and vinyl.

CLEANERS

Carburetor cleaner and choke cleaner is a strong solvent for gum, varnish and carbon. Most carburetor cleaners leave a dry-type lubricant film which will not harden or gum up. Because of this film it is not recommended for use on electrical components.

Brake system cleaner is used to remove brake dust, grease and brake fluid from the brake system, where clean surfaces are absolutely necessary. It leaves no residue and often eliminates brake squeal caused by contaminants.

Electrical cleaner removes oxidation, corrosion and carbon deposits from electrical contacts, restoring full current flow. It can also be used to clean spark plugs, carburetor jets, voltage regulators and other parts where an oil-free surface is desired.

Demoisturants remove water and moisture from electrical components such as alternators, voltage regulators, electrical connectors and fuse blocks. They are non-conductive and non-corrosive.

Degreasers are heavy-duty solvents used to remove grease from the outside of the engine and from chassis components. They can be sprayed or brushed on and, depending on the type, are rinsed off either with water or solvent.

LUBRICANTS

Motor oil is the lubricant formulated for use in engines. It normally contains a wide variety of additives to prevent corrosion and reduce foaming and wear. Motor oil comes in various weights (viscosity ratings) from 0 to 50. The recommended weight of the oil depends on the season, temperature and the demands on the engine. Light oil is used in cold climates and under light load conditions. Heavy oil is used in hot climates and where high loads are encountered. Multi-viscosity oils are designed to have characteristics of both light and heavy oils and are available in a number of weights from 5W-20 to 20W-50.

Gear oil is designed to be used in differentials, manual transmissions and other areas where high-temperature lubrication is required.

Chassis and wheel bearing grease is a heavy grease used where increased loads and friction are encountered, such as for wheel bearings, balljoints, tie-rod ends and universal joints.

High-temperature wheel bearing grease is designed to withstand the extreme temperatures encountered by wheel bearings in disc brake equipped vehicles. It usually contains molybdenum disulfide (moly), which is a dry-type lubricant.

White grease is a heavy grease for metal-to-metal applications where water is a problem. White grease stays soft under both low and high temperatures (usually from -100 to +190-degrees F), and will not wash off or dilute in the presence of water.

Assembly lube is a special extreme pressure lubricant, usually containing moly, used to lubricate high-load parts (such as main and rod bearings and cam lobes) for initial start-up of a new engine. The assembly lube lubricates the parts without being squeezed out or washed away until the engine oiling system begins to function.

Silicone lubricants are used to protect rubber, plastic, vinyl and nylon parts.

Graphite lubricants are used where oils cannot be used due to contamination problems, such as in locks. The dry graphite will lubricate metal parts while remaining uncontaminated by dirt, water, oil or acids. It is electrically conductive and will not foul electrical contacts in locks such as the ignition switch.

Moly penetrants loosen and lubricate frozen, rusted and corroded fasteners and prevent future rusting or freezing.

Heat-sink grease is a special electrically non-conductive grease that is used for mounting electronic ignition modules where it is essential that heat is transferred away from the module.

SEALANTS

RTV sealant is one of the most widely used gasket compounds. Made from silicone, RTV is air curing, it seals, bonds, waterproofs, fills surface irregularities, remains flexible, doesn't shrink, is relatively easy to remove, and is used as a supplementary sealer with almost all low and medium temperature gaskets.

Anaerobic sealant is much like RTV in that it can be used either to seal gaskets or to form gaskets by itself. It remains flexible, is solvent resistant and fills surface imperfections. The difference between an anaerobic sealant and an RTV-type sealant is in the curing. RTV cures when exposed to air, while an anaerobic sealant cures only in the absence of air. This means that an anaerobic sealant cures only after the assembly of parts, sealing them together.

Thread and pipe sealant is used for sealing hydraulic and pneumatic fittings and vacuum lines. It is usually made from a Teflon compound, and comes in a spray, a paint-on liquid and as a wrap-around tape.

CHEMICALS

Anti-seize compound prevents seizing, galling, cold welding, rust and corrosion in fasteners. High-temperature anti-seize, usually made with copper and graphite lubricants, is used for exhaust system and exhaust manifold bolts.

Anaerobic locking compounds are used to keep fasteners from vibrating or working loose and cure only after installation, in the absence of air. Medium strength locking compound is used for small nuts, bolts and screws that may be removed later. High-strength locking compound is for large nuts, bolts and studs which aren't removed on a regular basis.

Oil additives range from viscosity index improvers to chemical treatments that claim to reduce internal engine friction. It should be noted that most oil manufacturers caution against using additives with their oils.

Gas additives perform several functions, depending on their chemical makeup. They usually contain solvents that help dissolve gum and varnish that build up on carburetor, fuel injection and intake parts. They also serve to break down carbon deposits that form on the inside surfaces of the combustion chambers. Some additives contain upper cylinder lubricants for valves and piston rings, and others contain chemicals to remove condensation from the gas tank.

MISCELLANEOUS

Brake fluid is specially formulated hydraulic fluid that can withstand the heat and pressure encountered in brake systems. Care must be taken so this fluid does not come in contact with painted surfaces or plastics. An opened container should always be resealed to prevent contamination by water or dirt.

Weatherstrip adhesive is used to bond weatherstripping around doors, windows and trunk lids. It is sometimes used to attach trim pieces.

Undercoating is a petroleum-based, tar-like substance that is designed to protect metal surfaces on the underside of the vehicle from corrosion. It also acts as a sound-deadening agent by insulating the bottom of the vehicle.

Waxes and polishes are used to help protect painted and plated surfaces from the weather. Different types of paint may require the use of different types of wax and polish. Some polishes utilize a chemical or abrasive cleaner to help remove the top layer of oxidized (dull) paint on older vehicles. In recent years many non-wax polishes that contain a wide variety of chemicals such as polymers and silicones have been introduced. These non-wax polishes are usually easier to apply and last longer than conventional waxes and polishes.

Safety first!

Regardless of how enthusiastic you may be about getting on with the job at hand, take the time to ensure that your safety is not jeopardized. A moment's lack of attention can result in an accident, as can failure to observe certain simple safety precautions. The possibility of an accident will always exist, and the following points should not be considered a comprehensive list of all dangers. Rather, they are intended to make you aware of the risks and to encourage a safety conscious approach to all work you carry out on your vehicle.

ESSENTIAL DOS AND DON'TS

DON'T rely on a jack when working under the vehicle. Always use approved jackstands to support the weight of the vehicle and place them under the recommended lift or support points.

DON'T attempt to loosen extremely tight fasteners (i.e. wheel lug nuts) while the vehicle is on a jack - it may fall.

DON'T start the engine without first making sure that the transmission is in Neutral (or Park where applicable) and the parking brake is set.

DON'T remove the radiator cap from a hot cooling system - let it cool or cover it with a cloth and release the pressure gradually.

DON'T attempt to drain the engine oil until you are sure it has cooled to the point that it will not burn you.

DON'T touch any part of the engine or exhaust system until it has cooled sufficiently to avoid burns.

DON'T siphon toxic liquids such as gasoline, antifreeze and brake fluid by mouth, or allow them to remain on your skin.

DON'T inhale brake lining dust - it is potentially hazardous (see *Asbestos* below).

DON'T allow spilled oil or grease to remain on the floor - wipe it up before someone slips on it.

DON'T use loose fitting wrenches or other tools which may slip and cause injury.

DON'T push on wrenches when loosening or tightening nuts or bolts. Always try to pull the wrench toward you. If the situation calls for pushing the wrench away, push with an open hand to avoid scraped knuckles if the wrench should slip.

DON'T attempt to lift a heavy component alone - get someone to help you.

DON'T rush or take unsafe shortcuts to finish a job.

DON'T allow children or animals in or around the vehicle while you are working on it.

DO wear eye protection when using power tools such as a drill, sander, bench grinder, etc. and when working under a vehicle.

DO keep loose clothing and long hair well out of the way of moving parts.

DO make sure that any hoist used has a safe working load rating adequate for the job.

DO get someone to check on you periodically when working alone on a vehicle.

DO carry out work in a logical sequence and make sure that everything is correctly assembled and tightened.

DO keep chemicals and fluids tightly capped and out of the reach of children and pets.

DO remember that your vehicle's safety affects that of yourself and others. If in doubt on any point, get professional advice.

ASBESTOS

Certain friction, insulating, sealing, and other products - such as brake linings, brake bands, clutch linings, torque converters, gaskets, etc. - may contain asbestos. Extreme care must be taken to avoid inhalation of dust from such products, since it is hazardous to health. If in doubt, assume that they do contain asbestos.

FIRE

Remember at all times that gasoline is highly flammable. Never smoke or have any kind of open flame around when working on a vehicle. But the risk does not end there. A spark caused by an electrical short circuit, by two metal surfaces contacting each other, or even by static electricity built up in your body under certain conditions, can ignite gasoline vapors, which in a confined space are highly explosive. Do not, under any circumstances, use gasoline for cleaning parts. Use an approved safety solvent.

Always disconnect the battery ground (-) cable at the battery before working on any part of the fuel system or electrical system. Never risk spilling fuel on a hot engine or exhaust component. It is strongly recommended that a fire extinguisher suitable for use on fuel and electrical fires be kept handy in the garage or workshop at all times. Never try to extinguish a fuel or electrical fire with water.

FUMES

Certain fumes are highly toxic and can quickly cause unconsciousness and even death if inhaled to any extent. Gasoline vapor falls into this category, as do the vapors from some cleaning solvents. Any draining or pouring of such volatile fluids should be done in a well ventilated area.

When using cleaning fluids and solvents, read the instructions on the container carefully. Never use materials from unmarked containers.

Never run the engine in an enclosed space, such as a garage. Exhaust fumes contain carbon monoxide, which is extremely poisonous. If you need to run the engine, always do so in the open air, or at least have the rear of the vehicle outside the work area.

If you are fortunate enough to have the use of an inspection pit, never drain or pour gasoline and never run the engine while the vehicle is over the pit. The fumes, being heavier than air, will concentrate in the pit with possibly lethal results.

THE BATTERY

Never create a spark or allow a bare light bulb near a battery. They normally give off a certain amount of hydrogen gas, which is highly explosive.

Always disconnect the battery ground (-) cable at the battery before working on the fuel or electrical systems.

If possible, loosen the filler caps or cover when charging the battery from an external source (this does not apply to sealed or maintenance-free batteries). Do not charge at an excessive rate or the battery may burst.

Take care when adding water to a non maintenance-free battery and when carrying a battery. The electrolyte, even when diluted, is very corrosive and should not be allowed to contact clothing or skin.

Always wear eye protection when cleaning the battery to prevent the caustic deposits from entering your eyes.

HOUSEHOLD CURRENT

When using an electric power tool, inspection light, etc., which operates on household current, always make sure that the tool is correctly connected to its plug and that, where necessary, it is properly grounded. Do not use such items in damp conditions and, again, do not create a spark or apply excessive heat in the vicinity of fuel or fuel vapor.

SECONDARY IGNITION SYSTEM VOLTAGE

A severe electric shock can result from touching certain parts of the ignition system (such as the spark plug wires) when the engine is running or being cranked, particularly if components are damp or the insulation is defective. In the case of an electronic ignition system, the secondary system voltage is much higher and could prove fatal.

Troubleshooting

CONTENTS

This section provides an easy reference guide to the more common problems which may occur during the operation of your vehicle. These problems and their possible causes are grouped under headings denoting various components or systems, such as Engine, Cooling system, etc. They also refer you to the chapter and/or section which deals with the problem.

Remember that successful troubleshooting is not a mysterious art practiced only by professional mechanics. It is simply the result of the right knowledge combined with an intelligent, systematic approach to the problem. Always work by a process of elimination, starting with the simplest solution and working through to the most complex - and never overlook the obvious. Anyone can run the gas tank dry or leave the lights on overnight, so don't assume that you are exempt from such oversights.

Finally, always establish a clear idea of why a problem has occurred and take steps to ensure that it doesn't happen again. If the electrical system fails because of a poor connection, check the other connections in the system to make sure that they don't fail as well. If a particular fuse continues to blow, find out why - don't just replace one fuse after another. Remember, failure of a small component can often be indicative of potential failure or incorrect functioning of a more important component or system.

ENGINE

1   Engine will not rotate when attempting to start

1   Battery terminal connections loose or corroded (Chapter 1).
2   Battery discharged or faulty (Chapter 1).
3   Automatic transaxle not completely engaged in Park (Chapter 7B) or clutch pedal not completely depressed (Chapter 8).
4   Broken, loose or disconnected wiring in the starting circuit (Chapters 5 and 12).
5   Starter motor pinion jammed in flywheel ring gear (Chapter 5).
6   Starter solenoid faulty (Chapter 5).
7   Starter motor faulty (Chapter 5).
8   Ignition switch faulty (Chapter 12).
9   Starter pinion or flywheel teeth worn or broken (Chapter 5).
10   Defective fusible link (see Chapter 12)

2   Engine rotates but will not start

1   Fuel tank empty.
2   Battery discharged (engine rotates slowly) (Chapter 5).
3   Battery terminal connections loose or corroded (Chapter 1).
4   Leaking fuel injector(s), faulty fuel pump, pressure regulator, etc. (Chapter 4).
5   Broken or stripped timing belt Chapter 2).
6   Ignition components damp or damaged (Chapter 5).
7   Worn, faulty or incorrectly gapped spark plugs (Chapter 1).
8   Broken, loose or disconnected wiring in the starting circuit (Chapter 5).
9   Broken, loose or disconnected wires at the ignition coils or faulty coils (Chapter 5).
10   Defective crankshaft sensor or PCM (see Chapter 6).

3   Engine hard to start when cold

1   Battery discharged or low (Chapter 1).
2   Malfunctioning fuel system (Chapter 4).

3   Faulty coolant temperature sensor or intake air temperature sensor (Chapter 6).
4   Fuel injector(s) leaking (Chapter 4).
5   Faulty ignition system (Chapter 5).
6   Defective MAP sensor (see Chapter 6).

4   Engine hard to start when hot

1   Air filter clogged (Chapter 1).
2   Fuel not reaching the fuel injection system (Chapter 4).
3   Corroded battery connections, especially ground (Chapter 1).
4   Faulty coolant temperature sensor or intake air temperature sensor (Chapter 6).

5   Starter motor noisy or excessively rough in engagement

1   Pinion or flywheel gear teeth worn or broken (Chapter 5).
2   Starter motor mounting bolts loose or missing (Chapter 5).

6   Engine starts but stops immediately

1   Loose or faulty electrical connections at ignition coil (Chapter 5).
2   Insufficient fuel reaching the fuel injector(s) (Chapter 4).
3   Vacuum leak at the gasket between the intake manifold/plenum and throttle body (Chapter 4).
4   Fault in the engine control system (Chapter 6).
5   Intake air leaks, broken vacuum lines (see Chapter 4)

7   Oil puddle under engine

1   Oil pan gasket and/or oil pan drain bolt washer leaking (Chapter 2).
2   Oil pressure sending unit leaking (Chapter 2).
3   Valve covers leaking (Chapter 2).
4   Engine oil seals leaking (Chapter 2).

8   Engine lopes while idling or idles erratically

1   Vacuum leakage (Chapters 2 and 4).
2   Leaking EGR valve (Chapter 6).
3   Air filter clogged (Chapter 1).
4   Fuel pump not delivering sufficient fuel to the fuel injection system (Chapter 4).
5   Leaking head gasket (Chapter 2).
6   Timing belt and/or pulleys worn (Chapter 2).
7   Camshaft lobes worn (Chapter 2).

9   Engine misses at idle speed

1   Spark plugs worn or not gapped properly (Chapter 1).
2   Faulty spark plug wires (Chapter 1).
3   Vacuum leaks (Chapters 2 and 4).
4   Faulty ignition coil (Chapter 5).
5   Uneven or low compression (Chapter 2).
6   Faulty fuel injector(s) (Chapter 4).

10   Engine misses throughout driving speed range

1   Fuel filter clogged and/or impurities in the fuel system (Chapter 1).
2   Low fuel output at the fuel injector(s) (Chapter 4).
3   Faulty or incorrectly gapped spark plugs (Chapter 1).

4   Leaking spark plug wires (Chapters 1 or 5).
5   Faulty emission system components (Chapter 6).
6   Low or uneven cylinder compression pressures (Chapter 2).
7   Burned valves (Chapter 2).
8   Weak or faulty ignition system (Chapter 5).
9   Vacuum leak in fuel injection system, throttle body, intake manifold or vacuum hoses (Chapter 4).

11   Engine stumbles on acceleration

1   Spark plugs fouled (Chapter 1).
2   Problem with fuel injection system (Chapter 4).
3   Fuel filter clogged (Chapter 4).
4   Fault in the engine control system (Chapter 6).
5   Intake manifold air leak (Chapters 2 and 4).
6   EGR system malfunction (Chapter 6).

12   Engine surges while holding accelerator steady

1   Intake air leak (Chapter 4).
2   Fuel pump or fuel pressure regulator faulty (Chapter 4).
3   Problem with fuel injection system (Chapter 4).
4   Problem with the emissions control system (Chapter 6).

13   Engine stalls

1   Fuel filter clogged and/or water and impurities in the fuel system (Chapter 4).
2   Ignition components damp or damaged (Chapter 5).
3   Faulty emissions system components (Chapter 6).
4   Faulty or incorrectly gapped spark plugs (Chapter 1).
5   Faulty spark plug wires (Chapter 1).
6   Vacuum leak in the fuel injection system, intake manifold or vacuum hoses (Chapters 2 and 4).

14   Engine lacks power

1   Worn camshaft lobes (Chapter 2).
2   Burned valves or incorrect valve timing (Chapter 2).
3   Faulty spark plug wires or faulty coil (Chapters 1 and 5).
4   Faulty or incorrectly gapped spark plugs (Chapter 1).
5   Problem with the fuel injection system (Chapter 4).
6   Plugged air filter (Chapter 1).
7   Brakes binding (Chapter 9).
8   Automatic transaxle fluid level incorrect (Chapter 1).
9   Clutch slipping (Chapter 8).
10 Fuel filter clogged and/or impurities in the fuel system (Chapter 4).
11 Emission control system not functioning properly (Chapter 6).
12 Low or uneven cylinder compression pressures (Chapter 2).
13 Restricted exhaust system (Chapters 4 and 6).

15   Engine backfires

1   Emission control system not functioning properly (Chapter 6).
2   Faulty spark plug wires or coil(s) (Chapter 5).
3   Problem with the fuel injection system (Chapter 4).
4   Vacuum leak at fuel injector(s), intake manifold or vacuum hoses (Chapters 2 and 4).
5   Burned valves or incorrect valve timing (Chapter 2).

16   Pinging or knocking engine sounds during acceleration or uphill

1   Incorrect grade of fuel.
2   Problem with the engine control system (Chapter 6).
3   Fuel injection system faulty (Chapter 4).
4   Improper or damaged spark plugs or wires (Chapter 1).
5   EGR valve not functioning (Chapter 6).
6   Vacuum leak (Chapters 2 and 4).

17   Engine runs with oil pressure light on

1   Low oil level (Chapter 1).
2   Idle rpm below specification (Chapter 1).
3   Short in wiring circuit (Chapter 12).
4   Faulty oil pressure sender (Chapter 2).
5   Worn engine bearings and/or oil pump (Chapter 2).

18   Engine diesels (continues to run) after switching off

1   Idle speed too high (Chapter 1).
2   Excessive engine operating temperature (Chapter 3).
3   Excessive carbon deposits on valves and pistons (see Chapter 2)

ENGINE ELECTRICAL SYSTEM

19   Battery will not hold a charge

1   Alternator drivebelt defective or not adjusted properly (Chapter 1).
2   Battery electrolyte level low (Chapter 1).
3   Battery terminals loose or corroded (Chapter 1).
4   Alternator not charging properly (Chapter 5).
5   Loose, broken or faulty wiring in the charging circuit (Chapter 5).
6   Short in vehicle wiring (Chapter 12).
7   Internally defective battery (Chapters 1 and 5).

20   Alternator light fails to go out

1   Faulty alternator or charging circuit (Chapter 5).
2   Alternator drivebelt defective or out of adjustment (Chapter 1).
3   Alternator voltage regulator inoperative (Chapter 5).

21   Alternator light fails to come on when key is turned on

1   Warning light bulb defective (Chapter 12).
2   Fault in the printed circuit, dash wiring or bulb holder (Chapter 12).

FUEL SYSTEM

22   Excessive fuel consumption

1   Dirty or clogged air filter element (Chapter 1).
2   Emissions system not functioning properly (Chapter 6).
3   Fuel injection system not functioning properly (Chapter 4).
4   Low tire pressure or incorrect tire size (Chapter 1).

23 Fuel leakage and/or fuel odor

1 Leaking fuel line (Chapters 1 and 4).
2 Tank overfilled.
3 Evaporative emissions control canister defective (Chapters 1 and 6).
4 Problem with fuel injection system (Chapter 4).

COOLING SYSTEM

24 Overheating

1 Insufficient coolant in system (Chapter 1).
2 Water pump defective (Chapter 3).
3 Radiator core blocked or grille restricted (Chapter 3).
4 Thermostat faulty (Chapter 3).
5 Electric coolant fan inoperative or blades broken (Chapter 3).
6 Cooling system pressure cap not maintaining proper pressure (Chapter 3).

25 Overcooling

1 Faulty thermostat (Chapter 3).
2 Inaccurate temperature gauge sending unit (Chapter 3)

26 External coolant leakage

1 Deteriorated/damaged hoses; loose clamps (Chapters 1 and 3).
2 Water pump defective (Chapter 3).
3 Leakage from radiator core or coolant reservoir bottle (Chapter 3).
4 Engine drain or water jacket core plugs leaking (Chapter 2).

27 Internal coolant leakage

1 Leaking cylinder head gasket (Chapter 2).
2 Cracked cylinder bore or cylinder head (Chapter 2).

28 Coolant loss

1 Too much coolant in system (Chapter 1).
2 Coolant boiling away because of overheating (Chapter 3).
3 Internal or external leakage (Chapter 3).
4 Faulty pressure cap (Chapter 3).

29 Poor coolant circulation

1 Inoperative water pump (Chapter 3).
2 Restriction in cooling system (Chapters 1 and 3).
3 Thermostat sticking (Chapter 3).

CLUTCH

30 Pedal travels to floor - no pressure or very little resistance

1 Broken or disconnected clutch cable or defective hydraulic release system (Chapter 8).
2 Broken release bearing or fork (Chapter 8).

31 Unable to select gears

1 Faulty transaxle (Chapter 7).
2 Faulty clutch disc or pressure plate (Chapter 8).
3 Faulty release lever or release bearing (Chapter 8).
4 Faulty shift lever assembly or rods (Chapter 8).

32 Clutch slips (engine speed increases with no increase in vehicle speed)

1 Clutch plate worn (Chapter 8).
2 Clutch plate is oil soaked by leaking rear main seal (Chapter 8).
3 Clutch plate not seated (Chapter 8).
4 Warped pressure plate or flywheel (Chapter 8).
5 Weak diaphragm springs (Chapter 8).
6 Clutch plate overheated. Allow to cool.
7 Faulty clutch self-adjusting mechanism (Chapter 8).

33 Grabbing (chattering) as clutch is engaged

1 Oil on clutch plate lining, burned or glazed facings (Chapter 8).
2 Worn or loose engine or transaxle mounts (Chapters 2 and 7).
3 Worn splines on clutch plate hub (Chapter 8).
4 Warped pressure plate or flywheel (Chapter 8).
5 Burned or smeared resin on flywheel or pressure plate (Chapter 8).

34 Transaxle rattling (clicking)

1 Release fork loose (Chapter 8).
2 Low engine idle speed (Chapter 1).

35 Noise in clutch area

Faulty release bearing (Chapter 8).

36 Clutch pedal stays on floor

1 Broken release bearing or fork (Chapter 8).
2 Broken or disconnected clutch cable (Chapter 8).
3 Broken or worn-out modular clutch assembly

37 High pedal effort

1 Binding clutch cable (Chapter 8).
2 Pressure plate faulty (Chapter 8).

MANUAL TRANSAXLE

38 Knocking noise at low speeds

1 Worn driveaxle constant velocity (CV) joints (Chapter 8).
2 Worn side gear shaft counterbore in differential case (Chapter 7A).*

39 Noise most pronounced when turning

Differential gear noise (Chapter 7A).*

40 Clunk on acceleration or deceleration

1 Loose engine or transaxle mounts (Chapters 2 and 7A).
2 Worn differential pinion shaft in case.*

3 Worn side gear shaft counterbore in differential case (Chapter 7A).*

4 Worn or damaged driveaxle inboard CV joints (Chapter 8).

41 Clicking noise in turns

Worn or damaged outboard CV joint (Chapter 8).

42 Vibration

1 Rough wheel bearing (Chapters 1 and 10).
2 Damaged driveaxle (Chapter 8).
3 Out of round tires (Chapter 1).
4 Tire out of balance (Chapters 1 and 10).
5 Worn CV joint (Chapter 8).

43 Noisy in neutral with engine running

1 Damaged input gear bearing (Chapter 7A).*
2 Damaged clutch release bearing (Chapter 8).

44 Noisy in one particular gear

1 Damaged or worn constant mesh gears (Chapter 7A).*
2 Damaged or worn synchronizers (Chapter 7A).*
3 Bent reverse fork (Chapter 7A).*
4 Damaged fourth speed gear or output gear (Chapter 7A).*
5 Worn or damaged reverse idler gear or idler bushing (Chapter 7A).*

45 Noisy in all gears

1 Insufficient lubricant (Chapter 7A).
2 Damaged or worn bearings (Chapter 7A).*
3 Worn or damaged input gear shaft and/or output gear shaft (Chapter 7A).*

46 Slips out of gear

1 Worn or improperly adjusted linkage (Chapter 7A).
2 Transaxle loose on engine (Chapter 7A).
3 Shift linkage does not work freely, binds (Chapter 7A).
4 Input gear bearing retainer broken or loose (Chapter 7A).*
5 Worn shift fork (Chapter 7A).*

47 Leaks lubricant

1 Driveshaft seals worn (Chapter 7A).
2 Excessive amount of lubricant in transaxle (Chapters 1 and 7A).
3 Loose or broken input gear shaft bearing retainer (Chapter 7A).*
4 Input gear bearing retainer O-ring and/or lip seal damaged (Chapter 7A).*
5 Vehicle speed sensor O-ring leaking (Chapter 7A).

48 Hard to shift

Shift linkage loose or worn (Chapter 7A).

Although the corrective action necessary to remedy the symptoms described is beyond the scope of this manual, the above information should be helpful in isolating the cause of the condition so that the owner can communicate clearly with a professional mechanic.

AUTOMATIC TRANSAXLE

➡Note: Due to the complexity of the automatic transaxle, it is difficult for the home mechanic to properly diagnose and service this component. For problems other than the following, the vehicle should be taken to a dealer or transaxle shop.

49 Fluid leakage

1 Automatic transaxle fluid is a deep red color. Fluid leaks should not be confused with engine oil, which can easily be blown onto the transaxle by air flow.

2 To pinpoint a leak, first remove all built-up dirt and grime from the transaxle housing with degreasing agents and/or steam cleaning. Then drive the vehicle at low speeds so air flow will not blow the leak far from its source. Raise the vehicle and determine where the leak is coming from. Common areas of leakage are:

a) *Pan (Chapters 1 and 7)*
b) *Dipstick tube (Chapters 1 and 7)*
c) *Transaxle oil lines (Chapter 7)*
d) *Speed sensor (Chapter 7)*
e) *Driveaxle oil seals (Chapter 7).*

50 Transaxle fluid brown or has a burned smell

Transaxle fluid overheated (Chapter 1).

51 General shift mechanism problems

1 Chapter 7, Part B, deals with checking and adjusting the shift linkage on automatic transaxles. Common problems which may be attributed to poorly adjusted linkage are:

a) *Engine starting in gears other than Park or Neutral.*
b) *Indicator on shifter pointing to a gear other than the one actually being used.*
c) *Vehicle moves when in Park.*

2 Refer to Chapter 7B for the shift linkage adjustment procedure.

52 Transaxle will not downshift with accelerator pedal pressed to the floor

2000 and 2001 models:

Misadjusted or broken Throttle Valve cable (Chapter 7B).

2002 and later models:

The transaxle on these models is electronically controlled. This type of problem - which is caused by a malfunction in the control unit, a sensor or solenoid, or the circuit itself - is beyond the scope of this book. Take the vehicle to a dealer service department or a competent automatic transmission shop.

53 Engine will start in gears other than Park or Neutral

Neutral start switch out of adjustment or malfunctioning (Chapter 7B).

54 Transaxle slips, shifts roughly, is noisy or has no drive in forward or reverse gears

There are many probable causes for the above problems, but the home mechanic should be concerned with only one possibility - fluid level. Before taking the vehicle to a repair shop, check the level and condition of the fluid as described in Chapter 1. Correct the fluid level as necessary or change the fluid and filter if needed. If the problem persists, have a professional diagnose the cause.

DRIVEAXLES

55   Clicking noise in turns

Worn or damaged outboard CV joint (Chapter 8).

56   Shudder or vibration during acceleration

1   Excessive toe-in (Chapter 10).
2   Incorrect spring heights (Chapter 10).
3   Worn or damaged inboard or outboard CV joints (Chapter 8).
4   Sticking inboard CV joint assembly (Chapter 8).

57   Vibration at highway speeds

1   Out of balance front wheels and/or tires (Chapters 1 and 10).
2   Out of round front tires (Chapters 1 and 10).
3   Worn CV joint(s) (Chapter 8).

BRAKES

➡Note: Before assuming that a brake problem exists, make sure that:

a) The tires are in good condition and properly inflated (Chapter 1).
b) The front end alignment is correct (Chapter 10).
c) The vehicle is not loaded with weight in an unequal manner.

58   Vehicle pulls to one side during braking

1   Incorrect tire pressures (Chapter 1).
2   Front end out of alignment (have the front end aligned).
3   Front, or rear, tire sizes not matched to one another.
4   Restricted brake lines or hoses (Chapter 9).
5   Malfunctioning drum brake or caliper assembly (Chapter 9).
6   Loose suspension parts (Chapter 10).
7   Excessive wear of brake shoe or pad material or disc/drum on one side.

59   Noise (high-pitched squeal when the brakes are applied)

Front and/or rear disc brake pads worn out. The noise comes from the wear sensor rubbing against the disc (does not apply to all vehicles). Replace pads with new ones immediately (Chapter 9).

60   Brake roughness or chatter (pedal pulsates)

1   Excessive lateral disc runout (Chapter 9).
2   Uneven pad wear (Chapter 9).
3   Defective disc (Chapter 9).

61   Excessive brake pedal effort required to stop vehicle

1   Malfunctioning power brake booster (Chapter 9).
2   Partial system failure (Chapter 9).
3   Excessively worn pads or shoes (Chapter 9).

4   Piston in caliper or wheel cylinder stuck or sluggish (Chapter 9).
5   Brake pads or shoes contaminated with oil or grease (Chapter 9).
6   Brake disc grooved and/or glazed (Chapter 1).
7   New pads or shoes installed and not yet seated. It will take a while for the new material to seat against the disc or drum.

62   Excessive brake pedal travel

1   Partial brake system failure (Chapter 9).
2   Insufficient fluid in master cylinder (Chapters 1 and 9).
3   Air trapped in system (Chapters 1 and 9).

63   Dragging brakes

1   Incorrect adjustment of brake light switch (Chapter 9).
2   Master cylinder pistons not returning correctly (Chapter 9).
3   Restricted brakes lines or hoses (Chapters 1 and 9).
4   Parking brake adjuster malfunctioning.

64   Grabbing or uneven braking action

1   Malfunction of proportioning valve (Chapter 9).
2   Malfunction of power brake booster unit (Chapter 9).
3   Binding brake pedal mechanism (Chapter 9).

65   Brake pedal feels spongy when depressed

1   Air in hydraulic lines (Chapter 9).
2   Master cylinder mounting bolts loose (Chapter 9).
3   Master cylinder defective (Chapter 9).

66   Brake pedal travels to the floor with little resistance

1   Little or no fluid in the master cylinder reservoir caused by leaking caliper piston(s) (Chapter 9).
2   Loose, damaged or disconnected brake lines (Chapter 9).
3   Master cylinder defective (Chapter 9).

67   Parking brake does not hold

Parking brake automatic adjuster or rear drum brake automatic adjuster not working properly (Chapter 9).

SUSPENSION AND STEERING SYSTEMS

➡Note: Before attempting to diagnose the suspension and steering systems, perform the following preliminary checks:

a) Tires for wrong pressure and uneven wear.
b) Steering universal joints from the column to the rack and pinion for loose connectors or wear.
c) Front and rear suspension and the rack- and-pinion assembly for loose or damaged parts.
d) Out-of-round or out-of-balance tires, bent rims and loose and/or rough wheel bearings.

68   Vehicle pulls to one side

1   Mismatched or uneven tires (Chapter 10).
2   Broken or sagging springs (Chapter 10).
3   Wheel alignment out-of-specifications (Chapter 10).
4   Front brake dragging (Chapter 9).

69 Abnormal or excessive tire wear

1 Wheel alignment out-of-specifications (Chapter 10).
2 Sagging or broken springs (Chapter 10).
3 Tire out-of-balance (Chapter 10).
4 Worn strut damper (Chapter 10).
5 Overloaded vehicle.
6 Tires not rotated regularly.

70 Wheel makes a thumping noise

1 Blister or bump on tire (Chapter 10).
2 Improper strut damper action (Chapter 10).

71 Shimmy, shake or vibration

1 Tire or wheel out-of-balance or out-of-round (Chapter 10).
2 Loose or worn wheel bearings (Chapters 1, 8 and 10).
3 Worn tie-rod ends (Chapter 10).
4 Worn lower balljoints (Chapters 1 and 10).
5 Excessive wheel runout (Chapter 10).
6 Blister or bump on tire (Chapter 10).

72 Hard steering

1 Lack of lubrication at balljoints, tie-rod ends and rack and pinion assembly (Chapter 10).
2 Front wheel alignment out-of-specifications (Chapter 10).
3 Low tire pressure(s) (Chapters 1 and 10).

73 Poor returnability of steering to center

1 Binding in balljoints or tie-rod ends (Chapter 10).
2 Binding in steering column (Chapter 10).
3 Lack of lubricant in steering gear assembly (Chapter 10).
4 Front wheel alignment out-of-specifications (Chapter 10).

74 Abnormal noise at the front end

1 Balljoints or tie-rod ends worn (Chapter 10).
2 Damaged strut mounting (Chapter 10).
3 Worn control arm bushings or tie-rod ends (Chapter 10).
4 Loose stabilizer bar (Chapter 10).
5 Loose wheel nuts (Chapters 1 and 10).
6 Loose suspension bolts (Chapter 10)

75 Wander or poor steering stability

1 Mismatched or uneven tires (Chapter 10).
2 Lack of lubrication at balljoints and tie-rod ends (Chapters 1 and 10).
3 Worn strut assemblies (Chapter 10).
4 Loose stabilizer bar (Chapter 10).
5 Broken or sagging springs (Chapter 10).
6 Wheels out of alignment (Chapter 10).

76 Erratic steering when braking

1 Wheel bearings worn (Chapter 10).
2 Broken or sagging springs (Chapter 10).
3 Leaking wheel cylinder or caliper (Chapter 10).
4 Warped discs or drums (Chapter 10).

77 Excessive pitching and/or rolling around corners or during braking

1 Loose stabilizer bar (Chapter 10).
2 Worn strut dampers or mountings (Chapter 10).
3 Broken or sagging springs (Chapter 10).
4 Overloaded vehicle.

78 Suspension bottoms

1 Overloaded vehicle.
2 Worn strut dampers (Chapter 10).
3 Incorrect, broken or sagging springs (Chapter 10).

79 Cupped tires

1 Front wheel or rear wheel alignment out-of-specifications (Chapter 10).
2 Worn strut dampers (Chapter 10).
3 Wheel bearings worn (Chapter 10).
4 Excessive tire or wheel runout (Chapter 10).
5 Worn balljoints (Chapter 10).

80 Excessive tire wear on outside edge

1 Inflation pressures incorrect (Chapter 1).
2 Excessive speed in turns.
3 Front end alignment incorrect (excessive toe-in). Have professionally aligned.
4 Suspension arm bent or twisted (Chapter 10).

81 Excessive tire wear on inside edge

1 Inflation pressures incorrect (Chapter 1).
2 Front end alignment incorrect (toe-out). Have professionally aligned.
3 Loose or damaged steering components (Chapter 10).

82 Tire tread worn in one place

1 Tires out-of-balance.
2 Damaged or buckled wheel. Inspect and replace if necessary.
3 Defective tire (Chapter 1).

83 Excessive play or looseness in steering system

1 Wheel bearing(s) worn (Chapter 10).
2 Tie-rod end loose (Chapter 10).
3 Steering gear loose (Chapter 10).
4 Worn or loose steering intermediate shaft (Chapter 10).

84 Rattling or clicking noise in steering gear

1 Steering gear loose (Chapter 10).
2 Steering gear defective.

Section

Reference to other Chapters

1

TUNE-UP AND ROUTINE MAINTENANCE

1 Dodge Neon Maintenance Schedule

The following maintenance intervals are based on the assumption that the vehicle owner will be doing the maintenance or service work, as opposed to having a dealer service department do the work. Although the time/mileage intervals are loosely based on factory recommendations, most have been shortened to ensure, for example, that such items as lubricants and fluids are checked/changed at intervals that promote maximum engine/driveline service life. Also, subject to the preference of the individual owner interested in keeping his or her vehicle in peak condition at all times, and with the vehicle's ultimate resale in mind, many of the maintenance procedures may be performed more often than recommended in the following schedule. We encourage such owner initiative.

When the vehicle is new it should be serviced initially by a factory authorized dealer service department to protect the factory warranty. In many cases the initial maintenance check is done at no cost to the owner (check with your dealer service department for more information).

EVERY 250 MILES OR WEEKLY, WHICHEVER COMES FIRST

Check the engine oil level (see Section 4)
Check the engine coolant level (see Section 4)
Check the windshield washer fluid level (see Section 4)
Check the brake and clutch fluid level (see Section 4)
Check the tires and tire pressures (see Section 5)
Check the automatic transaxle fluid level (see Section 6)
Check the power steering fluid level (see Section 7)
Check the operation of all lights
Check the horn operation

EVERY 3,000 MILES OR 3 MONTHS, WHICHEVER COMES FIRST

Change the engine oil and filter (see Section 8)*

EVERY 7,500 MILES OR 6 MONTHS, WHICHEVER COMES FIRST

Check and clean the battery and terminals (see Section 9)
Check the cooling system hoses and connections for leaks and damage (see Section 10)
Check the condition of all underhood hoses and connections (see Section 11)
Check the wiper blade condition (see Section 12)
Rotate the tires (see Section 13)
Check for freeplay in the steering linkage and balljoints (see Section 14)
Check the driveaxle boots and the suspension components (see Section 14)
Check the exhaust pipes and hangers (see Section 15)
Check the manual transaxle lubricant level (see Section 16)

EVERY 15,000 MILES OR 12 MONTHS, WHICHEVER COMES FIRST

All items listed above, plus:

Check the brake system (see Section 17)
Check the fuel system hoses and connections for leaks and damage (see Section 18)
Check the drivebelts and adjust if necessary (see Section 19)

EVERY 30,000 MILES OR 24 MONTHS, WHICHEVER COMES FIRST

All items listed above, plus:

Lubricate the body hinges and locks (see Section 20)*
Replace the air filter element and the PCV filter (see Section 21)*
Change the automatic transaxle fluid and filter (see Section 22)*
Adjust the bands on the automatic transaxle (2000 and 2001 models) (see Section 23)
Change the manual transaxle lubricant (see Section 24)*
Drain and replace the engine coolant (unless filled with Mopar 5 year/100,000 mile coolant) (see Section 25)
Check the fuel evaporative emission control system and hoses (see Section 26)
Replace the spark plugs (non-platinum type) (see Section 27)

EVERY 60,000 MILES OR 48 MONTHS, WHICHEVER COMES FIRST

All items listed above, plus:

Replace the drivebelts (see Section 19)
Replace the spark plug wires (see Section 28)
Check and replace, if necessary, the PCV valve (see Section 29)*

EVERY 90,000 MILES OR 72 MONTHS, WHICHEVER COMES FIRST

Replace the timing belt (see Chapter 2A)

EVERY 100,000 MILES OR 84 MONTHS, WHICHEVER COMES FIRST

Drain and replace the engine coolant if filled with Mopar 5 year/100,000 mile coolant (see Section 25)
Replace the spark plugs (platinum or iridium spark plugs) (see Section 27)*

*This item is affected by "severe" operating conditions as described below. If the vehicle in question is operated under "severe" conditions, perform all maintenance procedures marked with an asterisk (*) at the intervals specified by the mileage headings below.

Consider the conditions "severe" if most driving is done . . .
In dusty areas

Idling for extended periods and/or low-speed operation

When outside temperatures remain below freezing and most trips are less than four miles

In heavy city traffic where outside temperatures regularly reach 90-degrees F or higher

EVERY 15,000 MILES

Check and replace, if necessary, the air filter element (see Section 21)

Change the automatic transaxle fluid and filter (see Section 22)

Change the manual transaxle lubricant (see Section 24)

EVERY 30,000 MILES

Check and replace, if necessary, the PCV valve (see Section 29)

EVERY 48,000 MILES

Change the brake fluid by bleeding the complete system until the fluid being bled is clear (see Chapter 9 for the bleeding procedure).

EVERY 75,000 MILES

Replace the spark plugs (platinum or iridium-type plugs) (see Section 27)

Typical engine compartment layout

1	Windshield washer fluid reservoir	6	Battery
2	Engine oil filler cap	7	Power Distribution Center
3	Ignition coil pack	8	Air filter housing
4	Engine coolant reservoir	9	Automatic transaxle fluid dipstick
5	Brake fluid reservoir		
10	Spark plug boot		
11	Upper radiator hose		
12	Engine oil dipstick		
13	Cooling system pressure cap		
14	Power steering fluid reservoir		

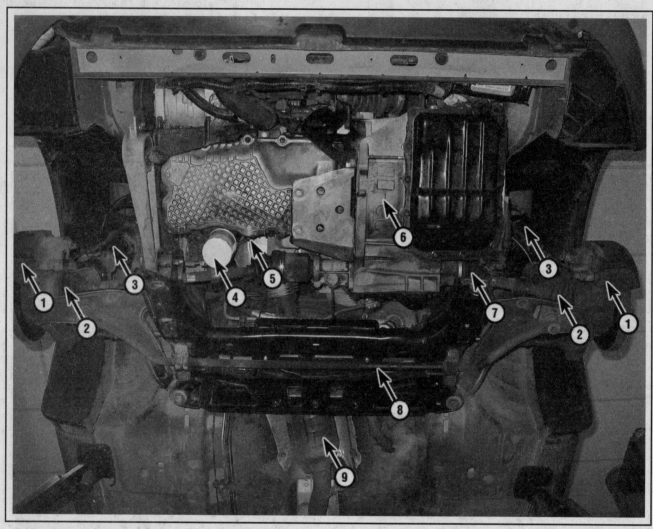

Typical engine compartment underside components

1	Brake caliper	4	Engine oil filter	7	Inner driveaxle boot
2	Outer driveaxle boot	5	Engine oil drain plug	8	Stabilizer bar
3	Spring and shock absorber strut	6	Transaxle	9	Catalytic converter

Typical rear underside components

1	Spring and shock absorber strut	3	Stabilizer bar	5	Trailing arm
2	Lateral arm	4	Muffler	6	Fuel tank

2 Introduction

This Chapter is designed to help the home mechanic maintain the Neon with the goals of maximum performance, economy, safety and reliability in mind.

Included is a master maintenance schedule, followed by procedures dealing specifically with each item on the schedule. Visual checks, adjustments, component replacement and other helpful items are included. Refer to the accompanying illustrations of the engine compartment and the underside of the vehicle for the locations of various components.

Adhering to the mileage/time maintenance schedule and following the step-by-step procedures, which is simply a preventive maintenance program, will result in maximum reliability and vehicle service life. Keep in mind that it's not possible for this comprehensive program to produce the same results if you maintain some items at the specified intervals but not others.

As you service the vehicle, you'll discover that many of the procedures can - and should - be grouped together because of the nature of the particular procedure you're performing or because of the close proximity of two otherwise unrelated components to one another.

For example, if the vehicle is raised, you should inspect the exhaust, suspension, steering and fuel systems while you're under the vehicle. When you're rotating the tires, it makes good sense to check the brakes, since the wheels are already removed. Finally, let's suppose you have to borrow or rent a torque wrench. Even if you only need it to tighten the spark plugs, you might as well check the torque of as many critical fasteners as time allows.

The first step in this maintenance program is to prepare before the actual work begins. Read through all the procedures you're planning, then gather together all the parts and tools needed. If it looks like you might run into problems during a particular job, seek advice from a mechanic or an experienced do-it-yourselfer.

OWNER'S MANUAL AND VECI LABEL INFORMATION

Your vehicle owner's manual was written for your year and model and contains very specific information on component locations, specifications, fuse ratings, part numbers, etc. The Owner's Manual is an important resource for the do-it-yourselfer to have; if one was not supplied with your vehicle, it can generally be ordered from a dealer parts department.

Among other important information, the Vehicle Emissions Control Information (VECI) label contains specifications and procedures for applicable tune-up adjustments and, in some instances, spark plugs (see Chapter 6 for more information on the VECI label). The information on this label is the exact maintenance data recommended by the manufacturer. This data often varies by intended operating altitude, local emissions regulations, month of manufacture, etc.

This Chapter contains procedural details, safety information and more ambitious maintenance intervals than you might find in manufacturer's literature. However, you may also find procedures or specifications in your Owner's Manual or VECI label that differ with what's printed here. In these cases, the Owner's Manual or VECI label can be considered correct, since it is specific to your particular vehicle.

3 Tune-up general information

The term "tune-up" is used in this manual to represent a combination of individual operations rather than one specific procedure.

The engine will be kept in relatively good running condition and the need for additional work will be minimized if the routine maintenance schedule is followed closely and frequent checks are made of fluid levels and high wear items, as suggested throughout this manual from the time the vehicle is new.

More likely than not, however, there will be times when the engine is running poorly due to lack of regular maintenance. This is even more likely if a used vehicle, which hasn't received regular and frequent maintenance checks, is purchased. In such cases, an engine tune-up will be needed outside of the regular routine maintenance intervals.

The first step in any tune-up or diagnostic procedure to help correct a poor running engine is a cylinder compression check. A compression check (see Chapter 2, Part B) will help determine the condition of internal engine components and should be used as a guide for tune-up and repair procedures. For instance, if a compression check indicates serious internal engine wear, a conventional tune-up will not improve the performance of the engine and would be a waste of time and money. Because of its importance, someone with the right equipment and the knowledge to use it properly should do the compression check.

The following procedures are those most often needed to bring a generally poor running engine back into a proper state of tune:

MINOR TUNE-UP

Check all engine related fluids (see Section 4)
Clean and inspect the battery (see Section 9)
Check all underhood hoses (see Section 11)
Check and adjust the drivebelts (see Section 19)
Check the air filter (see Section 21)
Service the cooling system (see Section 25)
Replace the spark plugs (see Section 27)
Check the PCV valve (see Section 29)

MAJOR TUNE-UP

All items listed under Minor tune-up plus . . .
Check the fuel system (see Section 18)
Replace the air filter (see Section 21)
Replace the spark plug wires (see Section 27)
Check the charging system (see Chapter 5)

4 Fluid level checks (every 250 miles or weekly)

➡**Note: The following are fluid level checks to be done on a 250 mile or weekly basis. Additional fluid level checks can be found in specific maintenance procedures that follow. Regardless of the intervals, develop the habit of checking under the vehicle periodically for evidence of fluid leaks.**

1 Fluids are an essential part of the lubrication, cooling, brake and window washer systems. Because the fluids gradually become depleted and/or contaminated during normal operation of the vehicle, they must be replenished periodically. See *Recommended lubricants and fluids* at

4.2 The engine oil dipstick is located at the front of the engine and is clearly marked

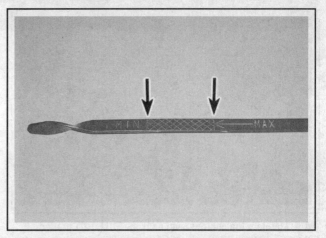

4.4 The oil level should be between the MIN and MAX marks, near the top of the cross-hatched area on the dipstick - if it isn't, add enough oil to bring the level up to or near the upper mark (do not overfill)

the end of this Chapter before adding fluid to any of the following components.

➡**Note: The vehicle must be on level ground when fluid levels are checked.**

ENGINE OIL

◗ **Refer to illustrations 4.2, 4.4 and 4.5**

2   Engine oil level is checked with a dipstick that is located on the side of the engine facing the front of the vehicle (see illustration). The dipstick extends through a tube and into the oil pan at thc bottom of the engine.

3   The oil level should be checked before the vehicle has been driven, or about 5 minutes after the engine has been shut off. If the oil is checked immediately after driving the vehicle, some of the oil will remain in the upper engine components, resulting in an inaccurate reading on the dipstick.

4   Pull the dipstick out of the tube and wipe all the oil off the end with a clean rag or paper towel. Insert the clean dipstick all the way back into the tube, then pull it out again. Note the oil level at the end of the dipstick. Add oil as necessary to bring the oil level to the top of the cross-hatched area, or MAX mark (see illustration).

5   Oil is added to the engine after removing a cap located on the

valve cover (see illustration). The cap will be marked "Engine oil." Use a funnel to prevent spills as the oil is added.

6   Don't allow the level to drop below the MIN mark on the dipstick or engine damage may occur. On the other hand, don't overfill the engine by adding too much oil - it may result in oil aeration and loss of oil pressure and also could result in oil fouled spark plugs, oil leaks or seal failures.

7   Checking the oil level is an important preventive maintenance step. A consistently low oil level indicates oil leakage through damaged seals, defective gaskets or past worn rings or valve guides. If the oil looks milky in color or has water droplets in it, the block or head may be cracked and leaking coolant is entering the crankcase. The engine should be checked immediately. The condition of the oil should also be checked. Each time you check the oil level, slide your thumb and index finger up the dipstick before wiping off the oil. If you see small dirt or metal particles clinging to the dipstick, the oil should be changed (see Section 8).

ENGINE COOLANT

◗ **Refer to illustration 4.9**

❋❋ WARNING:

Do not allow antifreeze to come in contact with your skin or painted surfaces of the vehicle. Flush contaminated areas immediately with plenty of water. Don't store new coolant or leave old coolant lying around where it's accessible to children or pets - they're attracted by its sweet smell. Ingestion of even a small amount of coolant can be fatal! Wipe up garage floor and drip pan spills immediately. Keep antifreeze containers covered and repair cooling system leaks as soon as they're noticed.

❋❋ CAUTION:

Never mix green colored ethylene glycol antifreeze with Mopar 5 year antifreeze because doing so will destroy the efficiency of the Mopar 5 year antifreeze. Check the coolant reservoir under the hood to determine what type coolant you have. Always refill with the correct coolant.

4.5 Turn the oil filler cap counterclockwise to remove it

➡**Note: Non-toxic antifreeze is now manufactured and available at local auto parts stores, but even this type must be disposed of properly.**

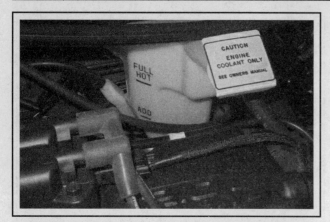

4.9 Maintain the coolant level near the FULL HOT mark on the reservoir

8   All vehicles covered by this manual are equipped with a pressurized coolant recovery system. A white plastic coolant reservoir is located at the rear of the engine compartment, on the center of the firewall.

9   The coolant level in the reservoir should be checked regularly.

❋❋ WARNING:

Do not remove the cooling system pressure cap to check the coolant level when the engine is warm!

The level in the reservoir varies with the temperature of the engine. When the engine is cold, the coolant level should be slightly above the ADD mark on the reservoir. Once the engine has warmed up, the level should be at or near the FULL HOT mark. If it isn't, allow the engine to cool, then remove the cap from the tank and add a 50/50 mixture of ethylene glycol based antifreeze and water (see illustration).

10 Drive the vehicle and recheck the coolant level. If only a small amount of coolant is required to bring the system up to the proper level, water can be used. However, repeated additions of water will dilute the antifreeze and water solution. In order to maintain the proper ratio of antifreeze and water, always top up the coolant level with the correct mixture. Don't use rust inhibitors or additives. An empty plastic milk jug or bleach bottle makes an excellent container for mixing coolant.

11 If the coolant level drops consistently, there may be a leak in the system. Inspect the radiator, hoses, filler cap, drain plugs and water pump (see Section 10). If no leaks are noted, have the pressure cap pressure tested by a service station.

4.17 Brake fluid level, indicated on the translucent white plastic brake fluid reservoir, should be kept at the upper (FULL) mark

4.14 Flip up the cap to add washer fluid

12 If you have to remove the pressure cap, wait until the engine has cooled completely, then wrap a thick cloth around the cap and turn it to the first stop. If coolant or steam escapes, or if you hear a hissing noise, let the engine cool down longer, then remove the cap.

13 Check the condition of the coolant as well. It should be relatively clear. If it's brown or rust colored, the system should be drained, flushed and refilled. Even if the coolant appears to be normal, the corrosion inhibitors wear out, so it must be replaced at the specified intervals.

WINDSHIELD WASHER FLUID

▶ **Refer to illustration 4.14**

14 The fluid for the windshield is stored in a plastic reservoir. The reservoir level should be maintained about one inch (25 mm) below the filler cap. The reservoir is accessible after opening the hood and is located on the right (passenger's) side of the engine compartment, in front of the strut tower (see illustration).

15 In milder climates, plain water can be used in the reservoir, but it should be kept no more than two-thirds full to allow for expansion if the water freezes. In colder climates, use windshield washer system antifreeze, available at any auto parts store, to lower the freezing point of the fluid. Mix the antifreeze with water in accordance with the manufacturer's directions on the container.

❋❋ CAUTION:

DO NOT use cooling system antifreeze - it will damage the vehicle's paint. To help prevent icing in cold weather, warm the windshield with the defroster before using the washer.

BRAKE AND CLUTCH FLUID

▶ **Refer to illustration 4.17**

16 The brake fluid reservoir is located on top of the brake master cylinder on the driver's side of the engine compartment near the firewall. The clutch fluid reservoir is mounted on the firewall, next to the brake master cylinder.

17 The fluid level should be maintained at the upper (FULL or MAX) mark on either reservoir (see illustration).

18 If additional fluid is necessary to bring the level up, use a rag to clean all dirt off the top of the reservoir to prevent contamination of the system. Also, make sure all painted surfaces around the reservoir are covered, since brake fluid will ruin paint. Carefully pour new, clean

brake fluid obtained from a sealed container into the reservoir. Be sure the specified fluid is used; mixing different types of brake fluid can cause damage to the system. See *Recommended lubricants and fluids* at the end of this Chapter or your owner's manual.

19 At this time the fluid and the master cylinder should be inspected for contamination. Normally the brake hydraulic system won't need periodic draining and refilling, but if rust deposits, dirt particles or water droplets are observed in the fluid, the system should be dismantled, cleaned and refilled with fresh fluid. Over time brake fluid will absorb moisture from the air. Moisture in the fluid lowers the fluid boiling point; if the fluid boils, the brakes will become ineffective. Normal brake fluid is clear in color. If the brake fluid is dark brown in color, it's a good idea to replace it (see Chapter 9).

20 Reinstall the fluid reservoir cap.

21 The brake fluid in the master cylinder will drop slightly as the brake lining material at each wheel wears down during normal operation. If the master cylinder requires repeated replenishing to maintain the correct level, there is a leak in the brake system that should be corrected immediately. Check all brake lines and connections, along with the calipers (disc brakes), wheel cylinders (drum brakes) and power brake booster (see Section 17 and Chapter 9 for more information).

22 If you discover that the reservoir is empty or nearly empty, the system should be thoroughly inspected, refilled and then bled (see Chapter 8 for clutch system bleeding and Chapter 9 for brake system bleeding).

5   Tire and tire pressure checks (every 250 miles or weekly)

▶ **Refer to illustrations 5.2, 5.3, 5.4a, 5.4b and 5.8**

1   Periodic inspection of the tires may spare you the inconvenience of being stranded with a flat tire. It can also provide you with vital information regarding possible problems in the steering and suspension systems before major damage occurs.

2   Original tires on this vehicle are equipped with 1/2-inch (13 mm) wide bands that will appear when tread depth reaches 1/16-inch (1.5 mm), at which point the tires can be considered worn out. Tread wear can be monitored with a simple, inexpensive device known as a tread depth indicator (see illustration).

3   Note any abnormal tread wear (see illustration). Tread pattern irregularities such as cupping, flat spots and more wear on one side than the other are indications of front end alignment and/or balance problems. If any of these conditions are noted, take the vehicle to a tire shop or service station to correct the problem.

4   Look closely for cuts, punctures and embedded nails or tacks. Sometimes a tire will hold air pressure for a short time or leak down

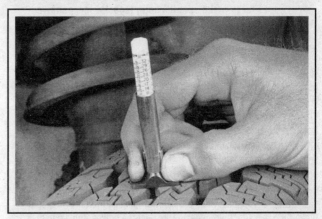

5.2 Use a tire tread depth indicator to monitor tire wear - they are available at auto parts stores or service stations and are relatively inexpensive

UNDERINFLATION

CUPPING

Cupping may be caused by:
- Underinflation and/or mechanical irregularities such as out-of-balance condition of wheel and/or tire, and bent or damaged wheel.
- Loose or worn steering tie-rod or steering idler arm.
- Loose, damaged or worn front suspension parts.

OVERINFLATION

INCORRECT TOE-IN OR EXTREME CAMBER

FEATHERING DUE TO MISALIGNMENT

5.3 This chart will help you determine the condition of the tires, the probable cause(s) of abnormal wear and the corrective action necessary

5.4a If a tire loses air on a steady basis, check the valve stem core first to make sure it's snug (special inexpensive wrenches are commonly available at auto parts stores)

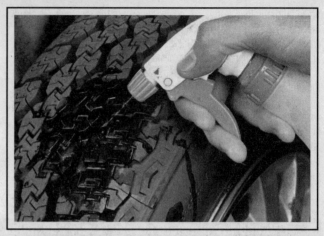

5.4b If the valve stem core is tight, raise the corner of the vehicle with the low tire and spray a soapy water solution onto the tread as the tire is turned slowly - leaks will cause small bubbles to appear

very slowly after a nail has embedded itself in the tread. If a slow leak persists, check the valve stem core to make sure it's tight (see illustration). Examine the tread for an object that may have embedded itself in the tire or for a "plug" that may have begun to leak (radial tire punctures are repaired with a plug that's installed in a puncture). If a puncture is suspected, it can be easily verified by spraying a solution of soapy water onto the puncture area (see illustration). The soapy solution will bubble if there's a leak. Unless the puncture is unusually large, a tire shop or service station can usually repair the tire.

5 Carefully inspect the inner sidewall of each tire for evidence of brake fluid leakage. If you see any, inspect the brakes immediately.

6 Correct air pressure adds miles to the lifespan of the tires, improves mileage and enhances overall ride quality. Tire pressure cannot be accurately estimated by looking at a tire, especially if it's a radial. A tire pressure gauge is essential. Keep an accurate gauge in the vehicle. The pressure gauges attached to the nozzles of air hoses at gas stations are often inaccurate.

7 Always check tire pressure when the tires are cold. Cold, in this case, means the vehicle has not been driven over a mile in the three hours preceding a tire pressure check. A pressure rise of four to eight pounds is not uncommon once the tires are warm.

8 Unscrew the valve cap protruding from the wheel or hubcap and push the gauge firmly onto the valve stem (see illustration). Compare the reading on the gauge to the recommended tire pressure shown on the placard on the end of the driver's side door. Be sure to reinstall the valve cap to keep dirt and moisture out of the valve stem mechanism.

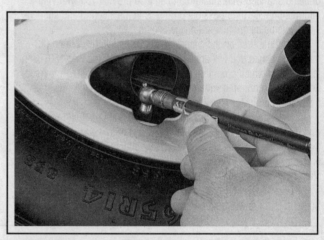

5.8 To extend the life of the tires, check the air pressure at least once a week with an accurate gauge (don't forget the spare!)

Check all four tires and, if necessary, add enough air to bring them up to the recommended pressure.

9 Don't forget to keep the spare tire inflated to the specified pressure (refer to your owner's manual or the tire sidewall). Note that the pressure recommended for the compact spare is higher than for the tires on the vehicle.

6 Automatic transaxle fluid level check (every 250 miles or weekly)

▶ **Refer to illustrations 6.3 and 6.4**

1 Fluid inside the transaxle should be at normal operating temperature to get an accurate reading on the dipstick. This is done by driving the vehicle for several miles, making frequent starts and stops to allow the transaxle to shift through all gears.

2 Park the vehicle on a level surface and apply the parking brake. With the engine running, apply the brakes and place the gear selector lever momentarily in Reverse, then Drive and repeat the sequence again ending with the gear selector in the Park position.

3 With the engine still running, locate the transaxle fluid dipstick near the air filter housing. The dipstick marked "TRANS FLUID" (see illustration). Remove the dipstick and wipe the fluid from the end with a clean rag.

4 Insert the dipstick back into the transaxle until the cap seats completely. Remove the dipstick again and note the fluid level on the end. The level should be in the area marked HOT (see illustration). If the fluid isn't hot (temperature approximately 100-degrees F), the level should be in the area marked WARM.

5 If the fluid level is at or below the ADD mark on the dipstick, add just enough of the specified fluid (see *Recommended lubricants and*

6.3 The automatic transaxle dipstick is located at the left end of the engine, near the air filter housing, and is clearly marked

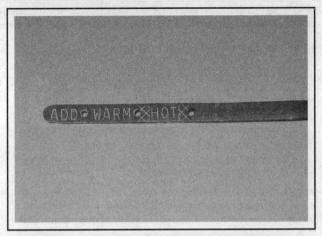

6.4 Check the fluid with the transaxle at normal operating temperature - the level should be kept in the HOT range

fluids at the end of this Chapter) to raise the level to within the marks indicated for the appropriate temperature. Fluid should be slowly added into the dipstick tube, using a funnel to prevent spills.

6   DO NOT overfill the transaxle. Never allow the fluid level to go above the upper end of the cross-hatched area on the dipstick - it could cause internal transaxle damage. The best way to prevent overfilling is to add fluid a little at a time, driving the vehicle and checking the level between additions.

7   Use only transaxle fluid specified by the manufacturer. This information can be found in the *Recommended lubricants and fluids* Section at the end of this Chapter or in your owner's manual.

8   The condition of the fluid should also be checked along with the level. If it's a dark reddish-brown color, or if it smells burned, it should be changed. If you're in doubt about the condition of the fluid, purchase some new fluid and compare the two for color and odor.

7   Power steering fluid level check (every 250 miles or weekly)

♦ **Refer to illustrations 7.2 and 7.5**

1   Unlike manual steering, the power steering system relies on hydraulic fluid that may, over a period of time, require replenishing.

2   The fluid reservoir for the power steering pump is located on the right (passenger's) side of the engine compartment (see illustration).

3   The power steering fluid level can be checked with the engine either hot or cold.

4   With the engine off, use a rag to clean the reservoir cap and the area around the cap. This will help prevent dirt from falling into the reservoir when the cap is removed.

5   Turn and pull out the reservoir cap, which has a dipstick attached

to it. Wipe the fluid off the dipstick with a clean rag. Reinstall the cap, then remove it again and note the fluid level. It should be at the appropriate mark on the dipstick in relation to the fluid temperature (see illustration).

6   If additional fluid is required, pour the specified type fluid (see *Recommended lubricants and fluids* at the end of this Chapter or your owner's manual) directly into the reservoir using a funnel to prevent spills.

7   If the reservoir requires frequent topping up, all power steering hoses, hose connections, the power steering pump and the steering gear should be carefully examined for leaks.

7.2 The power steering fluid reservoir is located on the right (passenger's) side of the engine compartment

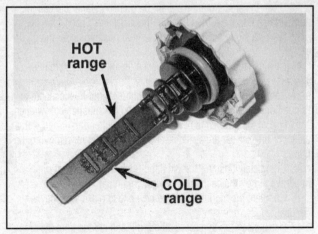

7.5 The power steering fluid dipstick is marked for checking the fluid level cold or hot

8 Engine oil and filter change (every 3,000 miles or 3 months)

▶ **Refer to illustrations 8.3, 8.8, 8.13 and 8.18**

1 Frequent oil changes are the most important preventive maintenance procedures that can be performed by the home mechanic. When engine oil ages, it gets diluted and contaminated, which ultimately leads to premature engine wear.

2 Although some sources recommend oil filter changes every other oil change, a new filter should be installed every time the oil is changed.

3 Gather together all necessary tools and materials before beginning this procedure (see illustration).

➡ **Note: To avoid rounding off the corners of the drain plug, use a box-end type wrench or socket. In addition, you should have plenty of clean rags and newspapers handy to mop up any spills.**

4 Raise the front of the vehicle and support it securely on jackstands.

✳✳ WARNING:

Never work under a vehicle that is supported only by a jack!

5 If this is your first oil change on the vehicle, familiarize yourself with the locations of the oil drain plug and the oil filter. Since the engine and exhaust components will be warm during the actual work, it's a good idea to figure out any potential problems beforehand.

6 Allow the engine to warm up to normal operating temperature. If oil or tools are needed, use the warm-up time to gather everything necessary for the job. The correct type of oil to buy for your application can be found in the *Recommended lubricants and fluids* Section at the end of this Chapter or your owner's manual.

7 Move all necessary tools, rags and newspapers under the vehicle. Place a drain pan capable of holding at least 5 quarts under the drain plug. Keep in mind that the oil will initially flow from the engine with some force, so position the pan accordingly.

8 Being careful not to touch any of the hot exhaust components, use the breaker bar and socket or box-end wrench to remove the drain plug (see illustration). Depending on how hot the oil is, you may want to wear gloves while unscrewing the plug the final few turns.

9 Allow the oil to drain into the pan. It may be necessary to move the pan further under the engine when the oil flow slows to a trickle.

10 After all the oil has completely drained, clean the plug thoroughly with a rag. Small metal particles may cling to it and would immediately contaminate the new oil.

11 Clean the area around the drain plug opening and reinstall the plug. Tighten it to the torque listed in this Chapter's Specifications.

12 Next, carefully move the drain pan into position under the oil filter.

13 Now use the filter wrench to loosen the oil filter in a counterclockwise direction (see illustration).

14 Sometimes the oil filter is on so tight it cannot be loosened, or it's positioned in an area inaccessible with a conventional filter wrench. Other type of tools which fit over the end of the filter and turned with a ratchet or breaker bar are available and may be better for removing the filter.

15 Completely unscrew the old filter. Be careful, it's full of oil. Empty the old oil inside the filter into the drain pan.

16 Compare the old filter with the new one to make sure they're identical.

17 Use a clean rag to remove all oil, dirt and sludge from the area where the oil filter seals on the engine. Check the old filter to make sure the rubber gasket isn't stuck to the engine mounting surface.

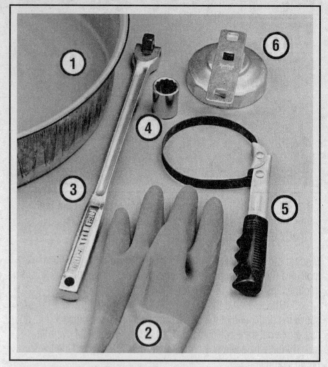

8.3 These tools are required when changing the engine oil and filter

1 **Drain pan** - *It should be fairly shallow in depth, but wide to prevent spills and capable of holding at least 5 quarts*

2 **Rubber gloves** - *When removing the drain plug and filter, you will get oil on your hands (the gloves will prevent burns)*

3 **Breaker bar** - *Sometimes the oil drain plug is tight, and a long breaker bar is needed to loosen it*

4 **Socket** - *To be used with the breaker bar or a ratchet (must be the correct size to fit the drain plug)*

5 **Filter wrench** - *This is a metal band-type wrench, which requires clearance around the filter to be effective*

6 **Filter wrench** - *This type fits on the bottom of the filter and can be turned with a ratchet or breaker bar (different size wrenches are available for different types of filters)*

18 Apply a light coat of clean engine oil to the rubber gasket on the new oil filter (see illustration).

19 Attach the new filter to the engine, following the tightening directions printed on the filter canister or packing box. Most filter manufacturers recommend against using a filter wrench due to the possibility of overtightening and damage to the seal.

20 Remove all tools and materials from under the vehicle, being careful not to spill the oil in the drain pan. Lower the vehicle.

21 Working inside the engine compartment, locate and remove the oil filler cap from the engine valve cover (see illustration 4.5).

22 Using a funnel to prevent spills, pour the specified type and amount of new oil required (see the *Specifications* Section in the end of this Chapter) into the engine. Wait a few minutes to allow the oil to drain down to the pan, then check the level on the dipstick (see Section 4 if necessary). If the oil level is at or above the lower mark on the dipstick, start the engine and allow the new oil to circulate.

23 Run the engine for only about a minute, then shut it off. Immediately look under the vehicle and check for leaks at the oil pan drain

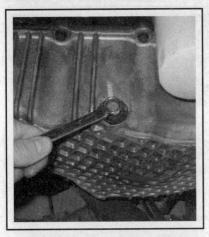

8.8 To avoid rounding off the corners, use the correct size box-end wrench or a socket to remove the engine oil drain plug

8.13 Since the oil filter is probably very tight, you'll need a special wrench for removal

8.18 Lubricate the oil filter gasket with clean engine oil before installing the filter on the engine

plug and around the oil filter. If either one is leaking, tighten it with a bit more force.

24 With the new oil circulated and the filter now completely full, wait a few minutes for the oil to drain back down into the pan then recheck the oil level on the dipstick. If necessary, add enough oil to bring the level to the upper mark on the dipstick. DO NOT overfill!

25 During the first few trips after an oil change, make it a point to check for leaks and keep a close watch on the oil level.

26 The old oil drained from the engine cannot be reused in its present state and should be disposed of. Check with your local auto parts store, disposal facility or environmental agency to see if they will accept the oil for recycling. After the oil has cooled it can be drained into a container (capped plastic jugs, topped bottles, milk cartons, etc.) for transport to one of these disposal sites. Don't dispose of the oil by pouring it on the ground or down a drain!

9 Battery check, maintenance and charging (every 7,500 miles or 6 months)

❊❊ WARNING:

Certain precautions must be followed when checking and servicing the battery. Hydrogen gas, which is highly explosive, is produced by the battery. Keep lighted tobacco, open flames, bare light bulbs or other possible sources of ignition away from the battery. Furthermore, the electrolyte inside the battery is sulfuric acid which is highly corrosive and can burn your skin and cause severe injury to your eyes. Always wear eye protection! It will also destroy clothing and ruin painted surfaces.

SERVICING

▶ **Refer to illustrations 9.1, 9.5, 9.6, 9.7a, 9.7b and 9.8**

1 A routine preventive maintenance program for the battery in your vehicle is the only way to ensure quick and reliable starts. But before performing any battery maintenance, make sure that you have the proper equipment necessary to work safely around the battery (see illustration).

2 Prior to servicing the battery, always turn the engine and all accessories off and disconnect the cable from the negative terminal of the battery.

3 The battery is located in the left side of the engine compartment.

4 Inspect the external condition of the battery. Check the battery case for cracks or other damage.

5 If corrosion, which looks like white, fluffy deposits (see illustra-

tion) is evident, particularly around the terminals, the battery should be removed for cleaning. Loosen the cable clamp nuts, being careful to remove the ground (negative) cable first, and slide them off the terminals. Then remove the battery (see Step 8).

6 Check the entire length of each cable for cracks, worn insulation and frayed conductors (see illustration).

7 Clean the cable clamps thoroughly with a battery brush (see illustration) and a solution of warm water and baking soda. Wash the terminals and the top of the battery case with the same solution but make sure that the solution doesn't get into the battery. When cleaning the cables, terminals and battery top, wear safety goggles and rubber gloves to prevent any solution from coming in contact with your eyes or hands. Wear old clothes too - even diluted, sulfuric acid splashed onto clothes will burn holes in them. If the terminals have been extensively corroded, clean them up with a terminal cleaner (see illustration). Thoroughly wash all cleaned areas with plain water.

8 If it's necessary to remove the battery, loosen the clamp bolt and remove the battery from the engine compartment (see illustration).

9 Inspect the battery carrier. If it's dirty or covered with corrosion, clean it with the same solution of warm water and baking soda and rinse it with clean water.

10 If the battery is a maintenance-type, it has removable cell caps which allow you to add water (use distilled water only) to the battery when the electrolyte level gets low.

11 If you are not sure what type of battery you have (some maintenance-types have recessed cell caps that resemble maintenance-free batteries), one simple way to confirm your type of battery is to look for a built-in hydrometer. Most maintenance-free batteries have built-in

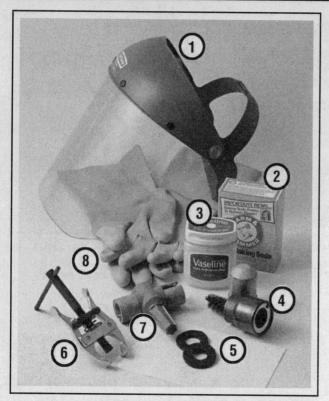

9.1 Tools and materials required for battery maintenance

1 *Face shield/safety goggles* - When removing corrosion with a brush, the acidic particles can easily fly up into your eyes
2 *Baking soda* - A solution of baking soda and water can be used to neutralize corrosion
3 *Petroleum jelly* - A layer of this on the battery posts will help prevent corrosion
4 *Battery post/cable cleaner* - This wire brush cleaning tool will remove all traces of corrosion from the battery posts and cable clamps
5 *Treated felt washers* - Placing one of these on each post, directly under the cable clamps, will help prevent corrosion
6 *Puller* - Sometimes the cable clamps are very difficult to pull off the posts, even after the nut/bolt has been completely loosened. This tool pulls the clamp straight up and off the post without damage
7 *Battery post/cable cleaner* - Here is another cleaning tool which is a slightly different version of Number 4 above, but it does the same thing
8 *Rubber gloves* - Another safety item to consider when servicing the battery; remember that's acid inside the battery!

hydrometers that indicate the state of charge by the color displayed in the hydrometer window, since measuring the specific gravity of the electrolyte is not possible. Also check for cut-outs near the cell caps - if the caps can be removed, cut-outs are usually provided to assist with prying off the caps.

12 If your battery is a maintenance-type, remove the cell caps and check the level of the electrolyte. It should be up to the split-ring inside the battery. If the level is low, add distilled water (distilled water is mineral-free, tap water contains minerals that will shorten the life of your battery) to bring the electrolyte up to the proper level.

13 Install the battery and tighten the hold-down clamp securely.

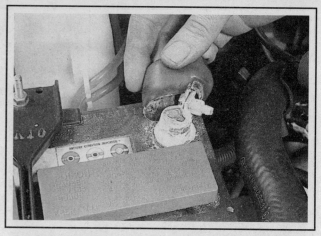

9.5 Battery terminal corrosion usually appears as a light, fluffy powder (typical)

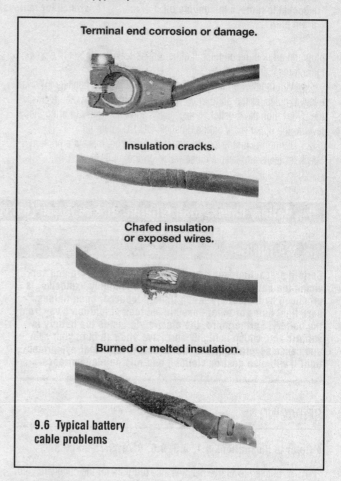

Terminal end corrosion or damage.

Insulation cracks.

Chafed insulation or exposed wires.

Burned or melted insulation.

9.6 Typical battery cable problems

CHARGING

✻ WARNING:

When batteries are being charged, hydrogen gas, which is very explosive and flammable, is produced. Do not smoke or allow open flames near a charging or a recently charged battery. Wear eye protection when near the battery during charging. Also, make sure the charger is unplugged before connecting or disconnecting the battery from the charger.

9.7a When cleaning the cable clamps, all corrosion must be removed

9.7b Regardless of the type of tool used on the battery post, a clean, shiny surface should be the result (the inside of the clamp is tapered to match the taper on the post, so don't remove too much material

14 Slow-rate charging is the best way to restore a battery that's discharged to the point where it will not start the engine. It's also a good way to maintain the battery charge in a vehicle that's only driven a few miles between starts. Maintaining the battery charge is particularly important in the winter when the battery must work harder to start the engine and electrical accessories that drain the battery are in greater use.

15 It's best to use a one or two-amp battery charger (sometimes called a "trickle" charger). They are the safest and put the least strain on the battery. They are also the least expensive. For a faster charge, you can use a higher amperage charger, but don't use one rated more than 1/10th the amp/hour rating of the battery. Rapid boost charges that claim to restore the power of the battery in one to two hours are hardest on the battery and can damage batteries not in good condition. This type of charging should only be used in emergency situations.

16 The average time necessary to charge a battery should be listed in the instructions that come with the charger. As a general rule, a trickle charger will charge a battery in 12 to 16 hours.

17 Detach the cables from the battery terminals (negative first, positive last).

18 On maintenance-type batteries, remove the cell caps. Make sure the electrolyte level is OK before beginning to charge the battery. Cover the holes with a clean cloth to prevent spattering electrolyte.

19 On batteries with the terminals located on the side, install bolts (with the appropriate thread size and pitch) in the terminals so the charger can be attached.

20 Connect the battery charger leads to the battery posts (positive to positive, negative to negative), then plug in the charger. Make sure it is set at 12 volts if it has a selector switch. If the battery charger does not have a built-in timer, it's a good idea to use one in case you forget - so you won't over charge the battery.

21 If you're using a charger with a rate higher than two amps, check the battery regularly during charging to make sure it doesn't overheat. If you're using a trickle charger, you can safely let the battery charge overnight after you've checked it regularly for the first couple of hours.

22 If the battery has removable cell caps, measure the specific gravity with a hydrometer every hour during the last few hours of the charging cycle. Hydrometers are available inexpensively from auto parts stores - follow the instructions that come with the hydrometer. Consider the battery charged when there's no change in the specific gravity read-

ing for two hours and the electrolyte in the cells is outgassing (bubbling) freely. The specific gravity reading from each cell should be very close to the others. If not, the battery probably has a bad cell(s).

23 Most batteries with sealed tops have built-in hydrometers on the top that indicate the state of charge by the color displayed in the hydrometer window. Normally, a bright-colored hydrometer indicates a full charge and a dark hydrometer indicates the battery still needs charging. Check the battery manufacturer's instructions to be sure you know what the colors mean.

➡**Note: It may be necessary to jiggle the battery to bring the test indicator fluid into view.**

24 If the battery has a sealed top and does not have a built-in hydrometer, you can hook up a voltmeter across the battery terminals to check the charge. A fully charged battery should read approximately 12.6 volts or higher.

25 Further information on the battery and jump starting can be found in Chapter 5 and at the front of this manual, respectively.

9.8 The battery is secured by a clamp at its base - to remove the battery, loosen the bolt, slide the clamp back and lift the battery out

10 Cooling system check (every 7,500 miles or 6 months)

▶ **Refer to illustration 10.4**

> ❊❊ **WARNING:**
>
> The electric cooling fan(s) on these models can activate at any time the ignition switch is in the ON position. Make sure the ignition is OFF when working in the vicinity of the fan(s).

1   Many major engine failures can be attributed to a faulty cooling system. If the vehicle is equipped with an automatic transaxle, a transmission fluid cooler is incorporated inside the radiator bottom tank.

2   The cooling system should be checked with the engine cold. Do this before the vehicle is driven for the day or after it has been shut off for three or four hours and the upper radiator hose feels cool to the touch.

3   Remove the cooling system pressure cap (see the underhood photo at the beginning of this Chapter) and thoroughly clean the cap with water. Also clean the filler neck. All traces of corrosion should be removed.

4   Carefully check the upper and lower radiator hoses along with the smaller diameter heater hoses. Inspect the entire length of each hose, replacing any that are cracked, swollen or deteriorated (see illustration). Cracks may become more apparent when a hose is squeezed.

5   Also check that all hose connections are tight. If the vehicle came equipped with spring-type hose clamps which lose their tension over time, replace them with the more reliable screw-type clamps when new hoses are installed. A leak in the cooling system will usually show up as white or rust-colored deposits on the areas adjoining the leak.

6   Use compressed air, water or a soft brush to remove bugs, leaves, and other debris from the front of the radiator or air conditioning condenser. Be careful not to damage the delicate cooling fins, or cut yourself on them.

7   Finally, have the cap and system pressure tested. If you do not have a pressure tester available, most gas stations and repair shops will do this for a minimal charge.

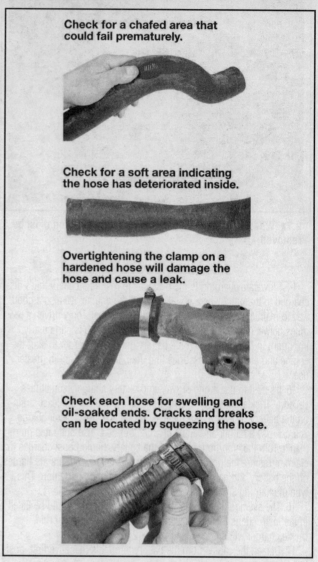

Check for a chafed area that could fail prematurely.

Check for a soft area indicating the hose has deteriorated inside.

Overtightening the clamp on a hardened hose will damage the hose and cause a leak.

Check each hose for swelling and oil-soaked ends. Cracks and breaks can be located by squeezing the hose.

10.4 Hoses, like drivebelts, have a habit of failing at the worst possible time - to prevent the inconvenience of a blown radiator or heater hose, inspect them carefully as shown here

11 Underhood hose check and replacement (every 7,500 miles or 6 months)

> ❊❊ **WARNING:**
>
> Replacement of air conditioning hoses must be left to a dealer service department or air conditioning shop equipped to depressurize the system safely. Never remove air conditioning components or hoses until the system has been depressurized.

GENERAL

1   High temperatures under the hood can cause the deterioration of the rubber and plastic hoses used for engine, accessory and emission systems operation. Periodic inspection should be made for cracks, loose clamps, material hardening and leaks.

2   Information specific to the cooling system hoses can be found in Section 10.

3   Some hoses use clamps to secure the hoses to fittings. Where clamps are used, check to be sure that they haven't lost their tension, allowing the hose to leak. Where clamps are not used, make sure the hose hasn't expanded and/or hardened where it slips over the fitting, allowing it to leak.

VACUUM HOSES

4   It's quite common for vacuum hoses, especially those in the emissions system, to be color coded or identified by colored stripes

molded into the hose. Various systems require hoses with different wall thickness, collapse resistance and temperature resistance. When replacing hoses, make sure the new ones are made of the same material as the original.

5   Often the only effective way to check a hose is to remove it completely from the vehicle. Where more than one hose is removed, be sure to label the hoses and their attaching points to insure proper reattachment.

6   Include plastic T-fittings in the check of vacuum hoses. Check the fittings for cracks and the hose where it fits over the fitting for enlargement, which could cause leakage.

7   A small piece of vacuum hose (1/4-inch [6 mm] inside diameter) can be used as a stethoscope to detect vacuum leaks. Hold one end of the hose to your ear and probe around vacuum hoses and fittings, listening for the "hissing" sound characteristic of a vacuum leak.

✳ WARNING:

When probing with the vacuum hose stethoscope, be careful not to allow your body or the hose to come into contact with moving engine components such as the drivebelt, cooling fan, etc.

FUEL HOSE

✳ WARNING:

Gasoline is extremely flammable, so take extra precautions when you work on any part of the fuel system. Don't smoke or allow open flames or bare light bulbs near the work area, and don't work in a garage where a gas-type appliance (such as a water heater or clothes dryer) is present. If you spill any fuel on your skin, rinse it off immediately with soap and water. When you perform any kind of work on the fuel system, wear fuel-resistant gloves and safety glasses, and have a Class B type fire extinguisher on hand. Before working on any part of the fuel system, relieve the fuel system pressure (see Chapter 4).

8   Check all rubber fuel hoses for damage and deterioration. Check especially for cracks in areas where the hose bends and just before quick-connect fittings (see Chapter 4).

9   High quality fuel line, specifically designed for fuel injection systems, must be used for fuel line replacement.

✳ WARNING:

Never use vacuum line, clear plastic tubing or water hose for fuel lines.

BRAKE HOSES

10   The hoses used to connect the brake calipers or wheel cylinders to the metal lines are subject to extreme working conditions. They must endure high hydraulic pressures, heat and still maintain flexibility. The brake hoses typically can be inspected without removing the wheels. Carefully examine each hose for leakage, cracks, bulging, delaminating and damage. If any damage is found, the hose must be replaced immediately (see Chapter 9).

FUEL AND BRAKE SYSTEM METAL LINES

11   Sections of metal line are often used for fuel line between the fuel tank and fuel injection system. Carefully check to be sure the line has not been bent and crimped and that no cracks have started in the line.

12   If a section of metal fuel line must be replaced, only seamless steel tubing should be used, since copper and aluminum tubing do not have the strength necessary to withstand normal engine operating vibration.

13   Check the metal brake lines where they enter the master cylinder and ABS unit (if equipped) for cracks in the lines or loose fittings. Any sign of brake fluid leakage calls for an immediate thorough inspection of the brake system.

12   Windshield wiper blade inspection and replacement (every 7,500 miles or 6 months)

▶ **Refer to illustrations 12.3, 12.5a and 12.5b**

1   The windshield wiper blade elements should be checked periodically for cracks and deterioration.

2   Road film can build up on the wiper blades and affect their efficiency, so they should be washed regularly with a mild detergent solution.

3   The action of the wiping mechanism can loosen the wiper arm retaining nuts, so they should be checked and tightened at the same time the wiper blades are checked (see illustration).

4   Lift the wiper blade away from the windshield.

12.3 Remove the cap and check the wiper arm mounting nut for tightness

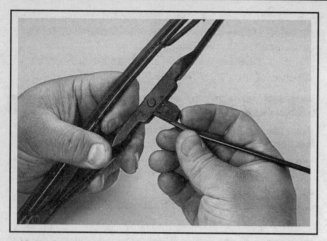

12.5a Depress the release lever . . .

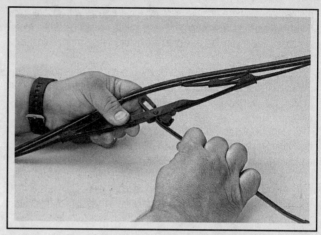

12.5b . . . then slide the wiper element down and out of the hook in the end of the arm

5   Press the release lever and slide the blade assembly out of the hook in the end of the wiper arm (see illustrations). Carefully rest the wiper arm on the windshield.

6   The rubber wiper element is secured to the blade assembly at one end of the blade element channel. Compress the locking feature on the element so it clears the tangs on the blade assembly channel claw and then slide the element out of the frame.

7   Installation is the reverse of removal. Make sure the rubber element and blade assembly is securely attached.

13 Tire rotation (every 7,500 miles or 6 months)

▶ **Refer to illustrations 13.2a and 13.2b**

1   The tires should be rotated at the specified intervals and whenever uneven wear is noticed. Since the vehicle will be raised and the tires removed, this is a good time to check the brakes also (see Section 17).

2   Radial tires must be rotated in a specific pattern (see illustrations).

➡**Note: Most vehicles are sold with non-directional radial tires, but some performance tires are directional, and have an arrow on the sidewall indicating the direction they must turn when mounted on the vehicle.**

3   See the information in *Jacking and towing* at the front of this manual for the proper procedures to follow when raising the vehicle and changing a tire; however, if the brakes are to be checked, don't apply the parking brake as stated. Make sure the tires are blocked to prevent the vehicle from rolling.

➡**Note: Prior to raising the vehicle, loosen all lug nuts a quarter turn.**

4   Preferably, the entire vehicle should be raised at the same time. This can be done on a hoist or by jacking up each corner of the vehicle and lowering it onto jackstands. Always use jackstands and make sure the vehicle is safely supported.

❊❊ **WARNING:**

Never work under a vehicle that is supported only by a jack!

5   After the tire rotation, check and adjust the tire pressures as necessary and tighten the wheel lug nuts to the torque listed in this Chapter's Specifications.

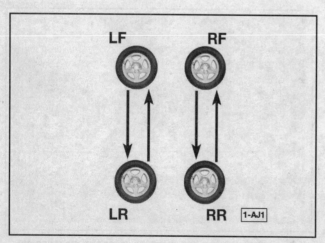

13.2a The recommended rotation pattern for directional radial tires

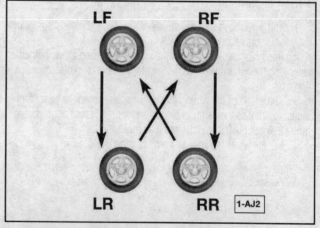

13.2b The recommended rotation pattern for non-directional radial tires

14 Steering and suspension check (every 7,500 miles or 6 months)

➡Note: The steering linkage and suspension components should be checked periodically. Worn or damaged suspension and steering components can result in excessive and abnormal tire wear, poor ride quality and vehicle handling and reduced fuel economy. For detailed illustrations of the steering and suspension components, refer to Chapter 10.

SHOCK ABSORBER/STRUT CHECK

1 Park the vehicle on level ground, turn the engine off and set the parking brake. Check the tire pressures.

2 Push down at one corner of the vehicle, then release it while noting the movement of the body. It should stop moving and come to rest in a level position within one or two bounces.

3 If the vehicle continues to move up-and-down or if it fails to return to its original position, a worn or weak strut assembly is probably the reason.

4 Repeat the above check at each of the three remaining corners of the vehicle.

5 Raise the vehicle and support it securely on jackstands.

6 Check the struts for evidence of fluid leakage. A light film of fluid is no cause for concern. Make sure that any fluid noted is from the struts and not from some other source. If leakage is noted, replace the struts as a set.

7 Check the struts to be sure that they are securely mounted and undamaged. Check the upper mounts for damage and wear. If damage or wear is noted, replace the shocks/struts as a set (front or rear).

8 If the struts must be replaced, refer to Chapter 10 for the procedure.

SUSPENSION AND STEERING CHECK

9 Raise the vehicle and support it securely on jackstands.

⁂ WARNING:

Never work under a vehicle that is supported only by a jack! Visually inspect the steering and suspension components (front and rear) for damage and distortion. Look for damaged seals, boots and bushings and leaks of any kind. Examine the bushings where the lower control arm meets the chassis and on the stabilizer bar connections.

10 Grasp each front tire at the front and rear edges, push in at the front, pull out at the rear and feel for play in the steering system components. If any freeplay is noted, check the tie-rod ends for looseness.

11 Check the balljoints using the procedure described in Chapter 10.

12 Additional steering and suspension system information and illustrations can be found in Chapter 10.

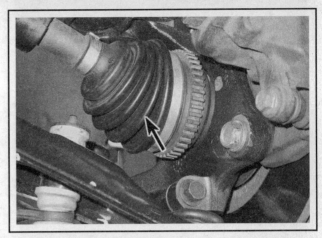

14.16 Check the inner and outer boot on each driveaxle for cracks and/or leaking grease

13 Inspect the steering gear boots for cracks as well as loose clamps. If you notice lubricant leaking from the boots, the rack seals have failed, in which case the steering gear will have to be replaced with a new or rebuilt unit, or overhauled by an automotive service technician.

DRIVEAXLE BOOT CHECK

▶ **Refer to illustration 14.16**

14 If the driveaxle boots are damaged, letting grease out and water and dirt in, serious (not to mention costly) damage can occur to the CV joints. The boots should be inspected very carefully at the recommended intervals or anytime the vehicle is raised.

15 Raise the front of the vehicle and support it securely on jackstands.

⁂ WARNING:

Never work under a vehicle that is supported only by a jack!

16 Place the transaxle in Neutral. While rotating the wheels, inspect the four driveaxle boots (two on each driveaxle) very carefully for cracks, tears, holes, deteriorated rubber and loose or missing clamps (see illustration). If the boots are dirty, wipe them clean before beginning the inspection.

17 If damage or deterioration is evident, replace the boots and check the CV joints for damage (see Chapter 8).

18 Place the transaxle in Park or in gear as applicable and lower the vehicle.

15 Exhaust system check (every 7,500 miles or 6 months)

▶ **Refer to illustrations 15.3a and 15.3b**

➡Note: Perform the following procedure with the engine cold.

1 Raise the vehicle and support it securely on jackstands.

⁂ WARNING:

Never work under a vehicle that is supported only by a jack!

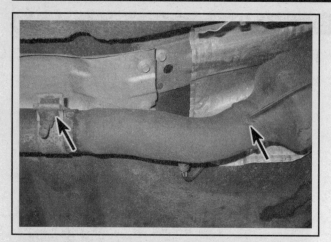

15.3a Check the exhaust system connections, clamps and welds for damage

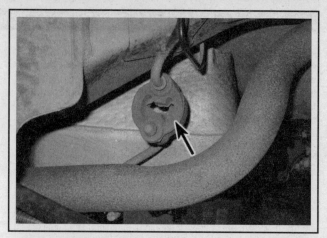

15.3b Also check the exhaust system rubber hangers for cracks and deterioration

2   With the engine cold (at least three hours after the vehicle has been driven), check the complete exhaust system from its starting point at the engine to the end of the tailpipe.

3   Check the pipes and connections for signs of leakage and/or corrosion indicating a potential failure. Make sure that all brackets and hangers are in good condition and tight (see illustrations).

4   At the same time, inspect the underside of the body for holes, corrosion and open seams which may allow exhaust gases to enter the passenger compartment. Seal all body openings with silicone sealant or body putty.

5   Rattles and other noises can often be traced to the exhaust system, especially the mounts and hangers. Try to move the pipes, muffler and catalytic converter. If the components can come into contact with the body, secure the exhaust system with new mounts.

6   This is also an ideal time to check the running condition of the engine by inspecting the very end of the tailpipe. The exhaust deposits here are an indication of the engine's state-of-tune. If the pipe is black and sooty or coated with white deposits, the engine may be in need of a tune-up (including a thorough fuel injection system inspection).

16   Manual transaxle lubricant level check (every 7,500 miles or 6 months)

▶ **Refer to illustration 16.3**

1   Manual transaxles do not have a fluid dipstick. The lubricant level is checked by removing the plug from the side of the transaxle case. The lubricant level should be checked with the engine cold and the vehicle level.

2   Raise the vehicle and support it securely on jackstands in a level position.

❋❋ WARNING:

Never work under a vehicle that is supported only by a jack!

3   Locate the rubber plug on the side of the transaxle differential near the driveaxle shaft (see illustration). Use a rag to clean it and the surrounding area.

4   Use a pliers or a screwdriver to remove the plug. If oil begins to run out, let it find its own level (presuming the vehicle is relatively level). If oil does not run out, insert your finger to feel the lubricant level. It should be within 3/16-inch (5 mm) of the bottom of the plug hole.

5   If the transaxle requires additional lubricant, use a funnel with a rubber tube or a syringe to pour or squeeze the recommended lubricant into the plug hole to restore the level. If you overfill It, let the fluid run out until it is level with the plug hole.

❋❋ CAUTION:

Use only the specified transaxle lubricant - see *Recommended*

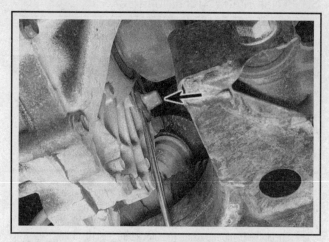

16.3 The manual transaxle check/fill plug is located on the side of the transaxle case

lubricants and fluids at the end of this Chapter or your owner's manual.

➡**Note: Most auto parts stores sell pumps that screw into the oil containers which make this job much easier and less messy.**

6   Install the plug and lower the vehicle. Test drive it and check for leaks.

17 Brake system check (every 15,000 miles or 12 months)

✵✵ WARNING:

Dust created by the brake system is harmful to your health. Never blow it out with compressed air and don't inhale any of it. An approved filtering mask should be worn when working on the brakes. Do not, under any circumstances, use petroleum-based solvents to clean brake parts. Use brake system cleaner only!

➡Note: For detailed photographs of the brake system, refer to Chapter 9.

1   In addition to the specified intervals, the brakes should be inspected every time the wheels are removed or whenever a defect is suspected. Any of the following symptoms could indicate a potential brake system defect: The vehicle pulls to one side when the brake pedal is depressed; the brakes make squealing or dragging noises when applied; brake travel is excessive; the pedal pulsates; brake fluid leaks, usually onto the inside of the tire or wheel.

2   The front disc brake pads have built-in wear indicators that will make a high-pitched squealing or scraping noise when they are worn to the replacement point. When you hear this noise, replace the pads immediately or expensive damage to the discs can result.

3   Loosen the wheel lug nuts.

4   Raise the vehicle and place it securely on jackstands.

5   Remove the wheels (see *Jacking and towing* at the front of this book, or your owner's manual, if necessary).

DISC BRAKES

▶ **Refer to illustration 17.6**

6   There are two pads - an outer and an inner - in each caliper. The pads are visible through small inspection holes in each caliper (see illustration).

7   Check the pad thickness by looking at each end of the caliper and through the inspection hole in the caliper body. If the lining material is less than the thickness listed in this Chapter's Specifications, replace the pads.

➡Note: Keep in mind that the lining material is riveted or bonded to a metal backing plate and the metal portion is not included in this measurement.

8   If it is difficult to determine the exact thickness of the remaining pad material by the above method, or if you are at all concerned about the condition of the pads, remove the caliper(s), then remove the pads from the calipers for further inspection (refer to Chapter 9).

9   Once the pads are removed from the calipers, clean them with brake cleaner and re-measure them.

10 Measure the disc thickness with a micrometer to make sure that it still has service life remaining. If any disc is thinner than the specified minimum thickness, replace it (refer to Chapter 9). Even if the disc has service life remaining, check its condition. Look for scoring, gouging and burned spots. If these conditions exist, remove the disc and have it resurfaced (see Chapter 9).

11 Before installing the wheels, check all brake lines and hoses for damage, wear, deformation, cracks, corrosion, leakage, bends and twists, particularly in the vicinity of the rubber hoses at the calipers.

12 Check the clamps for tightness and the connections for leakage. Make sure that all hoses and lines are clear of sharp edges, moving parts and the exhaust system. If any of the above conditions are noted, repair, reroute or replace the lines and/or fittings as necessary (see Chapter 9).

REAR DRUM BRAKES

▶ **Refer to illustrations 17.14 and 17.16**

13 Refer to Chapter 9 and remove the rear brake drums.

14 Note the thickness of the lining material on the rear brake shoes (see illustration) and look for signs of contamination by brake fluid and grease. If the lining material is within 1/16-inch (1.5 mm) of the recessed rivets or metal shoes, replace the brake shoes with new ones. The shoes should also be replaced if they are cracked, glazed (shiny lining surfaces) or contaminated with brake fluid or grease. See Chapter 9 for the replacement procedure.

15 Check the shoe return and hold-down springs and the adjusting mechanism to make sure they're installed correctly and in good condition. Deteriorated or distorted springs, if not replaced, could allow the

17.6 With the wheel off, check the thickness of the pads through the inspection hole

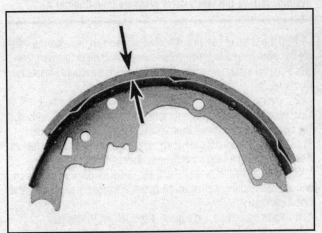

17.14 If the lining is bonded to the brake shoe, measure the lining thickness from the outer surface to the metal shoe; if the lining is riveted to the shoe, measure from the lining outer surface to the rivet head

linings to drag and wear prematurely.

16 Check the wheel cylinders for leakage by carefully peeling back the rubber boots (see illustration). If brake fluid is noted behind the boots, the wheel cylinders must be replaced (see Chapter 9).

17 Check the drums for cracks, score marks, deep scratches and hard spots, which will appear as small discolored areas. If imperfections cannot be removed with emery cloth, the drums must be resurfaced by an automotive machine shop (see Chapter 9 for more detailed information).

18 Install the brake drums.

19 Install the wheels and lug nuts.

20 Remove the jackstands and lower the vehicle.

21 Tighten the wheel lug nuts to the torque listed in this Chapter's Specifications.

BRAKE BOOSTER CHECK

22 Sit in the driver's seat and perform the following sequence of tests.

23 With the brake fully depressed, start the engine - the pedal should move down a little when the engine starts.

24 With the engine running, depress the brake pedal several times - the travel distance should not change.

25 Depress the brake, stop the engine and hold the pedal in for about 30 seconds - the pedal should neither sink nor rise.

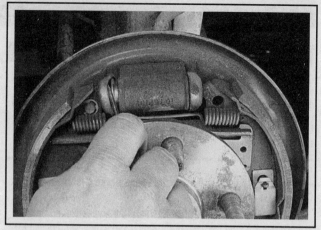

17.16 Pull the boot away from the cylinder and check for fluid leakage

26 Restart the engine, run it for about a minute and turn it off. Then firmly depress the brake several times - the pedal travel should decrease with each application.

27 If your brakes do not operate as described above when the preceding tests are performed, the brake booster has failed. Refer to Chapter 9 for the replacement procedure.

18 Fuel system hoses and connections check (every 15,000 miles or 12 months)

▶ Refer to illustration 18.5

❈❈ WARNING:

Gasoline is extremely flammable, so take extra precautions when you work on any part of the fuel system. Don't smoke or allow open flames or bare light bulbs near the work area, and don't work in a garage where a gas-type appliance (such as a water heater or clothes dryer) is present. If you spill any fuel on your skin, rinse it off immediately with soap and water. When you perform any kind of work on the fuel system, wear fuel-resistant gloves and safety glasses, and have a Class B type fire extinguisher on hand. Before working on any part of the fuel system, relieve the fuel system pressure (see Chapter 4).

1 If the smell of gasoline is noticed while driving, or after the vehicle has been parked in the sun, the fuel system and evaporative emissions control system (see Section 26) should be thoroughly inspected immediately.

2 The fuel system is under pressure even when the engine is off. Consequently, the fuel system must be depressurized before servicing the system (see Chapter 4). Even after depressurization, if any fuel lines are disconnected for servicing, be prepared to catch some fuel as it spills out. Plug all disconnected fuel lines immediately.

3 Remove the fuel tank filler cap and check for damage, corrosion and a proper sealing imprint on the gasket. Replace the cap with a new one if necessary.

4 Raise the vehicle and support it securely on jackstands.

❈❈ WARNING:

Never work under a vehicle that is supported only by a jack!

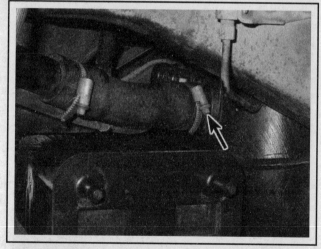

18.5 Check the fuel filler neck-to-tank hose and clamp

5 Inspect the fuel tank and filler neck for punctures, cracks and other damage. The hose connection between the filler neck and the tank is especially critical (see illustration). Sometimes the filler neck hose will leak due to loose clamps or deteriorated rubber; problems a home mechanic can usually rectify.

6 Carefully inspect all rubber hoses and metal lines leading to and-from the fuel tank. Check for loose connections, deteriorated hoses, crimped lines and damage of any kind. Follow the lines up to the front of the vehicle, carefully inspecting them all the way. Repair or replace damaged sections as necessary (see Chapter 4).

19 Drivebelt check, adjustment and replacement (every 15,000 miles or 12 months)

※※ WARNING:

The electric cooling fan(s) on these models can activate at any time the ignition switch is in the ON position. Make sure the ignition is OFF when working in the vicinity of the fan(s).

CHECK

▸ **Refer to illustrations 19.2, 19.3 and 19.5**

1   The drivebelts are located at the front of the engine and play an important role in the operation of the vehicle and its components. Due to their function and material makeup, the belts are prone to failure after a period of time and should be inspected periodically to prevent major damage. The alternator drivebelt is adjustable, while the power steering pump and air conditioning compressor (if equipped) drivebelt is adjusted by an automatic tensioner.

2   The drivebelts are very difficult to see from above. Apply the parking brake, loosen the right (passenger's side) front wheel lug nuts, raise the front of the vehicle and support it securely on jackstands. Remove the wheel, then remove the drivebelt splash shield (see illustration).

3   Use a flashlight to carefully check each belt. Check for a severed core, separation of the adhesive rubber on both sides of the core and for core separation from the belt side. Inspect the ribs for separation from the adhesive rubber and for cracking or separation of the ribs, torn or worn ribs or cracks in the inner ridges of the ribs (see illustration). Also check for fraying and glazing, which gives the belt a shiny appearance. Inspect both sides of the belt by twisting the belt to check the underside. Use your fingers to feel the belt where you can't see it. If any of the above conditions are evident, replace the belt(s).

4   The tension of the alternator drivebelt is checked by pushing on it at a distance halfway between the pulleys. Apply about 10 pounds of force with your thumb and see how much the belt moves (deflects); the belt should deflect about 1/4-inch (6 mm).

5   The tension of the power steering pump/air conditioning compressor drivebelt is adjusted by an automatic tensioner. Look at the

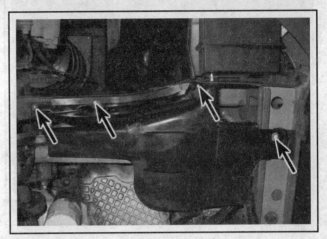

19.2 Remove these fasteners and detach the drivebelt splash shield

wear indicator on the tensioner (see illustration). The marks should be within the specified range; if not, the belt will have to be replaced.

ADJUSTMENT - ALTERNATOR BELT

▸ **Refer to illustrations 19.6a and 19.6b**

6   Loosen the pivot bolt and lock nut, then turn the adjusting bolt clockwise to tighten the belt or counterclockwise to loosen the belt (see illustrations). When you have obtained the desired tension, tighten the lock nut and pivot bolt securely.

REPLACEMENT

➡**Note: Since belts tend to wear out more or less at the same time, it's a good idea to replace both of them at the same time.**

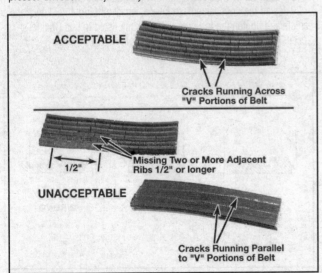

ACCEPTABLE

Cracks Running Across "V" Portions of Belt

1/2"

Missing Two or More Adjacent Ribs 1/2" or longer

UNACCEPTABLE

Cracks Running Parallel to "V" Portions of Belt

19.3 Here are some of the more common problems associated with drivebelts (check the belts very carefully to prevent an untimely breakdown)

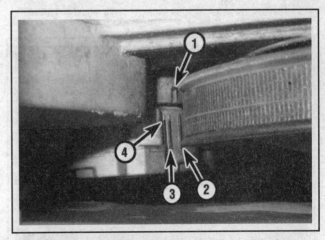

19.5 Details of the power steering pump/air conditioning compressor drivebelt tensioner

1	Belt length indicator	3	Normal length
2	Maximum length (belt worn out)	4	Minimum length

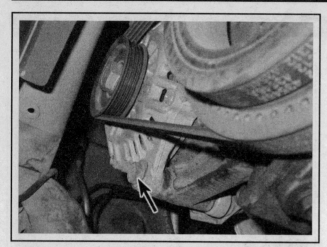

19.6a First, loosen the pivot bolt located at the bottom of the alternator

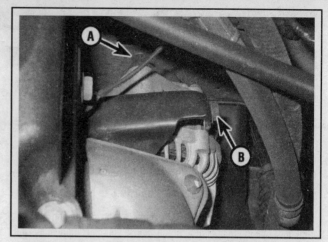

19.6b Next, loosen the lock nut (A) and rotate the adjusting bolt (B) counterclockwise to release tension on the drivebelt

7   Apply the parking brake, loosen the right (passenger's side) front wheel lug nuts, raise the front of the vehicle and support it securely on jackstands. Remove the wheel, then remove the drivebelt splash shield (see illustration 19.2).

Power steering pump/air conditioning compressor belt

♦ **Refer to illustrations 19.8 and 19.9**

8   The automatic tensioner must be released to allow drivebelt replacement. Place a 17mm wrench on the tensioner pulley tang and rotate it clockwise until the belt can be removed (see illustration). Remove the belt and slowly release the tensioner. Install the new belt then rotate the tensioner clockwise to allow the belt to slip over it, then release the tensioner slowly until it contacts the drivebelt.

9   When installing the belt, make sure the belt is centered on the pulleys (see illustration).

10 Install the drivebelt splash shield, wheel and lug nuts. Lower the vehicle and tighten the lug nuts to the torque listed in this Chapter's Specifications.

Alternator belt

11 Remove the power steering pump/air conditioning compressor

drivebelt (see Steps 7 and 8).

12 Follow Step 6 for drivebelt adjustment, but slip the belt off the pulleys and remove it.

13 When installing the belt, make sure the belt is centered on the pulleys (see illustration 19.9).

14 Adjust the belt as described in Step 6.

15 Install the drivebelt splash shield, wheel and lug nuts. Lower the vehicle and tighten the lug nuts to the torque listed in this Chapter's Specifications.

AUTOMATIC TENSIONER REPLACEMENT

16 Remove the power steering/air conditioning compressor drivebelt (see Steps 7 and 8).

17 Unscrew the tensioner mounting bolt and remove the tensioner.

18 Install the tensioner assembly by reversing the removal procedure. Tighten the mounting bolt to the torque listed in this Chapter's Specifications.

19 Install the drivebelt as described previously in this Section.

20 Install the drivebelt splash shield, wheel and lug nuts. Lower the vehicle and tighten the lug nuts to the torque listed in this Chapter's Specifications.

19.8 Use an open end wrench to rotate the drivebelt tensioner clockwise to release tension on the belt

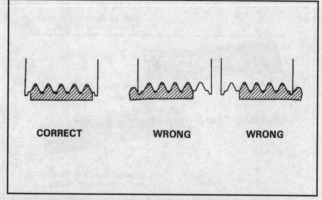

CORRECT   WRONG   WRONG

19.9 When installing a drivebelt, make sure it is centered - it must not overlap either edge of the pulley

20 Body hinge and lock lubrication (every 30,000 miles or 24 months)

1   Regular lubrication of the hinges and locks will keep them working smoothly and from wearing out prematurely. A container of multi-purpose grease, graphite spray, silicone spray and an oil can filled with engine oil will be required to lubricate the hinges, latches and locks.

2   Open the hood and smear a little multi-purpose grease on the hood latch mechanism and striker. Have an assistant pull the hood release lever from inside the vehicle as you lubricate the cable at the latch.

3   Lubricate all the hinges (door, hood, liftgate, etc.) with the recommended lubricant (see *Recommended lubricants and fluids* at the end of this Chapter) to keep them in proper working order.

4   The key lock cylinders can be lubricated with spray-type graphite or silicone lubricant which is available at auto parts stores.

5   Lubricate the door weather-stripping with silicone spray. This will reduce chafing and retard wear.

6   Some components should not be lubricated for the following reasons. Some are permanently lubricated, some lubricants will cause component failure or the lubricants will be detrimental to the component's operating characteristics. Do not lubricate the following: alternator bearings, drivebelts, drivebelt idler pulley, wheel bearings, rubber bushings, starter motor bearings, suspension strut bearings, or accelerator or cruise control cables.

21 Air filter and PCV filter replacement (every 30,000 miles or 24 months)

AIR FILTER

▶ **Refer to illustrations 21.2a and 21.2b**

1   The air filter element is located in the air filter housing on the driver's side of the engine compartment.

2   Remove the screws securing the top cover of the air filter housing, raise the cover, then lift the filter element out (see illustrations).

3   Inspect the inside of the air cleaner housing, top and bottom, for dirt, debris or damage. If necessary, clean the inside of the housing with a rag or shop vacuum as applicable. If the air filter housing is damaged and requires replacement, refer to Chapter 4.

4   Install the air filter, making sure it properly engages to the throttle body flange first and then snaps down tightly in front of the lip in the housing, then lower the cover and install the mounting screws.

PCV FILTER

▶ **Refer to illustration 21.6**

5   The PCV filter is located in the air filter housing on the driver's side of the engine compartment.

6   Remove the screws securing the top cover of the air filter housing, raise the cover, then lift the PCV filter out (see illustration).

7   Install the PCV filter, making sure it properly engages onto the PCV hose rubber grommet, then lower the cover and install the mounting screws.

21.2a To remove the air filter, remove the air filter cover screws . . .

21.2b . . . then raise the cover and remove the air filter element from the housing

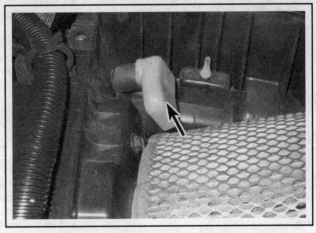

21.6 Remove the PCV filter from the PCV filter rubber tube

22 Automatic transaxle fluid and filter change (every 30,000 miles or 24 months)

▶ **Refer to illustrations 22.4a, 22.4b and 22.5**

1   The automatic transaxle fluid and filter should be changed and the magnet (if equipped) cleaned at the recommended intervals.

2   Raise the front of the vehicle and support it securely on jack-stands.

✳✳ WARNING:

Never work under a vehicle that is supported only by a jack!

3   Position a container capable of holding at least 5 quarts under the transaxle oil pan.

4   Since the transaxle drain pan does not have a drain plug, this procedure can get a bit messy, so you should have plenty of clean rags and newspapers handy to mop up any spills that may occur. If the drain pan you're using isn't very large in diameter, place it on a piece of plastic such as a trash bag to catch any splashing oil. Loosen, but don't remove, the pan bolts. Now remove the bolts on each side of the pan leaving two bolts loosely in place at the front and rear of the pan (see illustrations). Tap the corners of the pan using a soft-faced mallet to break the seal and allow the fluid to drain into the container (the remaining bolts will prevent the pan from completely separating from the transaxle). Remove the 2 bolts from the lower side of the pan and let it hang down to drain further. After the pan has finished draining, remove the remaining bolts and detach the pan.

5   On 2000 and 2001 models, remove the filter bolts (see illustration). On all models separate the filter by lowering it straight down. Also remove the gasket (2000 and 2001 models) or O-ring seal (2002 and later models).

6   Carefully remove all traces of old sealant from the pan, transaxle body (be careful not to nick or gouge the sealing surfaces) and the bolts.

7   Clean the pan and the magnet located inside the pan with a clean, lint-free cloth moistened with solvent. Don't forget to place the magnet back in its proper location at the bottom of the pan.

8   Fit the new filter in place, with a new O-ring installed, on the transaxle valve body. On 2000 and 2001 models, install the bolts and tighten them to the torque listed in this Chapter's Specifications.

9   Apply a 1/8-inch (3 mm) bead of RTV sealant to the pan sealing surface (stay on the inboard side of the bolt holes) and to the underside of each bolt head.

➡**Note: Some aftermarket replacement filters come with a fluid pan gasket. If a gasket is used, RTV sealant won't be necessary.**

10   Position the pan on the transaxle and install the bolts. Tighten them to the torque listed in this Chapter's Specifications following a criss-cross pattern. Work up to the final torque in three or four steps. Allow the RTV sealant time to dry according to the manufacturer's instructions.

11   Lower the vehicle and add three quarts of the specified fluid (see *Recommended lubricants and fluids* at the end of this Chapter) to the transaxle (see Section 6 if necessary). Start the engine and allow it to idle for at least two minutes while checking for leakage around the pan.

12   With the engine running and the brakes applied, move the shift lever through each of the gear positions and ending in Park. Check the fluid level on the dipstick. The level should be just up to the ADD mark. If necessary, add more fluid (a little at a time) until the level is just at the ADD mark (be careful not to overfill it).

13   Drive the vehicle until it reaches normal operating temperature. Recheck the fluid level and add as necessary until the fluid reaches the HOT range on the dipstick (see Section 6).

14   The old fluid drained from the transaxle cannot be reused in its present state and should be disposed of. Check with your local auto parts store, disposal facility or environmental agency to see if they will accept the fluid for recycling. After the fluid has cooled it can be drained into a container (capped plastic jugs, topped bottles, milk cartons, etc.) for transport to one of these disposal sites. Don't dispose of the fluid by pouring it on the ground or down a drain!

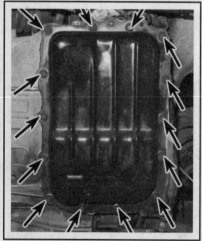

22.4a To drain the transaxle fluid, first loosen the bolts . . .

22.4b . . . then remove all the bolts except two at the front and two at the rear - after breaking the pan seal, remove the two bolts at the rear side and let the pan hang down to drain further

22.5 Remove the transaxle fluid filter bolts (2000 and 2001 models only)

23 Automatic transaxle band adjustment (2000 and 2001 models) (every 30,000 miles or 24 months)

1   The transaxle bands should be adjusted when specified in the maintenance schedule or at the time of a fluid and filter change.

KICKDOWN BAND

▶ **Refer to illustration 23.3**

2   The kickdown band adjustment screw and locknut is located at the top left side of the transaxle case. To access the adjustment screw, remove the air filter assembly (see Chapter 4).

3   Loosen the locknut approximately five turns and make sure the adjusting screw turns freely (see illustration).

4   Tighten the adjusting screw to the torque listed in this Chapter's Specifications.

5   Back the adjusting screw off 2-1/4 turns.

6   Hold the adjusting screw in position and tighten the locknut to the torque listed in this Chapter's Specifications.

23.3 Location of the kickdown band adjustment screw

LOW-REVERSE BAND

▶ **Refer to illustration 23.8**

7   To gain access to the Low-Reverse band, the transaxle pan must be removed (see Section 23).

8   Loosen the locknut approximately five turns (see illustration).

9   Tighten the adjusting screw to the torque listed in this Chapter's Specifications.

10 Back the adjusting screw off 3-1/2 turns.

11 Hold the adjusting screw in position and tighten the locknut to the torque listed in this Chapter's Specifications.

12 Install the pan and refill the transaxle (see Section 22).

23.8 Low/Reverse band locknut (A) and adjusting screw (B)

24 Manual transaxle lubricant change (every 30,000 miles or 24 months)

▶ **Refer to illustration 24.2**

1   Raise the vehicle and support it securely on jackstands in a level position.

❋❋ WARNING:

Never work under a vehicle that is supported only by a jack!

2   Remove the drain plug (see illustration). Drain the fluid into a suitable container capable of holding at least four quarts.

3   After the fluid has completely drained, install the drain plug and tighten it to the torque given in this Chapter's Specifications.

4   Fill the transaxle with the recommended lubricant (see Section 16).

5   The old oil drained from the transaxle cannot be reused in its present state and should be disposed of. Check with your local auto parts store, disposal facility or environmental agency to see if they will accept the oil for recycling. After the oil has cooled it can be drained

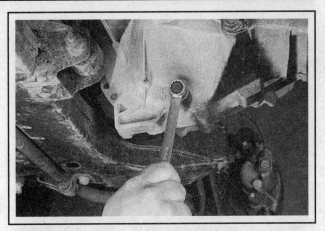

24.2 Location of the manual transaxle drain plug

into a container (capped plastic jugs, topped bottles, milk cartons, etc.) for transport to one of these disposal sites. Don't dispose of the oil by pouring it on the ground or down a drain!

---

25 Cooling system servicing (draining, flushing and refilling) (refer to the Maintenance schedule for service interval)

⁑ WARNING 1:

Do not allow coolant (antifreeze) to come in contact with your skin or painted surfaces of the vehicle. Flush contaminated areas immediately with plenty of water. Do not store new coolant or leave old coolant lying around where it's accessible to children or pets - they're attracted by its sweet smell. Ingestion of even a small amount of coolant can be fatal! Wipe up garage floor and drip pan spills immediately. Keep antifreeze containers covered and repair cooling system leaks as soon as they're noticed. Check with local authorities about the disposal of used antifreeze. Many communities have collection centers which will see that antifreeze is disposed of properly.

⁑ WARNING 2:

The electric cooling fan(s) on these models can activate at any time the ignition switch is in the ON position. Make sure the ignition is OFF when working in the vicinity of the fan(s).

⁑ CAUTION:

Never mix green colored ethylene glycol antifreeze with Mopar 5 year antifreeze because doing so will destroy the efficiency of the Mopar 5 year antifreeze. Check the coolant reservoir under the hood to determine what type coolant you have. Always refill with the correct coolant.

1 Periodically, the cooling system should be drained, flushed and refilled to replenish the coolant (antifreeze) mixture and prevent formation of rust and corrosion, which can impair the performance of the cooling system and cause engine damage.

2 At the same time the cooling system is serviced, all hoses and the radiator (pressure) cap should be inspected, tested and replaced if faulty (see Section 10).

DRAINING

▶ **Refer to illustration 25.5**

⁑ WARNING:

Wait until the engine is completely cool before beginning this procedure.

3 With the engine cold, remove the pressure cap and set the heater control to maximum heat.

4 Move a large container capable of holding at least eight quarts under the radiator drain fitting to catch the coolant mixture as it's drained. Attach a length of 5/16-inch (8 mm) diameter hose to the drain fitting, located at the bottom of the radiator on the right (passenger's) side, and direct it into the pan.

5 Open the drain fitting located at the bottom of the radiator (see illustration). Allow the coolant to completely drain out, then close the drain fitting.

FLUSHING

▶ **Refer to illustration 25.7**

6 Once the system is completely drained, remove the thermostat from the engine (see Chapter 3). Then reinstall the thermostat housing without the thermostat. This will allow the system to be thoroughly flushed.

7 Disconnect the upper radiator hose from the radiator, then place a garden hose in the upper radiator inlet and flush the system until the water runs clear at the upper radiator hose (see illustration).

8 Severe cases of radiator contamination or clogging will require removing the radiator (see Chapter 3) and reverse flushing it. This involves inserting the hose in the bottom radiator outlet to allow the clean water to run against the normal flow, draining out through the

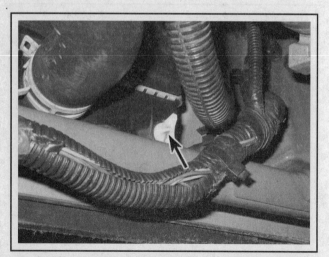

25.5 The drain fitting is located at the bottom of the radiator on the right hand (passenger's) side

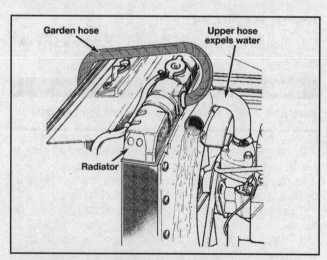

25.7 With the thermostat removed, disconnect the upper radiator hose and flush the radiator and engine block with a garden hose

top. A radiator repair shop should be consulted if further cleaning or repair is necessary.

9   When the coolant is regularly drained and the system refilled with the correct coolant mixture there should be no need to employ chemical cleaners or descalers.

10 Disconnect the coolant reservoir hose, remove the reservoir from the vehicle and flush it with clean water (see Chapter 3). Inspect it for damage and replace if necessary.

REFILLING

11 Install the thermostat, the thermostat housing and reconnect the radiator hose (see Chapter 3).

12 Install the coolant reservoir, reconnect the hose and close the radiator drain fitting.

13 Remove the cooling system pressure cap. Add the correct mixture of the proper type of antifreeze/coolant and water in the ratio specified on the antifreeze container or in this Chapter's Specifications through the filler neck until it reaches the pressure cap seat.

14 Add the same coolant mixture to the reservoir until the level is between the FULL HOT and ADD marks. Install the pressure cap.

15 Run the engine until normal operating temperature is reached (the fans will cycle on, then off), then allow the engine to cool. With the engine cold, add coolant as necessary to bring it up to the correct level.

16 Keep a close watch on the coolant level and the various cooling system hoses during the first few miles of driving and check for any coolant leaks. Tighten the hose clamps and add more coolant mixture as necessary.

26 Evaporative emissions control system check (every 30,000 miles or 24 months)

1   The function of the evaporative emissions control system is to prevent fuel vapors from escaping the fuel system and being released into the atmosphere. Vapors from the fuel tank are temporarily stored in a charcoal canister. The Powertrain Control Module (PCM) monitors the system and allows the vapors to be drawn into the intake manifold when the engine reaches normal operating temperature.

2   The charcoal canister is located above the fuel tank. The canister is maintenance-free and should last the life of the vehicle.

3   The most common symptom of a fault in the evaporative emis-

sions system is a strong fuel odor around or inside of the vehicle, or raw fuel leaking from the canister. These indications are usually more prevalent in hot temperatures. All systems are pressurized by a Leak Detection Pump (LDP). If normal system pressure cannot be achieved by the LDP, which indicates a leak, the PCM will store the appropriate fault code and illuminate the CHECK ENGINE light on the instrument panel. The most common cause of system pressure loss is a loose or poor-sealing gas cap.

4   For more information and replacement procedures see Chapter 6.

27 Spark plug check and replacement (see Maintenance schedule for intervals)

♦ **Refer to illustrations 27.2, 27.5a, 27.5b, 27.10, 27.12 and 27.13**

1   The spark plugs are located in the center of the cylinder head.

2   In most cases the tools necessary for spark plug replacement include a spark plug socket which fits onto a ratchet (this special socket is padded inside to protect the porcelain insulators on the new plugs and hold them in place), various extensions and a feeler gauge to check and adjust the spark plug gap (see illustration). Since these engines are equipped with an aluminum cylinder head, a torque wrench should be used when tightening the spark plugs.

3   The best approach when replacing the spark plugs is to purchase the new spark plugs beforehand, adjust them to the proper gap and then replace each plug one at a time. When buying the new spark plugs, be sure to obtain the correct plug for your specific engine. This information can be found in the *Specifications* Section at the end of this Chapter, in your owner's manual or on the Vehicle Emissions Control Information (VECI) label located under the hood. If differences exist between the sources, purchase the spark plug type specified on the VECI label as it was printed for your specific engine.

4   Allow the engine to cool completely before attempting to remove any of the plugs. During this cooling off time, each of the new spark plugs can be inspected for defects and the gaps can be checked.

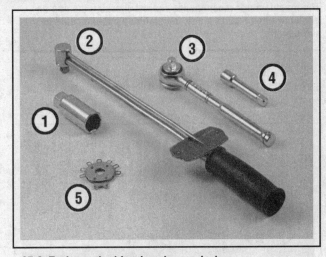

27.2 Tools required for changing spark plugs

1   **Spark plug socket** - This will have special padding inside to protect the spark plug's porcelain insulator

2   **Torque wrench** - Although not mandatory, using this tool is the best way to ensure the plugs are tightened properly

3   **Ratchet** - Standard hand tool to fit the spark plug socket

4   **Extension** - Depending on model and accessories, you may need special extensions and universal joints to reach one or more of the plugs

5   **Spark plug gap gauge** - This gauge for checking the gap comes in a variety of styles. Make sure the gap for your engine is included

27.5a Spark plug manufacturers recommend using a wire-type gauge when checking the gap - if the wire does not slide between the electrodes with a slight drag, adjustment is required

5   The gap is checked by inserting the proper thickness gauge between the electrodes at the tip of the plug (see illustration). The gap between the electrodes should be as listed in this Chapter's Specifications or in your owner's manual. The wire should touch each of the electrodes. If the gap is incorrect, use the adjuster on the thickness gauge body to bend the curved side electrode slightly until the proper gap is obtained (see illustration).

✳✳ CAUTION:

The manufacturer recommends against checking the gap on platinum-tipped spark plugs; the platinum coating could be scraped off.

6   Also, at this time check for cracks in the spark plug body (if any are found, the plug must not be used). If the side electrode is not exactly over the center one, use the adjuster to align the two.

27.10 Use a ratchet and extension to remove the spark plugs

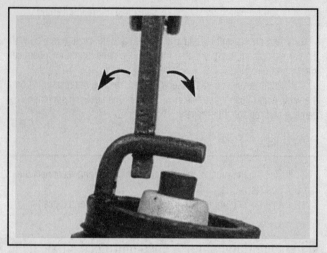

27.5b To change the gap, bend the side electrode only, as indicated by the arrows, and be very careful not to crack or chip the porcelain insulator surrounding the center electrode

7   Cover the fender to prevent damage to the paint. Fender covers are available from auto parts stores but an old blanket will work just fine.
8   Using a twisting motion, detach one of the spark plug boots/wires from the ignition coil, then withdraw the boot/wire from the valve cover. Pull only on the boot at the end of the wire; don't pull on the wire.

➡**Note: Due to the short length of the spark plug wire, always disconnect the spark plug wire from the ignition coil pack first.**

9   If compressed air is available, use it to blow any dirt or foreign material away from the spark plug area.

✳✳ WARNING:

Wear eye protection!

The idea here is to eliminate the possibility of material falling into the cylinder through the spark plug hole as the spark plug is removed.
10   Place the spark plug socket over the plug and remove it from the engine by turning it in a counterclockwise direction (see illustration).
11   Compare the spark plug with this chart (see illustration) to get an indication of the overall running condition of the engine.
12   It's a good idea to lightly coat the threads of the spark plugs with an anti-seize compound (see illustration) to insure that the spark plugs do not seize in the aluminum cylinder head.
13   It's often difficult to insert spark plugs into their holes without cross-threading them. To avoid this possibility, fit a piece of rubber hose over the end of the spark plug (see illustration). The flexible hose acts as a universal joint to help align the plug with the plug hole. Should the plug begin to cross-thread, the hose will slip on the spark plug, preventing thread damage. Install the spark plug and tighten it to the torque listed in this Chapter's Specifications.
14   Attach the plug wire to the new spark plug, again using a twisting motion on the boot until it is firmly seated on the end of the spark plug. Attach the other end to the ignition coil pack.
15   Follow the above procedure for the remaining spark plugs, replacing them one at a time to prevent mixing up the spark plug wires.

A normally worn spark plug should have light tan or gray deposits on the firing tip.

A carbon fouled plug, identified by soft, sooty, black deposits, may indicate an improperly tuned vehicle. Check the air cleaner, ignition components and engine control system.

An oil fouled spark plug indicates an engine with worn piston rings and/or bad valve seals allowing excessive oil to enter the chamber.

This spark plug has been left in the engine too long, as evidenced by the extreme gap. Plugs with such an extreme gap can cause misfiring and stumbling accompanied by a noticeable lack of power.

A physically damaged spark plug may be evidence of severe detonation in that cylinder. Watch that cylinder carefully between services, as a continued detonation will not only damage the plug, but could also damage the engine.

A bridged or almost bridged spark plug, identified by a build-up between the electrodes caused by excessive carbon or oil build-up on the plug.

27.11 Inspect the spark plug to determine engine running conditions

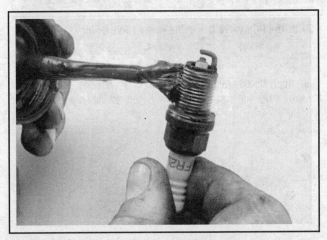
27.12 Apply a thin coat of anti-seize compound to the spark plug threads - DO NOT get any on the electrodes!

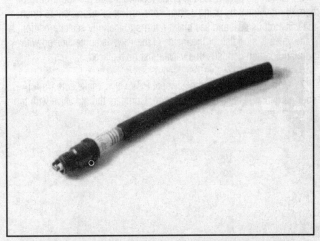
27.13 A length of snug-fitting rubber hose will save time and prevent damaged threads when installing the spark plugs

28 Spark plug wire check and replacement (every 60,000 miles or 48 months)

1   The spark plug wires should be replaced at the recommended intervals and/or checked when new spark plugs are installed.

2   Disconnect the spark plug wire from the ignition coil pack. Pull only on the boot at the end of the wire; don't pull on the wire itself. Use a twisting motion to free the boot/wire from the coil. Disconnect the same spark plug wire from the spark plug, using the same twisting method while pulling on the boot.

3   Check inside the boot for corrosion, which will look like a white, crusty powder (don't mistake the white dielectric grease used on some plug wire boots for corrosion protection).

4   Now push the wire and boot back onto the end of the spark plug. It should be a tight fit on the plug end. If not, remove the wire and use a pair of pliers to carefully crimp the metal connector inside the wire boot until the fit is snug.

5   Now push the wire and boot back into the end of the ignition coil terminal. It should be a tight fit in the terminal. If not, remove the wire and use a pair of pliers to carefully crimp the metal connector inside the wire boot until the fit is snug.

6   Now, using a cloth, clean each wire along its entire length. Remove all built-up dirt and grease. As this is done, inspect for burned areas, cracks and any other form of damage. Bend the wires in several places (but not too sharply) to ensure that the conductive material inside hasn't hardened. Repeat the procedure for the remaining wires.

7   If you are replacing the spark plug wires, purchase a complete set for your particular engine. The terminals and rubber boots should already be installed on the wires. Replace the wires one at a time to avoid mixing up the firing order and make sure the terminals are securely seated on the coil pack and the spark plugs.

8   Attach the plug wire to the new spark plug and to the ignition coil pack using a twisting motion on the boot until it is firmly seated.

29 Positive Crankcase Ventilation (PCV) valve check and replacement (every 60,000 miles or 48 months)

▶ **Refer to illustration 29.2**

1   The PCV valve controls the amount of crankcase vapors allowed to enter the intake manifold. Inside the PCV valve is a spring loaded valve that opens in relation to intake manifold vacuum, which allows crankcase vapors to be drawn from the valve cover back into the engine combustion chamber.

2   The PCV valve is located in the valve cover (see illustration).

3   To check the operation of the PCV valve, disconnect the hose from the valve, then unscrew the valve from the valve cover and reconnect the hose to it.

4   Start the engine and listen for a hissing sound coming from the PCV valve. Place your finger over the valve opening - you should feel vacuum. If there's no vacuum at the valve, check for a plugged hose, port or valve. Replace any plugged or deteriorated hoses.

5   Check the spring-loaded pintle located inside the valve for freedom of movement by using a small screwdriver or equivalent to push the valve off its seat and see that it returns to the fully seated position. If the valve is sluggish or the inside of the valve is contaminated with gum and carbon deposits, the valve must be replaced.

6   To remove the valve, see Step 3.

7   When purchasing a replacement PCV valve, make sure it's for your particular vehicle and engine size. Compare the old valve with the new one to make sure they're the same.

29.2 The PCV valve is located in the valve cover

8   Apply thread sealant to the threads of the valve, then install it in the valve cover and tighten it securely. Reconnect the hose to the valve.

Specifications

Recommended lubricants and fluids

➡Note: Listed here are manufacturer's recommendations at the time this manual was written. Manufacturers occasionally upgrade their fluid and lubricant specifications, so check with your local auto parts store for current recommendations.

Engine oil
    Type   API "Certified for gasoline engines"
    Viscosity   See accompanying chart
Manual transaxle lubricant
    2000 and 2001 models   Mopar manual transaxle fluid type MS 9417 or equivalent or equivalent
    2002 and later models   Mopar ATF +4 automatic transmission fluid (type MS 9602) or equivalent
Automatic transaxle fluid   Mopar ATF +4 automatic transmission fluid (type MS 9602), or equivalent
Power steering fluid
    2000 and 2001 models   Mopar power steering fluid, or equivalent
    2002 and later models   Mopar ATF +4 automatic transmission fluid (type 9602), or equivalent
Brake and clutch fluid   DOT type 3 brake fluid
Engine coolant   50/50 mixture of Mopar 5 year/100,000 mile Formula (MS-9769)
                                      antifreeze/coolant with HOAT (Hybrid Organic Additive Technology) and water*

Door latch   Multi-purpose grease
Fuel filler door remote control latch mechanism   Multi-purpose grease
Hood and door hinge lubricant   Engine oil
Key lock cylinder lubricant   Graphite spray
Parking brake mechanism grease   Mopar Spray White Lube or equivalent

* 2003 through 2005 vehicles are filled with a 50/50 mixture of Mopar 5 year/100,000 mile coolant that shouldn't be mixed with other coolants. Refer to the coolant reservoir label under the hood to determine what type coolant you have. Always refill with the correct coolant.

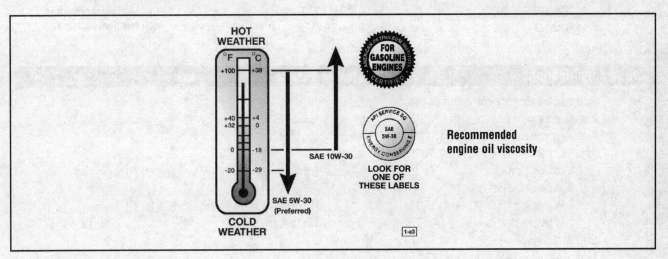

Recommended engine oil viscosity

Capacities*

Engine oil (including filter)   4.5 quarts (4.3 liters)
Automatic transaxle (drain and refill)**   4.0 quarts (3.8 liters)
Manual transaxle
    2000 models   2.0 to 2.3 quarts (1.9 to 2.2 liters)
    2001 and later models   2.5 to 2.8 quarts (2.4 to 2.7 liters)
Cooling system***   6.5 quarts (6.2 liters)

* All capacities approximate. Add as necessary to bring to appropriate level.

** The best way to determine the amount of fluid to add during a routine fluid change is to measure the amount drained. It's important to not overfill the transaxle.

*** Includes heater and coolant reservoir.

Capacities* (continued)

Brakes
Disc brake pad wear limit	1/8 inch (3 mm)
Drum brake shoe wear limit	1/16 inch (1.5 mm)
Parking brake lever travel	Adjusted automatically

Ignition system

Spark plug
    2000 through 2003 models
Type*	Champion RC 9YC or equivalent
Gap	0.033 to 0.038 inch (0.83 to 0.96 mm)

    2004 and 2005 models
Type	Champion 9001 or equivalent
Gap	0.044 inch (1.1 mm)

    Refer to the Vehicle Emission Control Information label in the engine compartment and follow the information on the label if it differs from that shown here.

Spark plug wire resistance (approximate)
    2000 and 2001 models
Wire numbers 1 and 4	3,500 to 4,900 ohms
Wire numbers 2 and 3	2,950 to 4,100 ohms

    2002 models
Wire numbers 1 and 4	2,970 to 8,910 ohms
Wire numbers 2 and 3	2,360 to 7,070 ohms

    2003 models
Wire numbers 1 and 4	2,900 to 9,000 ohms
Wire numbers 2 and 3	2,290 to 7,300 ohms

2004 and later models	2,280 to 7,290 ohms
Firing order	1-3-4-2

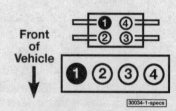

Cylinder numbering and coil terminal location

Torque specifications	Ft-lbs (unless otherwise indicated)	Nm
Automatic drivebelt tensioner mounting bolt	20	27
Automatic transaxle fluid pan bolts	165 in-lbs	19
Automatic transaxle fluid filter bolts (2000 and 2001 models)	40 in-lbs	5
Automatic transaxle band adjustment (2000 and 2001 models)		
Kickdown band	Tighten to 72 in-lbs (8 Nm) then back off 2-1/4 turns	
Kickdown band lock nut	35	47
Low reverse band	Tighten to 41 in-lbs (5 Nm) then back off 3-1/2 turns	
Low reverse band lock nut	125 in-lbs	14
Engine oil drain plug	20	27
Manual transaxle drain plug	20	28
Manual transaxle fill plug	29	39
Spark plugs		
2000 through 2002 models	20	28
2003 and later models	156 in-lbs	17.5
Wheel lug nuts	100	135

2A

ENGINE

Section

Reference to other Chapters
CHECK ENGINE light - See Chapter 6

1   General information

This Part of Chapter 2 is devoted to in-vehicle engine repair procedures. Information concerning engine removal and installation can be found in Part B of this Chapter.

The following repair procedures are based on the assumption that the engine is installed in the vehicle. If the engine has been removed from the vehicle and mounted on a stand, many of the steps outlined in this Part of Chapter 2 will not apply.

The Specifications included in this Part of Chapter 2 apply only to the procedures contained in this Part.

These models are equipped with the 2.0L SOHC four-cylinder engine. There are two different versions of this engine; the 2.0L 132 horsepower model or the 2.0L HO (high output) 150 horsepower model.

2   Repair operations possible with the engine in the vehicle

Many major repair operations can be accomplished without removing the engine from the vehicle.

Clean the engine compartment and the exterior of the engine with some type of degreaser before any work is done. It will make the job easier and help keep dirt out of the internal areas of the engine.

Depending on the components involved, it may be helpful to remove the hood to improve access to the engine as repairs are performed (refer to Chapter 11 if necessary). Cover the fenders to prevent damage to the paint. Special pads are available, but an old bedspread or blanket will also work.

If vacuum, exhaust, oil or coolant leaks develop, indicating a need for gasket or seal replacement, the repairs can generally be made with the engine in the vehicle. The intake and exhaust manifold gaskets, oil pan gasket, camshaft and crankshaft oil seals and cylinder head gasket are all accessible with the engine in place.

Exterior engine components, such as the intake and exhaust manifolds, the oil pan, the oil pump, the water pump, the starter motor, the alternator, the distributor and the fuel system components can be removed for repair with the engine in place.

Since the camshaft(s) and cylinder head can be removed without pulling the engine, valve component servicing can also be accomplished with the engine in the vehicle. Replacement of the timing belt and sprockets is also possible with the engine in the vehicle.

In extreme cases caused by a lack of necessary equipment, repair or replacement of piston rings, pistons, connecting rods and rod bearings is possible with the engine in the vehicle. However, this practice is not recommended because of the cleaning and preparation work that must be done to the components involved.

3   Top Dead Center (TDC) for number one piston - locating

▶ **Refer to illustration 3.9**

1   Top Dead Center (TDC) is the highest point in the cylinder that each piston reaches as it travels up-and-down when the crankshaft turns. Each piston reaches TDC on the compression stroke and again on the exhaust stroke, but TDC generally refers to piston position on the compression stroke. The cast-in timing mark arrow on the crankshaft timing belt pulley installed on the front of the crankshaft is referenced to the number one piston at TDC when the arrow is straight up, or at "12 o'clock", and aligned with the cast-in timing mark arrow on the oil pump housing (see Section 7).

2   Positioning a specific piston at TDC is an essential part of many procedures such as rocker arm removal, camshaft(s) removal, timing belt and sprocket replacement.

3   In order to bring any piston to TDC, the crankshaft must be turned using one of the methods outlined below. When looking at the front of the engine, normal crankshaft rotation is clockwise.

※※ WARNING:

Before beginning this procedure, be sure to set the emergency brake, place the transmission in Park or Neutral and disable the ignition system by disconnecting the primary electrical connector from the ignition coil pack.

a) *The preferred method is to turn the crankshaft with a large socket and breaker bar attached to the crankshaft balancer hub bolt that is threaded into the front of the crankshaft.*

b) *A remote starter switch, which may save some time, can also be used. Attach the switch leads to the S (switch) and B (battery) terminals on the starter solenoid. Once the piston is close to TDC, discontinue with the remote switch and use a socket and breaker bar as described in the previous paragraph.*

c) *If an assistant is available to turn the ignition switch to the Start position in short bursts, you can get the piston close to TDC without a remote starter switch. Use a socket and breaker bar as described in Paragraph a) to complete the procedure.*

4   Remove all spark plugs as this will make it easier to rotate the engine by hand.

5   Insert a compression gauge (screw-in type with a hose) in the number 1 spark plug hole. Place the gauge dial where you can see it while turning the crankshaft balancer hub bolt.

➡**Note: The number one cylinder is located at the front (timing belt end) of the engine.**

6   Turn the crankshaft clockwise until you see compression building up on the gauge - you are on the compression stroke for that cylinder. If you did not see compression build up, continue with one more complete revolution to achieve TDC for the number one cylinder.

7   Remove the compression gauge. Through the number one cylinder spark plug hole, insert a section of wooden dowel or plastic rod and slowly push it down until it reaches the top surface of the piston crown.

✳ CAUTION:

Don't insert a metal or sharp object into the spark plug hole as the piston crown may be damaged.

8   With the dowel or rod in place on top of the piston crown, slowly rotate the crankshaft clockwise until the dowel or rod is pushed upward, stops, and then starts to move back down. At this point, rotate the crankshaft slightly counterclockwise until the dowel or rod has reached it upper most travel. At this point, the number one piston is at the TDC position.

9   Check the alignment of the camshaft timing marks (see illustration). At this point the camshaft timing marks should be aligned. If not, repeat this procedure until alignment is correct.

10 After the number one piston has been positioned at TDC on the

3.9 Check the alignment of the camshaft timing marks

compression stroke, TDC for any of the remaining cylinders can be located by turning the crankshaft 180-degrees (1/2-turn) at a time and following the firing order (refer to the Specifications).

4   Valve cover - removal and installation

REMOVAL

▸ **Refer to illustrations 4.5 and 4.6**

1   Disconnect the cable from the negative battery terminal (see Chapter 5).

2   Remove the ignition coil pack from the valve cover (see Chapter 5).

3   Clearly label and then detach any electrical wiring harnesses which connect to or cross over the valve cover.

4   Disconnect the PCV valve hose and breather hose from the valve cover (see Chapter 6).

5   Remove the valve cover bolts (see illustration) and then lift off the cover. If the cover sticks to the cylinder head, tap on it with a soft-face hammer or place a wood block against the cover and tap on the wood with a hammer.

✳ CAUTION:

If you have to pry between the valve cover and the cylinder head, be extremely careful not to gouge or nick the gasket surfaces of either part. A leak could develop after reassembly.

➡**Note: Do not remove the baffle plate mounting screws. The screws are self tapping and can strip once they are removed and reinstalled.**

6   Remove the rubber gasket (see illustration). Thoroughly clean the valve cover and remove all traces of old gasket material. Gasket removal solvents are available from auto parts stores and may prove helpful. After cleaning the surfaces, degrease them with a rag soaked in lacquer thinner or acetone.

4.5 Gradually and evenly, loosen, and then remove, the valve cover mounting bolts

4.6 Install a new rubber gasket and new spark plug well seals onto the valve cover

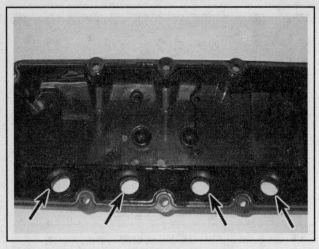

4.7 Install new tube seals into the valve cover

INSTALLATION

▶ **Refer to illustration 4.7**

7   Inspect the spark plug tube seals (see illustration) for deterioration and hardness. Replace them if necessary.

8   Install a new gasket on the cover, using RTV sealant to hold it in place. Place the cover on the engine and install the cover bolts.

9   Tighten the bolts to the torque listed in this Chapter's Specifications.

10 The remaining steps are the reverse of removal. When finished, run the engine and check for oil leaks.

5   Intake manifold - removal and installation

➡ **Note: High output models are equipped with a two-piece intake manifold. The upper and lower intake manifolds must be removed as a single assembly. The lower intake manifold houses the Manifold Tuning Valve assembly (see Chapter 6). If necessary, remove the lower intake manifold from the upper intake manifold after the assembly has been removed from the engine.**

1   Relieve the fuel system pressure (see Chapter 4).

2   Disconnect the cable from the negative battery terminal (see Chapter 5).

REMOVAL

All models

▶ **Refer to illustrations 5.9a, 5.9b and 5.14**

3   Remove the air inlet duct from the throttle body and the intake manifold (see Chapter 4).

4   Unplug the electrical connector from the Manifold Absolute Pressure (MAP) sensor (see Chapter 6).

5   Disconnect the fuel line from the fuel rail (see Chapter 4).

6   Remove the fuel rail (see Chapter 4).

7   Remove the brake booster hose (see Chapter 9) from the intake manifold.

8   Disconnect the knock sensor connector (see Chapter 6).

9   Remove the bolt from the intake manifold support bracket at the intake manifold (see illustration).

10 Disconnect the starter electrical connectors (see Chapter 5).

High output models

11 Disconnect the Manifold Tuning Valve connector (see Chapter 6).

12 Remove the engine oil dipstick from the housing.

13 Disconnect the electrical connectors by moving the intake manifold forward slightly.

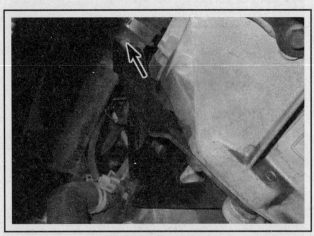

5.9a Working under the engine compartment, remove the upper intake manifold bracket bolt from the engine/transaxle/intake manifold bracket

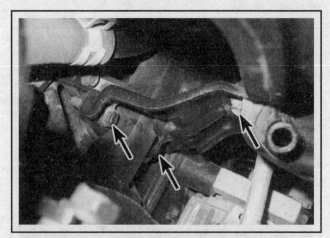

5.9b Remove the mounting bolts from the engine/intake manifold support bracket located near the alternator

All models

14 Remove the intake manifold bolts (see illustration) and remove the intake manifold from the cylinder head.

INSPECTION

15 Using a straightedge and feeler gauge, check the intake manifold mating surface for warpage. Check the intake manifold surface on the cylinder head also. If the warpage on either surface exceeds the limit listed in this Chapter's Specifications, the intake manifold and/or the cylinder head must be resurfaced at an automotive machine shop. If the warpage is too excessive for resurfacing, the intake manifold and/or the cylinder head must be replaced.

INSTALLATION

16 On high output models, install the lower intake manifold (Manifold Tuning Valve assembly) if it was removed. Tighten the lower manifold fasteners in several stages, working from the center out, to the torque listed in this Chapter's Specifications.

17 Install new seals and/or a new manifold gasket on the intake manifold or the lower intake manifold (high output models).

18 Install the intake manifold assembly and tighten the intake manifold fasteners in several stages (see illustration 5.14), working from the

5.14 Remove the upper intake manifold mounting bolts (non-HO model shown)

center out, to the torque listed in this Chapter's Specifications.

19 On high output models, connect the Manifold Tuning Valve connector (see Chapter 6) and all other harness connectors.

20 On high output models, install the oil dipstick tube.

21 The remaining installation is the reverse of removal.

6 Exhaust manifold - removal and installation

▶ **Refer to illustration 6.11**

⁂ WARNING:

Wait until the engine is completely cool before beginning this procedure.

REMOVAL

1 Disconnect the cable from the negative battery terminal (see Chapter 5).

2 Set the parking brake and block the rear wheels. Raise the vehicle and support it securely on jackstands.

3 Remove the engine protection cover from below the vehicle.

4 Remove the wire harness heat shield mounting bolts and remove the heat shield from the exhaust manifold support bracket.

➡**Note: On Ultra Low Emissions Vehicles (ULEV), unbolt the wire harness heat shield directly under the exhaust manifold.**

5 Remove the exhaust manifold support bracket.

6 Remove the exhaust manifold heat shield bolts and then remove the heat shield from the exhaust manifold.

7 Working under the vehicle, apply penetrating oil to the threads of the exhaust manifold-to-exhaust pipe studs and allow it to soak in for awhile then remove the exhaust pipe-to-exhaust manifold flange nuts and separate the exhaust pipe from the exhaust manifold.

8 On high output models, remove the cruise control vacuum reservoir, if equipped.

9 Remove the vacuum hose from the rear of the valve cover.

10 Remove the oxygen sensor (see Chapter 6).

11 Remove the exhaust manifold bolts and nuts (see illustration)

6.11 Locations of the exhaust manifold mounting bolts

and then remove the exhaust manifold.

12 Remove the exhaust manifold gasket.

INSPECTION

13 Inspect the exhaust manifold for cracks and any other obvious damage. If the manifold is cracked or damaged in any way, replace it.

14 Using a wire brush, clean up the threads of the exhaust manifold bolts and then inspect the threads for damage. Replace any bolts that have thread damage.

15 Using a scraper, remove all traces of gasket material from the mating surfaces and inspect them for wear and cracks.

✳✳ CAUTION:

When removing gasket material from any surface, especially aluminum or plastic, be very careful not to scratch or gouge the gasket surface. Any damage to the surface may leak after reassembly. Gasket removal solvents are available from auto parts stores and may prove helpful.

16 Using a straightedge and feeler gauge, inspect the exhaust manifold mating surface for warpage. Check the exhaust manifold surface on the cylinder head also. If the warpage on any surface exceeds the limits listed in this Chapter's Specifications, the exhaust manifold and/or cylinder head must be replaced or resurfaced at an automotive machine shop.

INSTALLATION

17 Coat the threads of the exhaust manifold bolts and studs with an anti-seize compound. Install a new gasket, install the manifold and install the fasteners. Tighten the bolts and nuts in several stages, working from the center out, to the torque listed in this Chapter's Specifications.

18 The remainder of installation is the reverse of removal. When you're done, be sure to run the engine and check for exhaust leaks.

---

7 Timing belt - removal, inspection and installation

REMOVAL

▶ **Refer to illustrations 7.8, 7.9, 7.17, 7.18, 7.19a, 7.19b and 7.19c**

✳✳ WARNING:

Wait until the engine is completely cool before beginning this procedure.

✳✳ CAUTION 1:

If the timing belt failed with the engine operating, damage to the valves may have occurred. Perform an engine compression check to confirm damage.

✳✳ CAUTION 2:

Do not try to turn the crankshaft with a camshaft sprocket bolt and do not rotate the crankshaft counterclockwise.

✳✳ CAUTION 3:

Do not turn the crankshaft or camshaft after the timing belt has been removed. Doing so will damage the valves from contact with the pistons.

1 Make sure the engine is cool before proceeding.
2 Position the number one piston at Top Dead Center (see Section 3).
3 Disconnect the cable from the negative battery terminal (see Chapter 5).
4 Remove the accessory drivebelts (see Chapter 1).
5 Remove the spark plugs (see Chapter 1).
6 Set the parking brake and block the rear wheels. Raise the front of the vehicle and support it securely on jackstands.
7 Remove the right (passenger-side) inner fender splash shield, if equipped.
8 Loosen the large bolt in the center of the crankshaft damper pulley. It might be very tight; to break it loose insert a large screwdriver or bar through the opening in the pulley to keep the pulley stationary and loosen the bolt with a socket and breaker bar (see illustration).
9 Install a 3-jaw puller onto the damper pulley and remove the pulley from the crankshaft (see illustration). Use the proper insert to keep the puller from damaging the crankshaft bolt threads. If the pulley is difficult to remove, tap the center bolt of the puller with a brass mallet to break it loose. Reinstall the bolt with a spacer so you can rotate the crankshaft later.
10 Remove the exhaust flange mounting bolts and separate the exhaust pipe from the exhaust manifold (see Chapter 4).
11 Drain the engine coolant (see Chapter 1).
12 Remove the ground strap from the engine mount bracket.

7.8 Insert a large screwdriver or bar through the opening in the pulley and wedge it against the engine block, then loosen the bolt with a socket and breaker bar

7.9 Install a 3-jaw puller onto the damper pulley, position the center post of the puller on the crankshaft end (use the proper insert to keep from damaging the crankshaft threads), tighten the puller and remove the pulley from the crankshaft

7.17 Remove the lower bolts that attach the timing belt outer cover to the engine

13 Remove the power steering pump and the power steering pump bracket (see Chapter 10) and position it off to the side without disconnecting the fluid lines.

14 Remove the lower torque strut (see Section 18).

15 Support the engine with a floor jack under the oil pan. Place a wood block on the jack head to prevent the floor jack from denting or damaging the oil pan.

16 Remove the right engine mount (see Section 18) and engine mount bracket.

17 Remove the lower and upper timing cover bolts (see illustration) and then remove the timing belt cover.

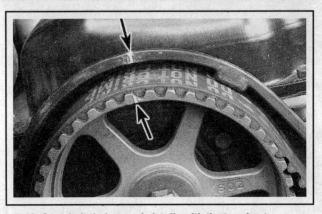

7.18 Camshaft timing mark details with the top, front cover removed

18 Before removing the timing belt, make sure that the camshaft sprocket timing marks are aligned (see illustration).

➡**Note: If you plan to reuse the timing belt, paint an arrow on it to indicate the direction of rotation (clockwise).**

19 Release tension on the timing belt:

a) *On 2000 models, insert a 6 mm Allen wrench into the belt tensioner and insert the long end of a 1/8-inch or a 3 mm Allen wrench into the small hole located on the front of the tensioner (see illustration). Then, using the 6 mm Allen wrench as a lever, rotate the tensioner COUNTERCLOCKWISE and simultaneously and lightly push in the 1/8-inch or 3 mm Allen wrench until it slides into the hole in the tensioner.*

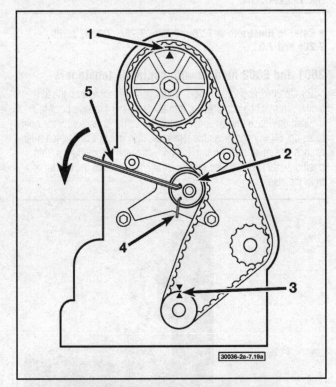

7.19a Timing belt alignment marks on 2000 models

1	*Camshaft timing marks*	*4*	*3 mm Allen wrench*
2	*Belt tensioner*	*5*	*8 mm Allen wrench*
3	*TDC marks on crankshaft sprocket*		

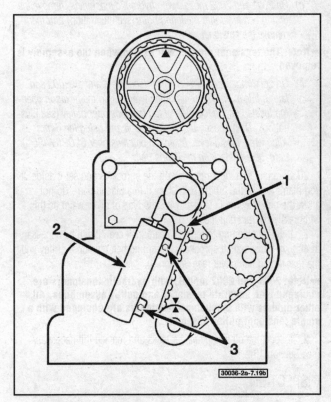

7.19b Timing belt alignment marks on 2001 and 2002 models with hydraulic tensioners

1	*Tensioner assembly*	*3*	*Tensioner bolts*
2	*Hydraulic tensioner*		

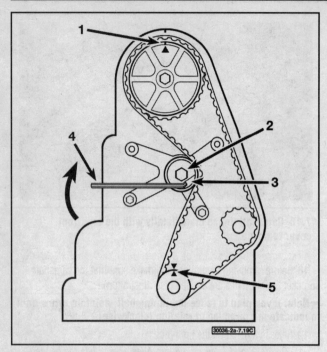

7.19c Timing belt alignment marks on 2002 models with mechanical tensioners and all 2003 and later models

1	Camshaft timing marks	4	6 mm Allen wrench
2	Lock nut	5	TDC marks on crankshaft
3	Top plate		sprocket

b) *On 2001 and 2002 models with hydraulic tensioners, remove the timing belt tensioner mounting bolts (see illustration) and then remove the tensioner.*

➡**Note: The tensioner plunger will extend when the assembly is removed.**

c) *On 2002 models with mechanical tensioners and all 2003 and later models, insert a 6 mm Allen wrench into the hexagon opening located on the top plate of the belt tensioner pulley (see illustration). Then, loosen the tensioner bolt and using the 6 mm Allen wrench as a lever, rotate the belt tensioner CLOCKWISE until there is slack on the timing belt.*

20 Carefully slip the timing belt off the sprockets and set it aside. If you plan to reuse the timing belt, place it in a plastic bag - do not allow the belt to come in contact with any type of oil or water as this will greatly shorten belt life.

21 If you're planning to replace a camshaft seal, it will be necessary to remove the camshaft sprockets, the timing belt tensioner pulley and the rear timing belt cover (see Section 9).

➡**Note: 2001 and 2002 models with hydraulic tensioners are equipped with separate tensioner and pulley assemblies. All other models with mechanical tensioners are equipped with a single tensioner/pulley assembly.**

22 Inspect the oil pump seal for leaks and replace it if necessary (see Section 16).

INSPECTION

▶ **Refer to illustration 7.24**

23 Rotate the tensioner pulley and idler pulley by hand and move them side-to-side to detect roughness and excess play. Visually

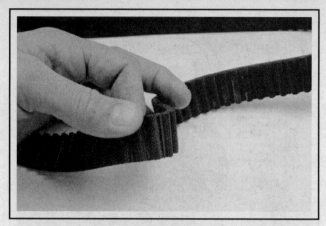

7.24 Carefully inspect the timing belt; bending it backwards will often make wear or damage more apparent

inspect the sprockets for any signs of damage and wear. Replace parts as necessary.

24 Inspect the timing belt for cracks, separation, wear, missing teeth and oil contamination. Replace the belt if it's in questionable condition (see illustration). If the timing belt is excessively worn or damaged on one side, it might be due to incorrect tracking (misalignment). If the belt looks like it was misaligned, be sure to replace the belt tensioner assembly.

25 Check the hydraulic tensioner for leaks or any obvious damage to the body.

INSTALLATION

▶ **Refer to illustrations 7.26a, 7.26b, 7.26c, 7.28a, 7.28b, 7.28c and 7.31**

2001 and 2002 models with hydraulic tensioners

26 On 2001 and 2002 models with hydraulic tensioners, the tensioner pin must be compressed into the tensioner housing prior to installation. Place the tensioner in the vise so the surface with the hole faces up. Slowly compress the tensioner, then install a 5/64-inch Allen wrench or similar tool through the body to retain the plunger in this position until it is installed (see illustrations). Remove the tensioner from the vise.

7.26a The tensioner pin must be compressed into the tensioner housing prior to installation

7.26b Place the tensioner in the vise so the hole faces up

7.26c Compress the pin with the vise and place a small Allen wrench, or something similar, through the hole to keep the pin retracted for reassembly on the engine

7.28a Using a box wrench, rotate the crankshaft timing sprocket until the TDC mark on the sprocket is aligned with the arrow on the oil pump housing . . .

7.28b . . . then back it off counterclockwise 3 notches BTDC

7.28c Rotate the crankshaft timing sprocket clockwise to 1/2-notch BTDC

All models

27 Confirm that the camshaft sprocket timing marks are aligned (see illustrations 7.19a, 7.19b and 7.19c).

28 Position the crankshaft timing sprocket as follows (see illustrations):

  a) *Initially align the TDC mark on the sprocket with the arrow on the oil pump housing.*

  b) *Back it off counterclockwise 3 notches BTDC.*

  c) *Rotate the crankshaft timing sprocket clockwise to 1/2-notch BTDC.*

29 Install the timing belt as follows; first place the belt onto the crankshaft sprocket, maintaining tension on the belt, wrap it around the water pump sprocket, idler pulley and camshaft sprocket, then slip the belt onto the tensioner pulley.

30 To take the slack out of the timing belt, rotate the crankshaft timing sprocket clockwise to align the marks (TDC), make sure the camshaft sprocket timing marks remain aligned.

31 Reset the tension on the timing belt:

  a) *On 2000 models, remove the Allen wrench from the tensioner (see illustration 7.19a). If you have installed a new tensioner*

assembly, remove the pull pin (new tensioners are locked in the "wound" position by a pull pin).

  b) *On 2001 and 2002 models with hydraulic tensioners, install the tensioner assembly - don't tighten the bolts at this time. Place a torque wrench on the center bolt of the tensioner pulley and apply 21 inch-lbs of torque. With the torque applied to the tensioner pulley, move the tensioner up against the tensioner pulley bracket and tighten the tensioner bolts to the torque listed in this Chapter's Specifications (see illustration 7.19b). Remove the torque wrench. Release the Allen wrench or pin from the tensioner. The timing belt tension is correct when the pin can be withdrawn and reinserted easily. Double check that the timing marks on both the camshaft sprocket(s) and crankshaft sprocket are still aligned at TDC.*

  c) *On 2002 models with mechanical tensioners and all 2003 and later models, insert a 6 mm Allen wrench into the hexagon opening located on the top plate of the belt tensioner pulley (see illustration 7.19c). Then, using the 6 mm Allen wrench as a lever, rotate the belt tensioner COUNTERCLOCKWISE until there is tension on the timing belt . Continue to rotate the tension until the setting notch is aligned with the spring tang (see illustration).*

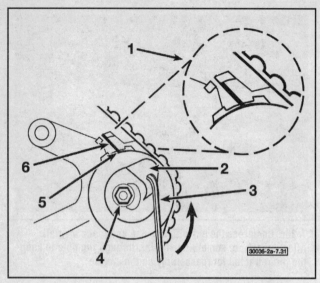

7.31 On 2003 and later models with mechanical tensioners, rotate the tensioner until the setting notch aligns with the spring tang

1	Setting notch and spring tang alignment details	4	Lock nut
2	Top plate	5	Setting notch
3	6 mm Allen wrench	6	Spring tang

Grip the Allen wrench to prevent the tensioner from rotating, and torque the tensioner bolt to 22 ft.-lbs. Recheck the alignment marks. If they are incorrect, loosen the bolt and repeat the procedure.

32 Using the bolt in the center of the crankshaft sprocket, turn the crankshaft clockwise through two complete revolutions.

✳✳ CAUTION:

If you feel resistance while turning the crankshaft - STOP, the valves may be hitting the pistons from incorrect valve timing. Stop and re-check the valve timing.

➡**Note: The camshaft and crankshaft sprocket marks will align every two revolutions of the crankshaft.**

33 On 2002 models with mechanical tensioners and all 2003 and later models, check the alignment of the spring tang and the alignment notch on the tensioner. If the alignment is incorrect, repeat the above procedure.

34 Recheck the alignment of the timing marks. If the marks do not align properly, loosen the tensioner, slip the belt off the camshaft sprocket, realign the marks, reinstall the belt, and check the alignment again.

35 Reinstall the remaining parts in the reverse order of removal.

36 Start the engine and road test the vehicle.

8   Crankshaft front oil seal - replacement

▸ Refer to illustrations 8.2, 8.3, 8.5 and 8.6

✳✳ CAUTION:

Do not rotate the camshaft(s) or crankshaft when the timing belt is removed or damage to the engine may occur.

1   Remove the timing belt cover(s) and then remove the timing belt(s) (see Section 7).

2   Pull the crankshaft sprocket from the crankshaft with a bolt-type gear puller (see illustration). Remove the Woodruff key.

3   Wrap the tip of a small screwdriver with tape. Working from below the right inner fender, use the screwdriver to pry the seal out of its bore (see illustration). Take care to prevent damaging the oil pump assembly, the crankshaft and the seal bore.

4   Thoroughly clean and inspect the seal bore and sealing surface on the crankshaft. Minor imperfections can be removed with emery cloth. If there is a groove worn in the crankshaft sealing surface (from contact with the seal), installing a new seal will probably not stop the leak.

5   Lubricate the new seal with engine oil and drive the seal into

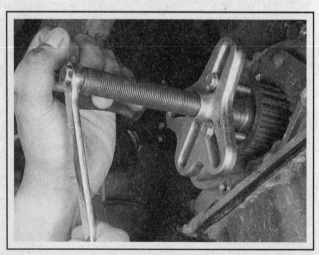

8.2 Attach a bolt-type gear puller to the crankshaft sprocket and remove the sprocket from the crankshaft

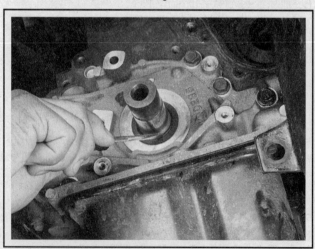

8.3 Working from below the inner fender, use the screwdriver to pry the seal out of its bore

8.5 Lubricate the new seal with engine oil and drive the seal into place with a hammer and socket

8.6 Position the crankshaft sprocket with the word FRONT facing out and install it onto the crankshaft

place with a hammer and an appropriate size socket (see illustration).

6   The remaining steps are the reverse of removal.

※※ CAUTION:

The crankshaft sprocket depth is set to a factory specification to insure correct timing belt alignment. Use a special tool with a built-in setting to install the crankshaft sprocket.

➡**Note: Position the crankshaft sprocket with the word FRONT facing out (see illustration).**

7   Reinstall the timing belt(s) and timing covers (see Section 7).

8   Run the engine and check for oil leaks.

9   Camshaft oil seal - replacement

▶ **Refer to illustrations 9.5, 9.7 and 9.9**

※※ CAUTION:

Do not rotate the camshaft(s) or crankshaft when the timing belt is removed or damage to the engine may occur.

1   Remove the timing belt cover and timing belt (see Section 7).

2   Rotate the crankshaft counterclockwise until the crankshaft sprocket is three notches BTDC (see illustration 7.28b). This will prevent engine damage if the camshaft sprocket is inadvertently rotated during removal.

3   Using a strap wrench or a chain wrench (use shop rags if you're using a chain wrench, to prevent damage to the sprocket teeth), remove

the camshaft sprocket bolt. Then, using two large screwdrivers, lever the sprocket off the camshaft.

4   Remove the timing belt tensioner assembly and the upper, rear timing belt cover.

➡**Note: 2001 and 2002 models with hydraulic tensioners are equipped with a timing belt pulley assembly and a separate tensioner.**

5   Note how far the seal is seated in the bore, then carefully pry it out with a small screwdriver (see illustration). Don't scratch the bore or damage the camshaft in the process (if the camshaft is damaged, the new seal will end up leaking).

6   Clean the bore and coat the outer edge of the new seal with engine oil or multi-purpose grease. Also lubricate the seal lip.

7   Using a socket with an outside diameter slightly smaller than the

9.5 Carefully pry the camshaft seal out of the bore - DO NOT nick or scratch the camshaft or seal bore

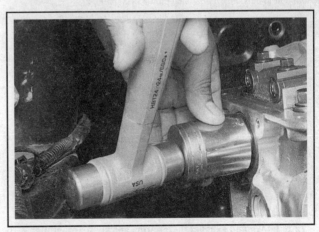

9.7 Gently tap the new seal into place with the spring side toward the engine

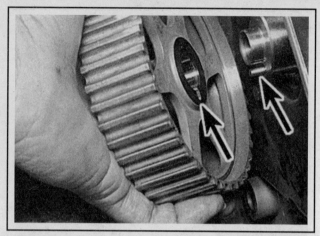

9.9 When installing a camshaft sprocket, make sure the pin in the camshaft is aligned with the hole in the sprocket

outside diameter of the seal (see illustration), carefully drive the new seal into place with a hammer. Make sure it's installed squarely and driven in to the same depth as the original. If a socket isn't available, a short section of pipe will also work.

8 Install the upper, rear timing belt cover. Install the timing belt tensioner assembly and torque the bolts to the Specifications listed in this Chapter.

9 Install the camshaft sprocket, aligning the pin in the camshaft with the hole in the sprocket (see illustration). Use an appropriate tool to hold the camshaft sprocket(s) while tightening the bolt(s) to the torque listed in this Chapter's Specifications.

10 Reinstall the timing belt (see Section 7).

11 Run the engine and check for oil leaks at the camshaft seal.

10 Rocker arm assembly - removal, inspection and installation

REMOVAL

▶ **Refer to illustration 10.6**

1 Position the number one piston at Top Dead Center (see Section 3).

2 Disconnect the cable from the negative battery terminal (see Chapter 5).

3 Remove the valve cover (see Section 4).

4 Prior to removing the rocker arm shafts, mark the front shaft as the intake rocker arm shaft and the rear shaft as the exhaust.

> ✳✳ **CAUTION:**
>
> **Do not interchange the rocker arms onto a different shaft as this could lead to premature wear.**

5 Loosen the rocker arm shaft bolts 1/4-turn at a time each, until the spring pressure is relieved, in the reverse of the TIGHTENING sequence (see illustration 10.15). Completely loosen the bolts, but do not remove them, since leaving them in place will prevent the assembly from falling apart when it is lifted off the cylinder head.

6 Lift the rocker arms and shaft assemblies from the cylinder head and set them on the workbench (see illustration).

7 Inspect the rocker arm assemblies.

INSPECTION

▶ **Refer to illustration 10.9**

8 Disassemble the rocker arm shaft components (see illustration 10.6).

> ✳✳ **CAUTION:**
>
> **Before disassembly, mark the rocker arm shafts, rocker arms, shaft retainers and plastic shaft spacers (intake only) so all the parts are reassembled in the same locations they were removed from. To keep the rocker arms and related parts in order, it's a good idea to remove them and put them onto two lengths of wire (such as unbent coat hangers) in the same order as they're removed, marking each wire (which simulates the rocker shaft) as to which end would be the front of the engine.**

9 Visually check the rocker arms for wear (see illustration).

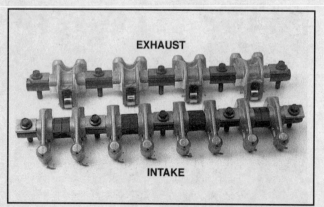

10.6 Intake and exhaust rocker arms and shaft assemblies are unique - don't intermix any of the parts

EXHAUST

INTAKE

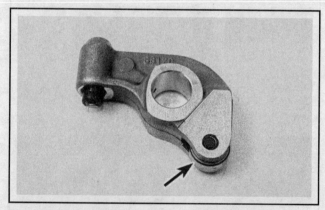

10.9 Visually check the rocker arm shaft bore and roller for score marks, pitting and evidence of overheating (blue discolored areas)

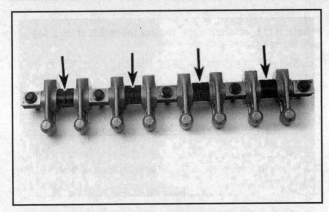

10.13 The intake rocker arm shaft and the plastic spacers must be installed in the correct locations

10.14 Both rocker arm shafts must be positioned with the notch facing UP

10.15 Camshaft rocker arm shaft bolt tightening sequence

Replace them if evidence of wear or damage is found.

10 Check all the rocker shaft components. Look for worn or scored shafts, etc. and replace any of the parts found to be damaged.

INSTALLATION

▶ **Refer to illustrations 10.13, 10.14 and 10.15**

11 Prior to installation, make sure each lash adjuster is at least partially full of oil (see Section 12).

12 When reassembling the parts, be sure they all go back on in the same locations they were removed from.

13 On the intake rocker arm shaft, make sure the plastic spacers are installed on the shaft in the correct location (see illustration).

14 Prior to installing the rocker arm shaft, to prevent internal damage, rotate the crankshaft to position the crankshaft timing sprocket at the 3 notches BTDC (see illustration 7.28b). Install the rocker arm assemblies with the notch in each rocker arm shaft located at the timing belt end of the engine and facing UP (see illustration).

15 Tighten the rocker arm bolts in sequence (see illustration) to the torque listed in this Chapter's Specifications.

16 The remainder of the reassembly is in the reverse order of disassembly. Run the engine and check for oil leaks and proper operation.

11 Camshaft - removal, inspection and installation

REMOVAL

➡**Note: The camshaft can't be removed with the cylinder head installed in the vehicle.**

1 Remove the cylinder head (see Section 13) and rocker arm assemblies (see Section 10).

2 Remove the Camshaft Position sensor (see Chapter 6).

3 Carefully withdraw the camshaft from the opening in the rear of the cylinder head.

✳ **CAUTION:**

Don't damage the camshaft lobes or bearing journals during removal and installation through the opening in the cylinder head.

INSPECTION

▶ **Refer to illustration 11.5**

4 Remove the seal(s) from the camshaft(s) and thoroughly clean the camshaft(s) and the gasket surface. Visually inspect the camshaft for wear and/or damage to the lobe surfaces, bearing journals and seal contact surfaces. Visually inspect the camshaft bearing surfaces in the cylinder head for scoring and other damage.

5 Measure the camshaft bearing journal diameters (see illustration). Measure the inside diameter of the camshaft bearing surfaces in the cylinder head, using a telescoping gauge. Subtract the journal measurement from the bearing measurement to obtain the camshaft bearing oil clearance. Compare this clearance with this Chapter's Specifications. Lubricate the camshaft journals with clean engine oil and install the camshaft into the cylinder head. Install the camshaft sensor. Set up

a dial indicator and measure the camshaft endplay. Compare your measurement with the value listed in this Chapter's Specifications.

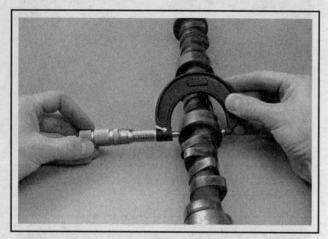

11.5 Measure the camshaft bearing journal diameters with a micrometer

6 Replace the camshaft if it fails any of the above inspections.

➡Note: If the lobes are worn, replace the rocker arms along with the camshaft.

Cylinder head replacement may be necessary if the camshaft bearing surfaces in the head are damaged or excessively worn or if the endplay is excessive.

INSTALLATION

7 Very carefully, clean the camshaft and bearing journals. Liberally coat the journals, lobes and thrust portions of the camshaft with assembly lube or engine oil.

8 Carefully install the camshaft in the cylinder head.

9 Install a new camshaft oil seal (see Section 9).

10 Install the Camshaft Position sensor (see Chapter 6).

11 Install the cylinder head (see Section 13).

12 Install the rocker arm shaft assembly (see Section 10).

13 Install the timing belt and covers (see Section 7).

14 Run the engine while checking for oil leaks.

12 Valve lash adjusters - removal, inspection and installation

➡Note: The valve lash adjuster is an integral part of each rocker arm and can't be replaced separately.

1 Remove the rocker arm shafts (see Section 10). Don't remove the rocker arms from the shafts.

2 Turn the rocker arm assembly upside down on the workbench. Inspect each lash adjuster carefully for signs of wear and damage, particularly on the surface that contacts the valve tip. Since the lash adjusters frequently become clogged, we recommend replacing the rocker arm/lash adjuster assembly if you're concerned about their condition or if the engine is exhibiting valve "tapping" noises.

3 If any are removed, assemble the rocker arms onto their shaft(s) (see Section 10).

4 The lash adjusters must be partially full of engine oil - indicated by little or no plunger action when the adjuster is depressed. If there's excessive plunger travel, place the rocker arm assembly into clean engine oil and pump the plunger until the plunger travel is eliminated.

➡Note: If the plunger still travels within the rocker arm when full of oil it's defective and the rocker arm assembly must be replaced.

5 When re-starting the engine after replacing the rocker arm/lash adjusters, the adjusters will normally make "tapping" noises. After warm-up, raise the speed of the engine from idle to 3,000 rpm for one minute. If the adjuster(s) do not become silent, replace the defective rocker arm/lash adjuster assembly(ies).

13 Cylinder head - removal and installation

13.4 Cover the intake ports with duct tape to keep out debris before removing the cylinder head

❊❊ **WARNING:**

Allow the engine to cool completely before beginning this procedure.

REMOVAL

▶ **Refer to illustrations 13.4, 13.10a and 13.10b**

1 Position the number one piston at Top Dead Center (see Section 3).

2 Disconnect the cable from the negative battery terminal (see Chapter 5).

3 Drain the cooling system and remove the spark plugs (see Chapter 1).

4 Remove the intake manifold (see Section 5). Cover the intake ports with duct tape to keep out debris (see illustration).

5 If necessary, remove the exhaust manifold (see Section 6).

13.10a Carefully lift the cylinder head straight up and place the head on wood blocks to prevent damage to the sealing surfaces

➡Note: On some models, the exhaust manifold is easier to remove after the cylinder head is removed.

6 Remove the ignition system components (see Chapter 5).
7 Remove the valve cover (see Section 4).
8 Remove the timing belt (see Section 7).
9 Loosen the cylinder head bolts, 1/4-turn at a time, in the reverse of the tightening sequence (see illustration 13.14) until they can be removed by hand.

➡Note: Write down the location of the different length bolts so they will be reinstalled in the correct location.

10 Carefully lift the cylinder head (see illustration) straight up and place the head on wood blocks to prevent damage to the sealing surfaces. If the head sticks to the engine block, dislodge it by placing a wood block against the head casting and tapping the wood with a hammer or by prying the head with a prybar placed carefully on a casting protrusion (see illustration).

❈❈ CAUTION:

The cylinder head is aluminum, so you must be very careful not to gouge the sealing surfaces.

➡Note: It's a good idea to have the head checked for warpage, even if you're just replacing the gasket.

11 Special gasket removal solvents that soften gaskets and make removal much easier are available at auto parts stores. Remove all traces of old gasket material from the block and head. Do not allow anything to fall into the engine. Clean and inspect all threaded fasteners and be sure the threaded holes in the block are clean and dry.

INSTALLATION

▶ **Refer to illustration 13.14**

12 Place a new gasket and the cylinder head in position on the engine block.
13 Apply clean engine oil to the cylinder head bolt threads prior to installation.

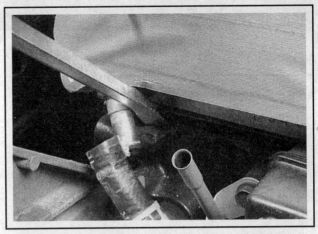

13.10b If the head sticks to the engine block, dislodge it by placing a wood block against the head casting and tapping the wood with a hammer or by prying the head with a prybar placed carefully on a casting protrusion

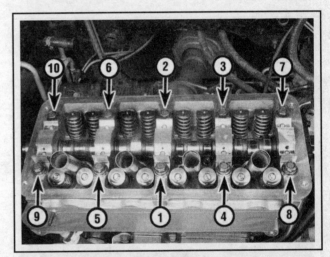

13.14 Cylinder head bolt TIGHTENING sequence

14 Tighten the cylinder head bolts in several stages in the recommended sequence (see illustration) to the torque listed in this Chapter's Specifications.

➡Note: The final step in the tightening procedure requires you to tighten the bolts a specific number of degrees. An angle-torque gauge, that fits on your torque wrench, is available at most auto parts stores and is highly recommended for this procedure. If the tool is not available, paint marks on the bolt heads and tighten them in sequence until the mark is the specified number of degrees from the starting point.

15 Reinstall the timing belt (see Section 7).
16 Reinstall the remaining parts in the reverse order of removal.
17 Be sure to refill the cooling system and check all fluid levels. Rotate the crankshaft clockwise slowly by hand through six complete revolutions. Recheck the camshaft timing marks (see Section 7).
18 Start the engine and run it until normal operating temperature is reached. Check for leaks and proper operation.

---

14 Oil pan - removal and installation

REMOVAL

▶ **Refer to illustrations 14.5a, 14.5b, 14.7a, 14.7b and 14.9**

1   Disconnect the cable from the negative battery terminal (see Chapter 5).

2   Raise the vehicle and support it securely on jackstands. Working under the vehicle, remove the engine protection cover.

3   Drain the engine oil (see Chapter 1).

4   Remove the oil filter adapter from the engine block.

5   Remove the structural collar and the intake manifold support bracket (see illustrations).

6   Remove the flywheel/driveplate inspection cover.

7   Remove the mounting bolts and lower the oil pan from the vehicle (see illustration). If the pan is stuck, tap it with a soft-face hammer (see illustration) or place a wood block against the pan and tap the wood block with a hammer.

8   Remove the oil pump pickup tube and screen assembly (see illustration) and clean both the tube and screen thoroughly. Install the pick-up tube and screen with a new seal (see illustration).

9   Thoroughly clean the oil pan and sealing surfaces on the block and pan (see illustration). Use a scraper to remove all traces of old gasket material. Gasket removal solvents are available at auto parts stores and may prove helpful. Check the oil pan sealing surface for distortion. Straighten or replace as necessary. After cleaning and straightening (if necessary), wipe the gasket surfaces of the pan and block clean with a rag soaked in lacquer thinner or acetone.

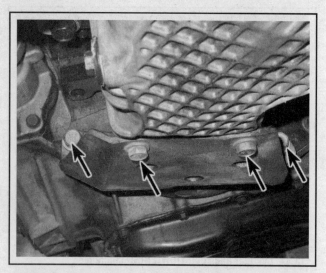

14.5a Structural collar mounting bolt locations

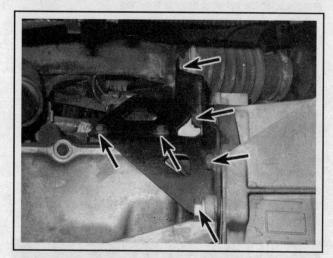

14.5b Intake manifold support bracket mounting bolt locations

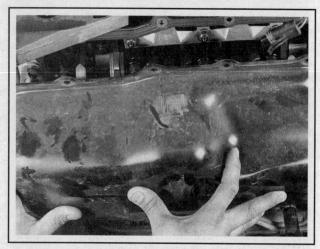

14.7a Lower the pan carefully (there might still be some residual oil in the pan)

14.7b If the pan is stuck, tap it with a soft-face hammer or place a wood block against the pan and tap the wood block with a hammer

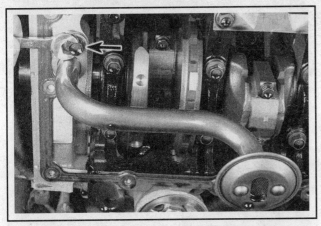

14.8a Remove the bolt and remove the oil pump pick-up tube and screen assembly - clean both the tube and screen thoroughly before reassembly

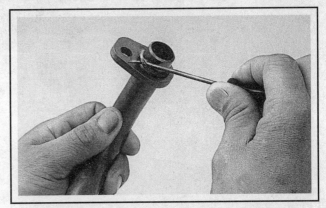

14.8b Install a new seal at the oil pump pick-up tube mounting flange

INSTALLATION

10 Apply a 1/8-inch bead of RTV sealant at the cylinder block-to-oil pump assembly joint at the oil pan flange. Install a new oil pan gasket.

11 Place the oil pan into position and install the bolts finger tight. Working side-to-side from the center out, tighten the bolts to the torque listed in this Chapter's Specifications.

12 Install the intake manifold support bracket (see illustration 14.5b). Tighten the bolts securely.

13 Install the structural collar as follows (see illustration 14.5a):

  a) Place the collar in position between the transaxle and the oil pan and then tighten the collar-to-oil pan bolts to about 30 in-lbs.

  b) Install the collar-to-transaxle bolts and then tighten them to 80 ft. lbs.

  c) Finally, torque the structural collar-to-oil pan bolts to 40 ft.-lbs.

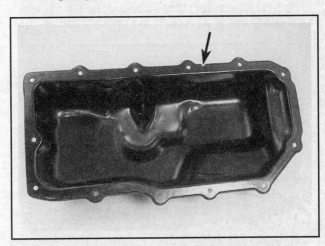

14.9 Thoroughly clean the oil pan and sealing surfaces on the engine block and oil pan with a scraper to remove all traces of old gasket material

15 Oil pump - removal, inspection and installation

REMOVAL

◆ **Refer to illustrations 15.5a, 15.5b, 15.6a, 15.6b and 15.6c**

1 Disconnect the cable from the negative battery terminal (see Chapter 5).

2 Raise the vehicle and then place it securely on jackstands.

3 Remove the oil pan and pick-up tube/strainer assembly (see Section 15).

4 Remove the timing belt and the crankshaft sprocket (see Section 7).

5 Remove the bolts and detach the oil pump assembly from the engine (see illustration).

✴ CAUTION:

If the pump doesn't come off by hand, tap it gently with a soft-faced hammer or pry on a casting boss (see illustration).

15.5a Remove the oil pump assembly mounting bolts and remove the assembly

15.5b If the pump doesn't come off by hand, tap it gently with a soft-faced hammer or pry gently on a casting protrusion

15.6a Remove the cover mounting screws . . .

15.6b . . . and the cover

6   Remove the mounting screws and remove the cover (see illustration). Remove the inner and outer rotor from the body (see illustrations).

✳✳ CAUTION:

Be very careful with these parts. Close tolerances are critical in creating the correct oil pressure. Any nicks or other damage will require replacement of the complete pump assembly.

7   If necessary, replace the crankshaft front seal within the oil pump body (see Section 8).

INSPECTION

▸ **Refer to illustrations 15.9a, 15.9b, 15.10a, 15.10b, 15.10c, 15.10d and 15.10e**

8   Clean all components including the block surfaces, with solvent, then inspect all surfaces for excessive wear and/or damage.
9   Disassemble the relief valve, unscrew the cap bolt and remove the bolt, washer, spring and relief valve (see illustrations). Check the oil pressure relief valve piston sliding surface and valve spring. If either the spring or the valve is damaged, they must be replaced as a set. If no damage is found reassemble the relief valve parts. Make sure to install the relief valve into the pump body with the grooved end going in first. Coating the parts with oil, and reinstall them in the oil pump body.

15.6c Arrangement of oil pump components

A Cover	C Inner rotor
B Outer rotor	D Oil pump body

Tighten the cap bolt securely.
10   Check the clearance of the oil pump components with a micrometer and a feeler gauge (see illustrations) and compare the results to this Chapter's Specifications.

15.9a Unscrew the oil pressure relief valve cap from the body

INSTALLATION

♦ Refer to illustration 15.12

11 Lubricate the housing and the inner and outer rotors with clean engine oil and install both rotors in the body. Install the cover and tighten the cover screws securely. Prime the oil pump with clean engine oil.

12 Install a new O-ring in the oil discharge passage (see illustration). Apply sealant to the oil pump body, and attach the pump assembly to the block. Tighten the bolts to the torque listed in this Chapter's Specifications.

13 Install the crankshaft sprocket (see Section 8) and the timing belt (see Section 7).

14 Install the pick-up tube/strainer assembly and oil pan (see Section 15).

15 Install a new oil filter and engine oil (see Chapter 1).

16 Start the engine and check for oil pressure and leaks.

17 Recheck the engine oil level.

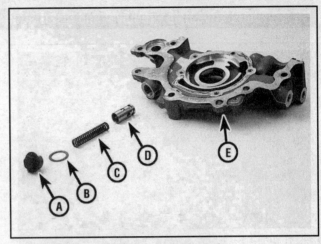

15.9b Oil pressure relief valve components

A	Cap	C	Spring	E	Oil pump
B	Gasket	D	Relief valve		body

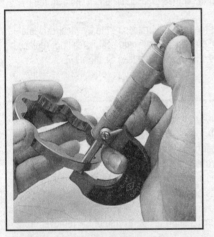

15.10a Measure the outer rotor thickness

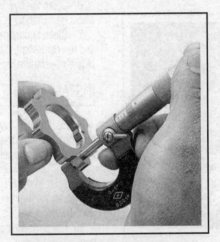

15.10b Measure the inner rotor thickness

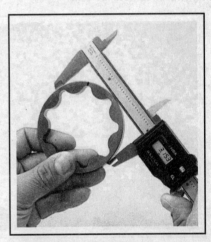

15.10c Use a dial caliper and measure the outer diameter of the outer rotor

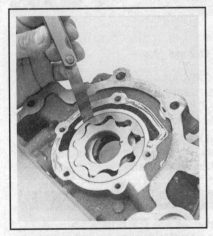

15.10d Use a flat feeler gauge and measure the outer rotor-to-case clearance

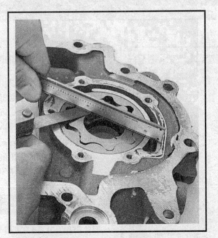

15.10e Place a precision straightedge over the rotors and measure the clearance between the rotors and the cover

15.12 Install a new O-ring seal in the oil pump body - apply clean engine oil to the seal

16 Flywheel/driveplate - removal and installation

16.4 Mark the relative position of the flywheel or driveplate to the crankshaft and, using an appropriate tool to hold the flywheel, remove the bolts

16.5 Remove the flywheel/driveplate from the crankshaft

REMOVAL

⬧ **Refer to illustrations 16.4 and 16.5**

1   Raise the vehicle and support it securely on jackstands, then refer to Chapter 7 and remove the transaxle assembly.

2   If the vehicle has a manual transaxle, remove the pressure plate and clutch disc (see Chapter 8). Now is a good time to check/replace the clutch components and the pilot bearing.

3   To ensure correct alignment during reinstallation, mark the position of the flywheel/driveplate to the crankshaft before removal.

4   Remove the bolts that secure the flywheel/driveplate to the crankshaft (see illustration). A tool is available at most auto parts stores to hold the flywheel/driveplate while loosening the bolts, if the tool is not available wedge a screwdriver in the ring gear teeth to jam the flywheel.

5   Remove the flywheel/driveplate from the crankshaft (see illustration). Since the flywheel is fairly heavy, be sure to support it while removing the last bolt.

6   Clean the flywheel to remove grease and oil. To inspect the flywheel, see Chapter 8.

7   Clean and inspect the mating surfaces of the flywheel/driveplate and the crankshaft. If the crankshaft rear main seal is leaking, replace it before reinstalling the flywheel/driveplate (see Section 17).

INSTALLATION

8   Position the flywheel/driveplate against the crankshaft. Align the previously applied match marks. Before installing the bolts, apply thread locking compound to the threads.

9   Hold the flywheel/driveplate with the holding tool, or wedge a screwdriver in the ring gear teeth to keep the flywheel/driveplate from turning as you tighten the bolts to the torque listed in this Chapter's Specifications.

10   The remainder of installation is the reverse of the removal procedure.

17 Rear main oil seal - replacement

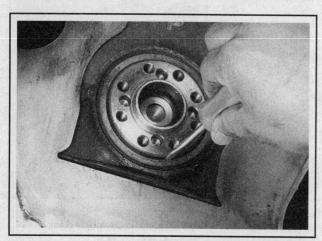

17.2 Carefully pry the crankshaft seal out of the bore - DO NOT nick or scratch the crankshaft or seal bore

⬧ **Refer to illustrations 17.2 and 17.4**

1   The one-piece rear main oil seal is pressed into a bore machined into the rear main bearing cap and engine block. Remove the transaxle (see Chapter 7), the clutch components, if equipped (see Chapter 8) and the flywheel or driveplate (see Section 17).

➥ **2 Note: Observe that the oil seal is installed flush with the outer surface of the block. Pry out the old seal with a 3/16-inch flat blade screwdriver (see illustration).**

⁑ **CAUTION:**

To prevent an oil leak after the new seal is installed, be very careful not to scratch or otherwise damage the crankshaft sealing surface or the bore in the engine block.

3   Clean the crankshaft and seal bore in the block thoroughly and de-grease these areas by wiping them with a rag soaked in lacquer thinner or acetone. Do not lubricate the lip or outer diameter of the new seal - it must be installed as it comes from the manufacturer.

4   Position the new seal onto the crankshaft.

➡Note: When installing the new seal, if so marked, the words THIS SIDE OUT on the seal must face out, toward the rear of the engine. Using an appropriate size driver and pilot tool, drive the seal into the cylinder block until it is flush with the outer surface of the block. If the seal is driven in past flush, there will be an oil leak. Check that the seal is flush (see illustration).

5   The remainder of installation is the reverse of removal.

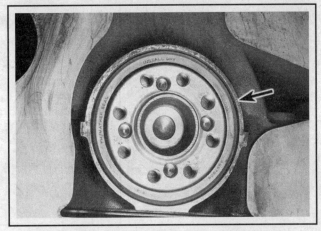

17.4 Position the new seal with the words THIS SIDE OUT facing out, toward the rear of the engine. Gently drive the seal into the cylinder block until it is flush with the outer surface of the block. Do not drive it past flush or there will be an oil leak - the seal must be flush

18 Engine mounts - check, replacement and torque strut adjustment

1   The engine mounting system on these vehicles consists of two load carrying mounts (right and left) and two torque struts (upper and lower). This four-point system requires proper adjustment in the event of replacement. The upper and lower torque struts need to be adjusted together to properly position the engine and load. The adjustment procedure must also be performed if a torque strut bolt is loosened for engine component replacement procedures or other reasons.

2   Engine mounts seldom require attention, but broken or deteriorated mounts should be replaced immediately or the added strain placed on the driveline components may cause damage or wear.

CHECK

3   During the check, the engine must be raised slightly to remove the weight from the mounts.

4   Raise the vehicle and support it securely on jackstands, then position a jack under the engine oil pan. Place a large wood block between the jack head and the oil pan to prevent oil pan damage, then carefully raise the engine just enough to take the weight off the mounts.

❄❄ WARNING:

DO NOT place any part of your body under the engine when it's supported only by a jack!

5   Check the mounts to see if the rubber is cracked, hardened or separated from the metal backing. Sometimes the rubber will split right down the center.

6   Check for relative movement between the mount plates and the engine or frame (use a large screwdriver or pry bar to attempt to move the mounts). If movement is noted, lower the engine and tighten the mount fasteners.

7   Rubber preservative may be applied to the mounts to slow deterioration.

REPLACEMENT

8   Disconnect the cable from the negative battery terminal (see Chapter 5), then raise the vehicle and support it securely on jackstands (if not already done).

9   Remove the front wheels and splash shields.

10 Remove the air filter housing (see Chapter 4).

Left engine mount

▶ **Refer to illustrations 18.13a, 18.13b and 18.15**

11 Remove the battery tray (see Chapter 5).

12 Place a floor jack under the transaxle (with a wood block between the jack head and the transaxle) and raise the transaxle slightly to relieve the weight from the mounts.

13 Remove the access plug and through bolt on the engine mount (see illustrations).

18.13a First, remove the access plug and . . .

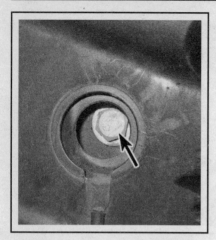

18.13b ... then remove the through bolt for the left engine mount

18.15 After the transaxle is lowered slightly, all the bolts for the left engine mount and bracket will be exposed

18.19 Location of the lower mounting bolts on the right engine mount

14 Remove the transaxle shift cable (see Chapter 7) and position it off to the side.

15 Remove the mounting bolts for the left engine mount and separate it from the left frame rail (see illustration).

16 Installation is the reverse of removal. Use thread locking compound on the mount bolts and be sure to tighten them securely.

Right engine mount

▶ Refer to illustration 18.19

✳✳ CAUTION:

The right engine mount is designed with large diameter mounting holes. Be sure to use paint to mark the exact position of the engine mount in relation to the frame rail. The engine mount has been preset for correct engine alignment.

17 Place a floor jack under the engine (with a wood block between the jack head and the oil pan) and raise the engine slightly to relieve the weight from the mounts.

18 Carefully mark the position of the right engine mount before removing any fasteners.

19 Remove the bolts and detach the mount from the frame and engine (see illustration)

20 Installation is the reverse of removal. Use thread locking compound on the mount bolts and be sure to tighten them securely.

Upper torque strut

▶ Refer to illustration 18.21

21 Remove the bolts that secure the upper torque strut to the shock tower bracket and engine mount bracket (see illustration).

22 Separate the upper torque strut from the vehicle.

23 Installation is the reverse of removal.

24 Be sure to perform the engine torque strut adjustment procedure (see Steps 29 through 37). Use thread locking compound on the torque strut bolts and be sure to tighten them securely.

Lower torque strut

▶ Refer to illustration 18.26

25 Remove the pencil strut, if equipped.

26 Remove the torque strut mounting bolts (see illustration) and separate the torque strut from the crossmember.

27 Installation is the reverse of removal.

28 Be sure to perform the engine torque strut adjustment procedure (see Steps 29 through 37). Use thread locking compound on the torque strut bolts and be sure to tighten them securely.

18.21 Remove the bolts from the upper torque strut and bracket

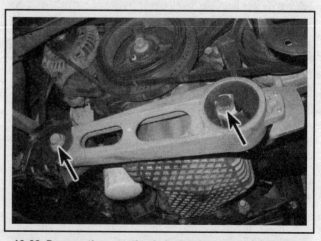

18.26 Remove the mounting bolts for the lower torque strut

TORQUE STRUT ADJUSTMENT

♦ **Refer to illustration 18.34**

29 Disconnect the cable from the negative battery terminal, then raise the vehicle and support it securely on jackstands.

30 Loosen the upper and lower torque strut attachment bolts at the crossmember and the shock tower bracket.

31 Install a floorjack on the forward edge of the transmission bell-housing.

32 Remove the upper and lower torque strut bolts (see above). Make sure the torque struts are free and not binding.

33 Reinstall the torque strut bolts hand tight.

34 Carefully apply upward force using the floorjack. Allow the engine to rotate to the rear and measure the distance between the engine mount bracket (point A) and the center of the hole on the shock tower bracket (point B) (see illustration). This distance should be 4.70 inches (119 mm).

35 With the engine in the correct position, tighten the upper and lower torque strut bolts to the Specifications listed in this Chapter.

36 Remove the floorjack.

37 Installation is the reverse of removal.

18.34 Measure the distance from the center of the rear engine mount bolt (point A) to the center of the hole on the shock tower bracket (point B); this distance should be 4.70 inches (119 mm)

General

Displacement	122 cubic inches (2.0 liters)
Bore	3.445 inches (87.5 mm)
Stroke	3.268 inches (83.0 mm)
Compression ratio	
2000 through 2002 models	9.8:1
2003 and later models	
High output models	9.8:1
All other models	9.3:1
Compression pressure	
Standard	170 to 225 psi (1,172 to 1,551 kPa)
Minimum	25 percent
Firing order	1-3-4-2
Oil pressure	
At curb idle (minimum)	4 psi (25 kPa)
At 3,000 rpm	25 to 80 psi (170 to 550 kPa)

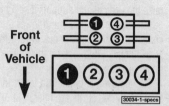

Front of Vehicle

Cylinder numbering and coil terminal location (2.0L SOHC engines)

Camshaft

Bearing journal diameter	
No. 1	1.6190 to 1.6199 inches (41.128 to 41.147 mm)
No. 2	1.634 to 1.635 inches (41.528 to 41.547 mm)
No. 3	1.650 to 1.651 inches (41.928 to 41.947 mm)
No. 4	1.666 to 1.668 inches (42.328 to 42.374 mm)
No. 5	1.6820 to 1.6829 inches (42.728 to 42.747 mm
Bearing bore diameter	
No. 1	1.6220 to 1.6228 inches (41.200 to 41.221 mm)
No. 2	1.637 to 1.638 inches (41.60 to 41.62 mm)
No. 3	1.653 to 1.654 inches (42.00 to 42.02 mm)
No. 4	1.669 to 1.670 inches (42.40 to 42.42 mm)
No. 5	1.6850 to 1.6858 inches (42.80 to 42.82 mm)
End play	0.002 to 0.015 inch (0.05 to 0.39 mm)
Bearing clearance	0.0021 to 0.0037 inch (0.053 to 0.093 mm)
Lobe lift	
High output models	
Intake	0.344 inch (8.5 mm)
Exhaust	0.314 inch (8.0 mm)
All other models	
Intake	0.335 inch (7.20 mm)
Exhaust	0.315 inch (7.03 mm)

Cylinder head

Cylinder head warpage (maximum)	
Head gasket surface	0.004 inch (0.101 mm)
Exhaust manifold mounting surfaces	0.006 inch (0.162 mm)

Intake and exhaust manifolds

Warpage limit	0.006 inch (0.152 mm)

Rocker arm shaft assemblies

Rocker arm shaft diameter	0.7861 to 0.7868 inch (19.966 to 19.984 mm)
Rocker arm inside diameter	0.787 to 0.788 inch (20.00 to 20.02 mm)
Rocker arm-to-shaft clearance	0.0006 to 0.0021 inch (0.016 to 0.054 mm)
Rocker arm shaft retainer width	
Intake	1.119 to 1.122 inches (28.43 to 28.49 mm)
Exhaust	
No. 1 and No. 5	1.147 to 1.152 inches (29.15 to 29.25 mm)
No. 2, No. 3 and No. 4	1.591 to 1.595 inches (40.40 to 40.50 mm)

Oil pump

Cover warpage limit	0.003 inch (0.076 mm)
Inner rotor thickness (minimum)	0.301 inch (7.64 mm)
Outer rotor thickness (minimum)	0.301 inch (7.64 mm)
Outer rotor diameter (minimum)	3.148 inch (79.95 mm)
Rotor-to-pump cover clearance	0.004 inch (0.102 mm)
Outer rotor-to-housing clearance	0.015 inch (0.039 mm)
Inner rotor-to-outer rotor lobe clearance	0.008 inch (0.203 mm)
Pressure relief spring free length (approximate)	2.39 inches (60.7 mm)

Torque specifications

	Ft-lbs (unless otherwise indicated)	Nm
Camshaft position sensor bolt	85 inch-lbs	10
Camshaft sprocket bolt	85	115
Crankshaft damper bolt	100	136
Cylinder head bolts (see illustration 13.14)		
2000 through 2002 models		
Step 1	25	34
Step 2	50	68
Step 3	50	68
Step 4	Tighten all bolts an additional 90-degrees (1/4 turn)	
2003 and later models		
Step 1	25	34
Step 2		
Bolts 1 through 6	50	68
Bolts 7 through 10	35	49
Step 3		
Bolts 1 through 6	50	68
Bolts 7 through 10	35	49
Step 4	Tighten all bolts an additional 90-degrees (1/4 turn)	
Exhaust manifold-to-cylinder head bolts	200 inch-lbs	23
Exhaust manifold-to-exhaust pipe bolts	20	26
Exhaust manifold heat shield bolts	95 inch-lbs	11
Flywheel/driveplate-to-crankshaft bolts	70	95
Intake manifold bolts	105 inch-lbs	12
Lower intake manifold bolts (high output models)	105 inch-lbs	12
Oil filter adapter fastener	60	80
Oil pan bolts	105 inch-lbs	12

Torque specifications (continued)   Ft-lbs (unless otherwise indicated)   Nm

	Ft-lbs (unless otherwise indicated)	Nm
Oil pump		
Attaching bolts	21	29
Cover screws	105 inch-lbs	12
Pick-up tube bolt	21	29
Relief valve cap	3041	
Rocker arm shaft bolts	21	29
Structural collar		
Step 1 (collar-to-oil pan bolts)	30 in-lbs	3
Step 2 (collar-to-transaxle bolts)	80	108
Step 3 (collar-to-oil pan bolts)	40	54
Thermostat housing bolts	See Chapter 3	
Timing belt cover bolts (front and rear)	105 inch-lbs	12
Tensioner components		
2000 models		
Tensioner assembly bolts	23	31
2001 and 2002 models (hydraulic tensioner systems)		
Tensioner bolts	23	31
Tensioner pulley assembly bolts	23	31
Tensioner pulley bolt	21	28
2002 and later models (mechanical tensioner systems)		
Tensioner pulley assembly plate bolts	23	31
Tensioner lock nut	22	30
Torque strut bolts (upper and lower)	87	118
Valve cover bolts	105 inch-lbs	12
Water pump mounting bolts	See Chapter 3	

Refer to Part B for additional torque specifications

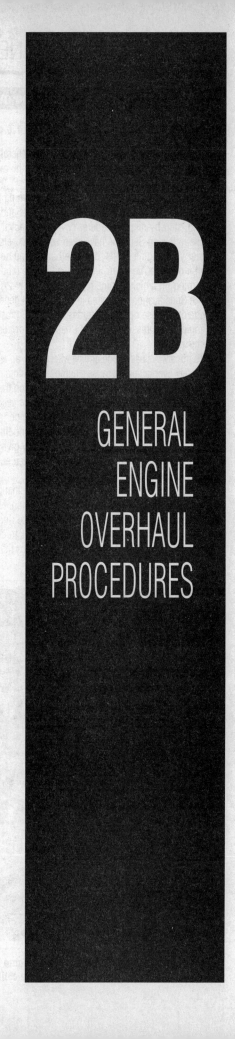

2B

GENERAL
ENGINE
OVERHAUL
PROCEDURES

Section

Reference to other Chapters

1 General information - engine overhaul

▶ **Refer to illustrations 1.2, 1.3, 1.4, 1.5, 1.6 and 1.7**

Included in this portion of Chapter 2 are general information and diagnostic testing procedures for determining the overall mechanical condition of your engine.

The information ranges from advice concerning preparation for an overhaul and the purchase of replacement parts and/or components to detailed, step-by-step procedures covering removal and installation.

The following Sections have been written to help you determine whether your engine needs to be overhauled and how to remove and install it once you've determined it needs to be rebuilt. For information concerning in-vehicle engine repair, see Chapter 2A.

The Specifications included in this Part are general in nature and include only those necessary for testing the oil pressure and checking the engine compression. Refer to Chapter 2A for additional engine Specifications.

It's not always easy to determine when, or if, an engine should be completely overhauled, because a number of factors must be considered.

High mileage is not necessarily an indication that an overhaul is needed, while low mileage doesn't preclude the need for an overhaul. Frequency of servicing is probably the most important consideration. An engine that's had regular and frequent oil and filter changes, as well as other required maintenance, will most likely give many thousands of miles of reliable service. Conversely, a neglected engine may require an overhaul very early in its service life.

Excessive oil consumption is an indication that piston rings, valve seals and/or valve guides are in need of attention. Make sure that oil leaks aren't responsible before deciding that the rings and/or guides are bad. Perform a cylinder compression check to determine the extent of the work required (see Section 3). Also check the vacuum readings under various conditions (see Section 4).

Check the oil pressure with a gauge installed in place of the oil pressure sending unit and compare it to this Chapter's Specifications (see Section 2). If it's extremely low, the bearings and/or oil pump are probably worn out.

Loss of power, rough running, knocking or metallic engine noises, excessive valve train noise and high fuel consumption rates may also point to the need for an overhaul, especially if they're all present at the same time. If a complete tune-up doesn't remedy the situation, major mechanical work is the only solution.

An engine overhaul involves restoring the internal parts to the specifications of a new engine. During an overhaul, the piston rings are replaced and the cylinder walls are reconditioned (rebored and/or honed) (see illustrations 1.2 and 1.3). If a rebore is done by an auto-motive machine shop, new oversize pistons will also be installed. The main bearings, connecting rod bearings and camshaft bearings are generally replaced with new ones and, if necessary, the crankshaft may be reground to restore the journals (see illustration 1.4). Generally, the valves are serviced as well, since they're usually in less-than-perfect condition at this point. While the engine is being overhauled, other components, such as the distributor, starter and alternator, can be rebuilt as well. The end result should be similar to a new engine that will give many trouble free miles.

➡**Note: Critical cooling system components such as the hoses, drivebelts, thermostat and water pump should be replaced with new parts when an engine is overhauled. The radiator should be checked carefully to ensure that it isn't clogged or leaking (see Chapter 3). If you purchase a rebuilt engine or short block, some rebuilders will not warranty their engines unless the radiator has been professionally flushed. Also, we don't recommend overhauling the oil pump - always install a new one when an engine is rebuilt.**

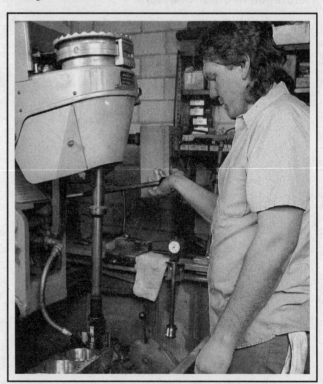

1.2 An engine block being bored. An engine rebuilder will use special machinery to recondition the cylinder bores

1.3 If the cylinders are bored, the machine shop will normally hone the engine on a machine like this

1.4 A crankshaft main bearing journal being ground

Overhauling the internal components on today's engines is a difficult and time-consuming task which requires a significant amount of specialty tools and is best left to a professional engine rebuilder (see illustrations 1.5, 1.6 and 1.7). A competent engine rebuilder will handle the inspection of your old parts and offer advice concerning the reconditioning or replacement of the original engine, never purchase parts or have machine work done on other components until the block has been thoroughly inspected by a professional machine shop. As a general rule, time is the primary cost of an overhaul, especially since the vehicle may be tied up for a minimum of two weeks or more. Be aware that some engine builders only have the capability to rebuild the engine you bring them while other rebuilders have a large inventory of rebuilt exchange engines in stock. Also be aware that many machine shops could take as much as two weeks time to completely rebuild your engine depending on shop workload. Sometimes it makes more sense to simply exchange your engine for another engine that's already rebuilt to save time.

1.5 A machinist checks for a bent connecting rod, using specialized equipment

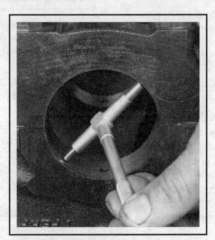

1.6 A bore gauge being used to check the main bearing bore

1.7 Uneven piston wear like this indicates a bent connecting rod

2   Oil pressure check

▶ **Refer to illustration 2.2**

1   Low engine oil pressure can be a sign of an engine in need of rebuilding. A "low oil pressure" indicator (often called an "idiot light") is not a test of the oiling system. Such indicators only come on when the oil pressure is dangerously low. Even a factory oil pressure gauge in the instrument panel is only a relative indication, although much better for driver information than a warning light. A better test is with a mechanical (not electrical) oil pressure gauge.

2   Locate the oil pressure indicator sending unit on the engine block. The oil pressure sending unit is located on the rear of the engine block, below the exhaust manifold (see illustration).

3   Unscrew and remove the oil pressure sending unit and then screw in the hose for your oil pressure gauge. If necessary, install an adapter fitting. Use Teflon tape or thread sealant on the threads of the adapter and/or the fitting on the end of your gauge's hose.

4   Connect an accurate tachometer to the engine, according to the tachometer manufacturer's instructions.

5   Check the oil pressure with the engine running (normal operating temperature) at the specified engine speed, and compare it to this

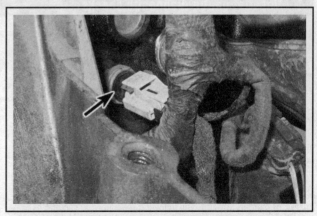

2.2 The oil pressure sending unit is located on the rear of the engine block, near the exhaust manifold

Chapter's Specifications. If it's extremely low, the bearings and/or oil pump are probably worn out.

3 Cylinder compression check

▶ **Refer to illustration 3.6**

1   A compression check will tell you what mechanical condition the upper end of your engine (pistons, rings, valves, head gaskets) is in. Specifically, it can tell you if the compression is down due to leakage caused by worn piston rings, defective valves and seats or a blown head gasket.

➡**Note: The engine must be at normal operating temperature and the battery must be fully charged for this check.**

2   Begin by cleaning the area around the spark plugs before you remove them (compressed air should be used, if available). The idea is to prevent dirt from getting into the cylinders as the compression check is being done.

3   Remove all of the spark plugs from the engine (see Chapter 1).

4   Block the throttle wide open.

5   Disable the ignition system by unplugging the wiring harness from the ignition coil pack (see Chapter 5) and by removing the fuel pump relay (see Chapter 4).

6   Install a compression gauge in the spark plug hole (see illustration).

7   Crank the engine over at least seven compression strokes and watch the gauge. The compression should build up quickly in a healthy engine. Low compression on the first stroke, followed by gradually increasing pressure on successive strokes, indicates worn piston rings. A low compression reading on the first stroke, which doesn't build up during successive strokes, indicates leaking valves or a blown head gasket (a cracked head could also be the cause). Deposits on the undersides of the valve heads can also cause low compression. Record the highest gauge reading obtained.

8   Repeat the procedure for the remaining cylinders and compare the results to this Chapter's Specifications.

9   Add some engine oil (about three squirts from a plunger-type oil can) to each cylinder, through the spark plug hole, and repeat the test.

10 If the compression increases after the oil is added, the piston rings are definitely worn. If the compression doesn't increase significantly, the leakage is occurring at the valves or head gasket. Leakage

3.6 Use a compression gauge with a threaded fitting for the spark plug hole, not the type that requires hand pressure to maintain the seal

past the valves may be caused by burned valve seats and/or faces or warped, cracked or bent valves.

11 If two adjacent cylinders have equally low compression, there's a strong possibility that the head gasket between them is blown. The appearance of coolant in the combustion chambers or the crankcase would verify this condition.

12 If one cylinder is slightly lower than the others, and the engine has a slightly rough idle, a worn lobe on the camshaft could be the cause.

13 If the compression is unusually high, the combustion chambers are probably coated with carbon deposits. If that's the case, the cylinder head(s) should be removed and decarbonized.

14 If compression is way down or varies greatly between cylinders, it would be a good idea to have a leak-down test performed by an automotive repair shop. This test will pinpoint exactly where the leakage is occurring and how severe it is.

4 Vacuum gauge diagnostic checks

▶ **Refer to illustrations 4.4 and 4.6**

A vacuum gauge provides inexpensive but valuable information about what is going on in the engine. You can check for worn rings or cylinder walls, leaking head or intake manifold gaskets, incorrect carburetor adjustments, restricted exhaust, stuck or burned valves, weak valve springs, improper ignition or valve timing and ignition problems.

Unfortunately, vacuum gauge readings are easy to misinterpret, so they should be used in conjunction with other tests to confirm the diagnosis.

Both the absolute readings and the rate of needle movement are important for accurate interpretation. Most gauges measure vacuum in inches of mercury (in-Hg). The following references to vacuum assume the diagnosis is being performed at sea level. As elevation increases (or atmospheric pressure decreases), the reading will decrease. For every 1,000 foot increase in elevation above approximately 2,000 feet, the gauge readings will decrease about one inch of mercury.

Connect the vacuum gauge directly to the intake manifold vacuum, not to ported (throttle body) vacuum (see illustration). Be sure no

4.4 A simple vacuum gauge can be handy in diagnosing engine condition and performance

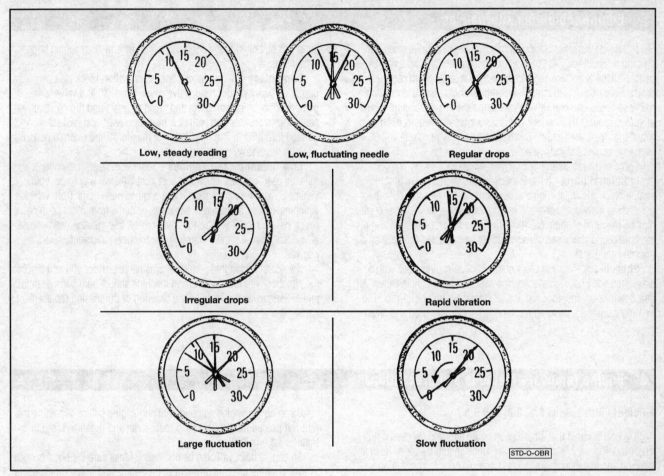

4.6 Typical vacuum gauge readings

hoses are left disconnected during the test or false readings will result.

Before you begin the test, allow the engine to warm up completely. Block the wheels and set the parking brake. With the transaxle in Park, start the engine and allow it to run at normal idle speed.

✳✳ WARNING:

Keep your hands and the vacuum gauge clear of the fans.

Read the vacuum gauge; an average, healthy engine should normally produce about 17 to 22 in-Hg with a fairly steady needle (see illustration). Refer to the following vacuum gauge readings and what they indicate about the engine's condition:

1   A low, steady reading usually indicates a leaking gasket between the intake manifold and cylinder head(s) or throttle body, a leaky vacuum hose, late ignition timing or incorrect camshaft timing. Check ignition timing with a timing light and eliminate all other possible causes, utilizing the tests provided in this Chapter before you remove the timing chain cover to check the timing marks.

2   If the reading is three to eight inches below normal and it fluctuates at that low reading, suspect an intake manifold gasket leak at an intake port or a faulty fuel injector.

3   If the needle has regular drops of about two-to-four inches at a steady rate, the valves are probably leaking. Perform a compression check or leak-down test to confirm this.

4   An irregular drop or down-flick of the needle can be caused by a sticking valve or an ignition misfire. Perform a compression check or leak-down test and read the spark plugs.

5   A rapid vibration of about four in-Hg vibration at idle combined with exhaust smoke indicates worn valve guides. Perform a leak-down test to confirm this. If the rapid vibration occurs with an increase in engine speed, check for a leaking intake manifold gasket or head gasket, weak valve springs, burned valves or ignition misfire.

6   A slight fluctuation, say one inch up and down, may mean ignition problems. Check all the usual tune-up items and, if necessary, run the engine on an ignition analyzer.

7   If there is a large fluctuation, perform a compression or leak-down test to look for a weak or dead cylinder or a blown head gasket.

8   If the needle moves slowly through a wide range, check for a clogged PCV system, incorrect idle fuel mixture, throttle body or intake manifold gasket leaks.

9   Check for a slow return after revving the engine by quickly snapping the throttle open until the engine reaches about 2,500 rpm and let it shut. Normally the reading should drop to near zero, rise above normal idle reading (about 5 in-Hg over) and then return to the previous idle reading. If the vacuum returns slowly and doesn't peak when the throttle is snapped shut, the rings may be worn. If there is a long delay, look for a restricted exhaust system (often the muffler or catalytic converter). An easy way to check this is to temporarily disconnect the exhaust ahead of the suspected part and redo the test.

5 Engine rebuilding alternatives

The do-it-yourselfer is faced with a number of options when purchasing a rebuilt engine. The major considerations are cost, warranty, parts availability and the time required for the rebuilder to complete the project. The decision to replace the engine block, piston/connecting rod assemblies and crankshaft depends on the final inspection results of your engine. Only then can you make a cost effective decision whether to have your engine overhauled or simply purchase an exchange engine for your vehicle.

Some of the rebuilding alternatives include:

Individual parts - If the inspection procedures reveal that the engine block and most engine components are in reusable condition, purchasing individual parts and having a rebuilder rebuild your engine may be the most economical alternative. The block, crankshaft and piston/connecting rod assemblies should all be inspected carefully by a machine shop first.

Short block - A short block consists of an engine block with a crankshaft and piston/connecting rod assemblies already installed. All new bearings are incorporated and all clearances will be correct. The existing camshafts, valve train components, cylinder head and external parts can be bolted to the short block with little or no machine shop work necessary.

Long block - A long block consists of a short block plus an oil pump, oil pan, cylinder head, valve cover, camshaft and valve train components, timing sprockets and chain or gears and timing cover. All components are installed with new bearings, seals and gaskets incorporated throughout. The installation of manifolds and external parts is all that's necessary.

Low mileage used engines - Some companies now offer low mileage used engines which is a very cost effective way to get your vehicle up and running again. These engines often come from vehicles which have been in totaled in accidents or come from other countries which have a higher vehicle turn over rate. A low mileage used engine also usually has a similar warranty like the newly remanufactured engines.

Give careful thought to which alternative is best for you and discuss the situation with local automotive machine shops, auto parts dealers and experienced rebuilders before ordering or purchasing replacement parts.

6 Engine removal - methods and precautions

▶ **Refer to illustrations 6.1, 6.2, and 6.3**

If you've decided that an engine must be removed for overhaul or major repair work, several preliminary steps should be taken. Read all removal and installation procedures carefully prior to committing to this job. These engines are removed by lowering the engine to the floor, along with the transaxle, and then raising the vehicle sufficiently to slide the assembly out; this will require a vehicle hoist as well as an engine hoist.

Locating a suitable place to work is extremely important. Adequate work space, along with storage space for the vehicle, will be needed. If a shop or garage isn't available, at the very least a flat, level, clean work surface made of concrete or asphalt is required.

Cleaning the engine compartment and engine before beginning the removal procedure will help keep tools clean and organized (see illustrations 6.1 and 6.2).

An engine hoist will also be necessary. Make sure the hoist is rated in excess of the combined weight of the engine and transaxle. Safety is of primary importance, considering the potential hazards involved in removing the engine from the vehicle.

If you're a novice at engine removal, get at least one helper. One person cannot easily do all the things you need to do to remove a big heavy engine and transaxle assembly from the engine compartment. Also helpful is to seek advice and assistance from someone who's experienced in engine removal.

Plan the operation ahead of time. Arrange for or obtain all of the

6.1 After tightly wrapping water-vulnerable components, use a spray cleaner on everything, with particular concentration on the greasiest areas, usually around the valve cover and lower edges of the block. If one section dries out, apply more cleaner

6.2 Depending on how dirty the engine is, let the cleaner soak in according to the directions and then hose off the grime and cleaner. Get the rinse water down into every area you can get at; then dry important components with a hair dryer or paper towels

tools and equipment you'll need prior to beginning the job (see illustration 6.3). Some of the equipment necessary to perform engine removal and installation safely and with relative ease are (in addition to a vehicle hoist and an engine hoist) a heavy duty floor jack (preferably fitted with a transmission jack head adapter), complete sets of wrenches and sockets as described in the front of this manual, wooden blocks, plenty of rags and cleaning solvent for mopping up spilled oil, coolant and gasoline.

Plan for the vehicle to be out of use for quite a while. A machine shop can do the work that is beyond the scope of the home mechanic. Machine shops often have a busy schedule, so before removing the engine, consult the shop for an estimate of how long it will take to rebuild or repair the components that may need work.

6.3 Get an engine stand sturdy enough to firmly support the engine while you're working on it. Stay away from three-wheeled models: they have a tendency to tip over more easily, so get a four-wheeled unit.

7   Engine - removal and installation

▶ Refer to illustration 7.8

❈❈ WARNING 1:

Gasoline is extremely flammable, so take extra precautions when you work on any part of the fuel system. Don't smoke or allow open flames or bare light bulbs near the work area, and don't work in a garage where a gas-type appliance (such as a water heater or clothes dryer) is present. Since gasoline is carcinogenic, wear fuel-resistant gloves when there's a possibility of being exposed to fuel, and, if you spill any fuel on your skin, rinse it off immediately with soap and water. Mop up any spills immediately and do not store fuel-soaked rags where they could ignite. The fuel system is under constant pressure, so, if any fuel lines are to be disconnected, the fuel pressure in the system must be relieved first (see Chapter 4 for more information). When you perform any kind of work on the fuel system, wear safety glasses and have a Class B type fire extinguisher on hand.

❈❈ WARNING 2:

The engine must be completely cool before beginning this procedure.

➡Note 1: Engine removal on these models is a difficult job, especially for the do-it-yourself mechanic working at home. Because of the vehicle's design, the manufacturer states that the engine and transaxle have to be removed as a unit from the bottom of the vehicle, not the top. With a floor jack and jack-stands, the vehicle can't be raised high enough and supported safely enough for the engine/transaxle assembly to slide out from underneath. The manufacturer recommends that removal of the engine transaxle assembly only be performed on a frame-contact type vehicle hoist.

➡Note 2: Read through the entire Section before beginning this procedure. The engine and transaxle are removed as a unit from below and then separated outside the vehicle.

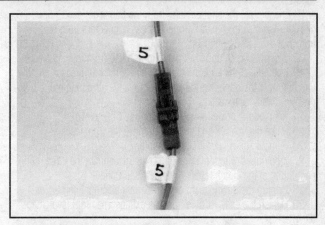

7.8 Label both ends of each wire and hose before disconnecting it

REMOVAL

1   Have the air conditioning system discharged by an automotive air conditioning technician.

2   Park the vehicle on a frame-contact type vehicle hoist, then engage the arms of the hoist with the jacking points of the vehicle. Raise the hoist arms until they contact the vehicle, but not so much that the wheels come off the ground.

3   Relieve the fuel system pressure (see Chapter 4).

4   Place protective covers on the fenders and cowl and remove the hood (see Chapter 11).

5   Remove the air filter housing (see Chapter 4).

6   Disconnect the accelerator cable (and cruise control cable, if equipped) and bracket from the engine and position them aside.

7   Remove the battery and the battery tray (see Chapter 5).

8   Clearly label and disconnect all vacuum lines, emissions hoses, wiring harness connectors, ground straps and fuel lines. Masking tape and/or a touch up paint applicator work well for marking items (see illustration). Take instant photos or sketch the locations of components and brackets.

9   Disconnect the electrical connectors from the PCM (see Chapter 6). Also detach any other electrical connectors between the engine and the vehicle.

10 Detach the positive cable from the engine compartment fuse/relay box and the ground cable from the vehicle. Unbolt the fuse/relay box and position it out of the way.

11 Detach the electrical connectors from the battery tray support. Also unbolt the ground cable from the right (passenger's) side strut tower.

12 Loosen the front wheel lug nuts, then raise the vehicle on the hoist.

➡**Note: Keep in mind that during this procedure you'll have to adjust the height of the vehicle to perform certain operations.**

13 Drain the cooling system and engine oil and remove the drivebelts (see Chapter 1).

14 Remove the alternator and its brackets (see Chapter 5).

15 Remove the power steering pump and bracket (see Chapter 10).

16 Detach the lower radiator hose from the engine.

17 Lower the vehicle and detach the heater hoses at the firewall.

18 Detach the upper radiator hose from the thermostat housing.

19 Remove the upper radiator support crossmember.

20 Remove the cooling fan(s), shroud(s) and radiator (see Chapter 3).

21 Disconnect the shift cable(s) from the transaxle (see Chapter 7A or 7B). Also disconnect any wiring harness connectors from the transaxle.

22 Disconnect the upper air conditioning line from the condenser.

23 Disconnect the air conditioning lines at the compressor and the junction near the upper torque strut. Remove the air conditioning compressor (see Chapter 3).

24 Remove the power steering pump (see Chapter 10).

25 Raise the vehicle on the hoist. Remove the front wheels.

26 Remove the driveaxles (see Chapter 8).

27 If equipped with a manual transaxle, disconnect the quick-connect fitting at the clutch release cylinder (see Chapter 8).

28 Unplug the downstream oxygen sensor electrical connector.

29 Detach the exhaust pipe from the exhaust manifold (see Chapter 4).

30 Detach the pressure hose from the power steering gear (see Chapter 10).

31 Remove the engine lower torque strut and the structural collar (see Chapter 2A).

32 If equipped with an automatic transaxle, remove the torque converter bolts (see Chapter 7B).

33 If equipped with a manual transaxle, remove the driveplate-to-modular clutch bolts (see Chapter 8).

34 Lower the vehicle.

35 Support the engine with a floor jack and block of wood. Remove the right (passenger's) side engine mount, including the portion that bolts to the engine. Using one of the mount-to-engine bolts, attach one end of an engine lifting sling or chain to the mount boss. Tighten the bolt securely. Attach the other end of the sling or chain to the other side of the engine, using one of the transaxle-to-engine bolts. Be sure the positioning of the chain or sling will support the engine and transaxle in a balanced attitude.

➡**Note: The sling or chain must be long enough to allow the engine hoist to lower the engine/transaxle assembly to the ground, without letting the hoist arm contact the vehicle.**

36 Roll the hoist into position and attach the sling or chain to it. Take up the slack until there is slight tension on the hoist, then remove the jack from under the engine. Remember that the transaxle end of the engine will be heavier, so position the chain on the hoist so it balances the engine and the transaxle level with the vehicle.

➡**Note: Depending on the design of the engine hoist, it may be helpful to position the hoist from the side of the vehicle, so that when the engine/transaxle assembly is lowered, it will fit between the legs of the hoist.**

37 Recheck to be sure nothing except the remaining mounts are still connecting the engine to the vehicle or to the transaxle. Disconnect and label anything still remaining.

38 Remove the through-bolt on the driver's side transaxle mount and the upper engine torque strut (see Chapter 2A).

39 Slowly lower the engine/transaxle to the ground.

40 Once the powertrain is on the floor, disconnect the engine lifting hoist and raise the vehicle until it clears the powertrain.

41 Reconnect the chain or sling and raise the engine and transaxle. Support the engine with blocks of wood or another floor jack, while leaving the sling or chain attached to the right-side mounting boss. Support the transaxle with another floor jack, preferably one with a transmission jack head adapter. At this point the transaxle can be unbolted and removed from the engine. Be very careful to ensure that the components are supported securely so they won't topple off their supports during disconnection.

42 Reconnect the lifting chain to the engine, then raise the engine and attach it to an engine stand.

INSTALLATION

43 Installation is the reverse of removal, noting the following points:

a) *Check the engine/transaxle mounts. If they're worn or damaged, replace them.*

b) *On manual transaxle equipped models, inspect the clutch components (see Chapter 8) and on automatic models inspect the converter seal and bushing.*

c) *Attach the transaxle to the engine following the procedure described in Chapter 7A or 7B.*

d) *Add coolant, oil, power steering and transmission fluids as needed (see Chapter 1).*

e) *Run the engine and check for proper operation and leaks. Shut off the engine and recheck fluid levels.*

---

8   Engine overhaul - disassembly sequence

1   It's much easier to remove the external components if it's mounted on a portable engine stand. A stand can often be rented quite cheaply from an equipment rental yard. Before the engine is mounted on a stand, the flywheel/driveplate should be removed from the engine.

2   If a stand isn't available, it's possible to remove the external engine components with it blocked up on the floor. Be extra careful not to tip or drop the engine when working without a stand.

3   If you're going to obtain a rebuilt engine, all external components must come off first, to be transferred to the replacement engine. These components include:

Clutch and flywheel (models with manual transaxle)
Driveplate (models with automatic transaxle)
Ignition system components
Emissions-related components
Engine mounts and mount brackets
Engine rear cover (spacer plate between flywheel/driveplate and engine block)
Intake/exhaust manifolds
Fuel injection components
Oil filter
Spark plug wires and spark plugs
Thermostat and housing assembly
Water pump

➡**Note: When removing the external components from the engine, pay close attention to details that may be helpful or important during installation. Note the installed position of gaskets, seals, spacers, pins, brackets, washers, bolts and other small items.**

4   If you're going to obtain a short block (assembled engine block, crankshaft, pistons and connecting rods), then remove the timing belt, cylinder head, oil pan, oil pump pick-up tube, oil pump and water pump from your engine so that you can turn in your old short block to the rebuilder as a core. See *Engine rebuilding alternatives* for additional information regarding the different possibilities to be considered.

9   Pistons and connecting rods - removal and installation

REMOVAL

▶ **Refer to illustrations 9.1 and 9.3**

➡**Note: Prior to removing the piston/connecting rod assemblies, remove the cylinder head and oil pan (see Chapter 2A).**

1   Use your fingernail to feel if a ridge has formed at the upper limit of ring travel (about 1/4-inch down from the top of each cylinder). If carbon deposits or cylinder wear have produced ridges, they must be completely removed with a special tool (see illustration). Follow the manufacturer's instructions provided with the tool. Failure to remove the ridges before attempting to remove the piston/connecting rod assemblies may result in piston breakage.

2   After the cylinder ridges have been removed, turn the engine so the crankshaft is facing up.

3   Before the main bearing cap assembly and connecting rods are removed, check the connecting rod endplay with feeler gauges. Slide them between the first connecting rod and the crankshaft throw until the play is removed (see illustration). Repeat this procedure for each connecting rod. The endplay is equal to the thickness of the feeler gauge(s). Check with an automotive machine shop for the endplay service limit (a typical end play limit should measure between 0.005 to 0.015 inch [0.127 to 0.369 mm]). If the play exceeds the service limit, new connecting rods will be required. If new rods (or a new

crankshaft) are installed, the endplay may fall under the minimum allowable. If it does, the rods will have to be machined to restore it. If necessary, consult an automotive machine shop for advice.

4   Check the connecting rods and caps for identification marks. If they aren't plainly marked, use paint or marker to clearly identify each rod and cap (1, 2, 3, etc., depending on the cylinder they're associated with).

✳✳ CAUTION:

Do not use a punch and hammer to mark the connecting rods or they may be damaged.

5   Loosen each of the connecting rod cap bolts 1/2-turn at a time until they can be removed by hand.

➡**Note: New connecting rod cap bolts must be used when reassembling the engine, but save the old bolts for use when checking the connecting rod bearing oil clearance.**

6   Remove the number one connecting rod cap and bearing insert. Don't drop the bearing insert out of the cap.

7   Remove the bearing insert and push the connecting rod/piston assembly out through the top of the engine. Use a wooden or plastic hammer handle to push on the upper bearing surface in the connecting rod. If resistance is felt, double-check to make sure that all of the ridge was removed from the cylinder.

9.1 Before you try to remove the pistons, use a ridge reamer to remove the raised material (ridge) from the top of the cylinders

9.3 Checking the connecting rod endplay (side clearance)

ENGINE BEARING ANALYSIS

Debris

Babbitt bearing embedded with debris from machinings

Microscopic detail of debris

Microscopic detail of gouges

Overplated copper alloy bearing gouged by cast iron debris

Aluminum bearing embedded with glass beads

Microscopic detail of glass beads

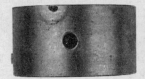

Damaged lining caused by dirt left on the bearing back

Misassembly

Result of a lower half assembled as an upper - blocking the oil flow

Excessive oil clearance is indicated by a short contact arc

Polished and oil-stained backs are a result of a poor fit in the housing bore

Result of a wrong, reversed, or shifted cap

Overloading

Damage from excessive idling which resulted in an oil film unable to support the load imposed

Damaged upper connecting rod bearings caused by engine lugging; the lower main bearings (not shown) were similarly affected

The damage shown in these upper and lower connecting rod bearings was caused by engine operation at a higher-than-rated speed under load

Misalignment

A warped crankshaft caused this pattern of severe wear in the center, diminishing toward the ends

A poorly finished crankshaft caused the equally spaced scoring shown

A bent connecting rod led to the damage in the "V" pattern

A tapered housing bore caused the damage along one edge of this pair

Lubrication

Result of dry start: The bearings on the left, farthest from the oil pump, show more damage

Result of a low oil supply or oil starvation

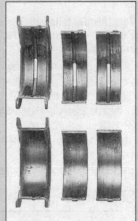

Severe wear as a result of inadequate oil clearance

Corrosion

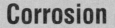

Microscopic detail of corrosion

Corrosion is an acid attack on the bearing lining generally caused by inadequate maintenance, extremely hot or cold operation, or interior oils or fuels

Microscopic detail of cavitation

Example of cavitation - a surface erosion caused by pressure changes in the oil film

Damage from excessive thrust or insufficient axial clearance

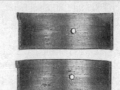

Bearing affected by oil dilution caused by excessive blow-by or a rich mixture

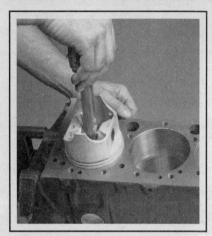

9.13 Install the piston ring into the cylinder then push it down into position using a piston so the ring will be square in the cylinder

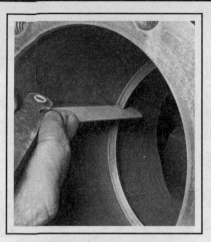

9.14 With the ring square in the cylinder, measure the ring end gap with a feeler gauge

9.15 If the ring end gap is too small, clamp a file in a vise as shown and file the piston ring ends - be sure to remove all raised material

8 Repeat the procedure for the remaining cylinders.

9 After removal, reassemble the connecting rod caps and bearing inserts in their respective connecting rods and install the cap bolts finger tight. Leaving the old bearing inserts in place until reassembly will help prevent the connecting rod bearing surfaces from being accidentally nicked or gouged.

10 The pistons and connecting rods are now ready for inspection and overhaul at an automotive machine shop.

PISTON RING INSTALLATION

♦ **Refer to illustrations 9.13, 9.14, 9.15, 9.19a, 9.19b and 9.22**

11 Before installing the new piston rings, the ring end gaps must be checked. It's assumed that the piston ring side clearance has been checked and verified correct.

12 Lay out the piston/connecting rod assemblies and the new ring sets so the ring sets will be matched with the same piston and cylinder during the end gap measurement and engine assembly.

13 Insert the top (number one) ring into the first cylinder and square it up with the cylinder walls by pushing it in with the top of the piston (see illustration). The ring should be near the bottom of the cylinder, at the lower limit of ring travel.

14 To measure the end gap, slip feeler gauges between the ends of the ring until a gauge equal to the gap width is found (see illustration). The feeler gauge should slide between the ring ends with a slight amount of drag. A typical ring gap should fall between 0.010 and 0.020 inch [0.25 to 0.50 mm] for compression rings and up to 0.030 inch [0.76 mm] for the oil ring steel rails. If the gap is larger or smaller than specified, double-check to make sure you have the correct rings before proceeding.

15 If the gap is too small, it must be enlarged or the ring ends may come in contact with each other during engine operation, which can cause serious damage to the engine. If necessary, increase the end gaps by filing the ring ends very carefully with a fine file. Mount the file in a vise equipped with soft jaws, slip the ring over the file with the ends contacting the file face and slowly move the ring to remove material from the ends. When performing this operation, file only by pushing the ring from the outside end of the file towards the vise (see illustration).

16 Excess end gap isn't critical unless it's greater than 0.040 inch (1.01 mm). Again, double-check to make sure you have the correct ring type.

17 Repeat the procedure for each ring that will be installed in the first cylinder and for each ring in the remaining cylinders. Remember to keep rings, pistons and cylinders matched up.

18 Once the ring end gaps have been checked/corrected, the rings can be installed on the pistons.

19 The oil control ring (lowest one on the piston) is usually installed first. It's composed of three separate components. Slip the spacer/expander into the groove (see illustration). If an anti-rotation tang is used, make sure it's inserted into the drilled hole in the ring groove. Next, install the upper side rail in the same manner (see illustration). Don't use a piston ring installation tool on the oil ring side rails, as they may be damaged. Instead, place one end of the side rail into the groove between the spacer/expander and the ring land, hold it firmly in place and slide a finger around the piston while pushing the rail into the groove. Finally, install the lower side rail.

20 After the three oil ring components have been installed, check to make sure that both the upper and lower side rails can be rotated smoothly inside the ring grooves.

21 The number two (middle) ring is installed next. It's usually stamped with a mark which must face up, toward the top of the piston. Do not mix up the top and middle rings, as they have different cross-sections.

➡**Note: Always follow the instructions printed on the ring package or box - different manufacturers may require different approaches.**

22 Use a piston ring installation tool and make sure the identification mark is facing the top of the piston, then slip the ring into the middle groove on the piston (see illustration). Don't expand the ring any more than necessary to slide it over the piston.

23 Install the number one (top) ring in the same manner. Make sure the mark is facing up. Be careful not to confuse the number one and number two rings.

24 Repeat the procedure for the remaining pistons and rings.

INSTALLATION

25 Before installing the piston/connecting rod assemblies, the cylinder walls must be perfectly clean, the top edge of each cylinder bore must be chamfered, and the crankshaft must be in place.

26 Remove the cap from the end of the number one connecting rod (refer to the marks made during removal). Remove the original bearing

9.19a Installing the spacer/expander in the oil ring groove

9.19b DO NOT use a piston ring installation tool when installing the oil control ring side rails

9.22 Use a piston ring installation tool to install the number 2 and the number 1 (top) rings - be sure the directional mark on the piston ring(s) is facing toward the top of the piston

inserts and wipe the bearing surfaces of the connecting rod and cap with a clean, lint-free cloth. They must be kept spotlessly clean.

CONNECTING ROD BEARING OIL CLEARANCE CHECK

▶ **Refer to illustrations 9.30, 9.35, 9.37 and 9.41**

27 Clean the back side of the new upper bearing insert, then lay it in place in the connecting rod.

28 Make sure the tab on the bearing fits into the recess in the rod. Don't hammer the bearing insert into place and be very careful not to nick or gouge the bearing face. Don't lubricate the bearing at this time.

29 Clean the back side of the other bearing insert and install it in the rod cap. Again, make sure the tab on the bearing fits into the recess in the cap, and don't apply any lubricant. It's critically important that the mating surfaces of the bearing and connecting rod are perfectly clean and oil free when they're assembled.

30 Position the piston ring gaps at 90-degree intervals around the piston as shown (see illustration).

31 Lubricate the piston and rings with clean engine oil and attach a piston ring compressor to the piston. Leave the skirt protruding

about 1/4-inch to guide the piston into the cylinder. The rings must be compressed until they're flush with the piston.

32 Rotate the crankshaft until the number one connecting rod journal is at BDC (bottom dead center) and apply a liberal coat of engine oil to the cylinder walls.

33 With the weight designation mark (L or H) on top of the piston facing the front (timing belt end) of the engine, gently insert the piston/connecting rod assembly into the number one cylinder bore and rest the bottom edge of the ring compressor on the engine block. Install the pistons with the cavity mark(s) facing toward the timing belt.

34 Tap the top edge of the ring compressor to make sure it's contacting the block around its entire circumference.

35 Gently tap on the top of the piston with the end of a wooden or plastic hammer handle (see illustration) while guiding the end of the connecting rod into place on the crankshaft journal. The piston rings may try to pop out of the ring compressor just before entering the cylinder bore, so keep some downward pressure on the ring compressor. Work slowly, and if any resistance is felt as the piston enters the cylinder, stop immediately. Find out what's hanging up and fix it before proceeding. Do not, for any reason, force the piston into the cylinder - you might break a ring and/or the piston.

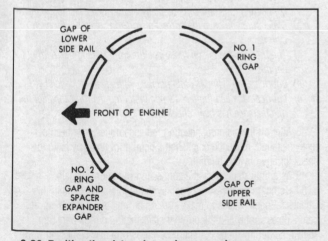

9.30 Position the piston ring end gaps as shown

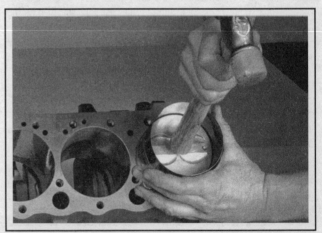

9.35 Use a plastic or wooden hammer handle to push the piston into the cylinder

9.37 Place Plastigage on each connecting rod bearing journal parallel to the crankshaft centerline

36 Once the piston/connecting rod assembly is installed, the connecting rod bearing oil clearance must be checked before the rod cap is permanently installed.

37 Cut a piece of the appropriate size Plastigage slightly shorter than the width of the connecting rod bearing and lay it in place on the number one connecting rod journal, parallel with the journal axis (see illustration).

38 Clean the connecting rod cap bearing face and install the rod cap. Make sure the mating mark on the cap is on the same side as the mark on the connecting rod (see illustration 9.4).

39 Install the old rod bolts, at this time, and tighten them to the torque listed in this Chapter's Specifications.

➡**Note: Use a thin-wall socket to avoid erroneous torque readings that can result if the socket is wedged between the rod cap and the bolt or nut. If the socket tends to wedge itself between the fastener and the cap, lift up on it slightly until it no longer contacts the cap. DO NOT rotate the crankshaft at any time during this operation.**

40 Remove the fasteners and detach the rod cap, being very careful not to disturb the Plastigage. Discard the cap bolts at this time as they cannot be reused.

➡**Note: You MUST use new connecting rod bolts.**

41 Compare the width of the crushed Plastigage to the scale printed on the Plastigage envelope to obtain the oil clearance (see illustration). The connecting rod oil clearance is usually about 0.001 to 0.002 inch. Consult an automotive machine shop for the clearance specified for the rod bearings on your engine.

42 If the clearance is not as specified, the bearing inserts may be the wrong size (which means different ones will be required). Before deciding that different inserts are needed, make sure that no dirt or oil was between the bearing inserts and the connecting rod or cap when the clearance was measured. Also, recheck the journal diameter. If the Plastigage was wider at one end than the other, the journal may be tapered. If the clearance still exceeds the limit specified, the bearing will have to be replaced with an undersize bearing.

✳✳ CAUTION:

When installing a new crankshaft always use a standard size bearing.

Final installation

43 Carefully scrape all traces of the Plastigage material off the rod

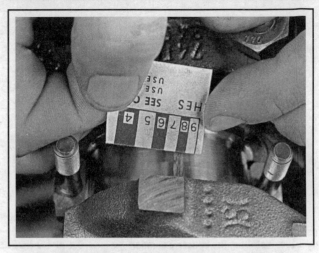

9.41 Use the scale on the Plastigage package to determine the bearing oil clearance - be sure to measure the widest part of the Plastigage and use the correct scale; it comes with both standard and metric scales

journal and/or bearing face. Be very careful not to scratch the bearing - use your fingernail or the edge of a plastic card.

44 Make sure the bearing faces are perfectly clean, then apply a uniform layer of clean moly-base grease or engine assembly lube to both of them. You'll have to push the piston into the cylinder to expose the face of the bearing insert in the connecting rod.

✳✳ CAUTION:

45 **If the connecting rod caps are secured to the rods with bolts (instead of nuts), install new connecting rod cap bolts. Do NOT reuse old bolts - they have stretched and cannot be reused. Slide the connecting rod back into place on the journal, install the rod cap, install the nuts or new bolts and tighten them to the torque listed in this Chapter's Specifications. Again, work up to the torque in three steps.**

46 Repeat the entire procedure for the remaining pistons/connecting rods.

47 The important points to remember are:

  a) *Keep the back sides of the bearing inserts and the insides of the connecting rods and caps perfectly clean when assembling them.*
  b) *Make sure you have the correct piston/rod assembly for each cylinder.*
  c) *The mark on the piston must face the front (timing belt end) of the engine.*
  d) *Lubricate the cylinder walls liberally with clean oil.*
  e) *Lubricate the bearing faces when installing the rod caps after the oil clearance has been checked.*

48 After all the piston/connecting rod assemblies have been correctly installed, rotate the crankshaft a number of times by hand to check for any obvious binding.

49 As a final step, check the connecting rod endplay again.

50 Compare the measured endplay to the tolerance listed in this Chapter's Specifications to make sure it's acceptable. If it was correct before disassembly and the original crankshaft and rods were reinstalled, it should still be correct. If new rods or a new crankshaft were installed, the endplay may be inadequate. If so, the rods will have to be removed and taken to an automotive machine shop for resizing.

10 Crankshaft - removal and installation

REMOVAL

▶ **Refer to illustrations 10.1 and 10.3**

➡**Note: The crankshaft can be removed only after the engine has been removed from the vehicle. It's assumed that the flywheel or driveplate, crankshaft pulley, timing belt, oil pan, oil pump body, oil filter and piston/connecting rod assemblies have already been removed. The rear main oil seal retainer must be unbolted and separated from the block before proceeding with crankshaft removal.**

1 Before the crankshaft is removed, measure the endplay. Mount a dial indicator with the indicator in line with the crankshaft and just touching the end of the crankshaft as shown (see illustration).

2 Pry the crankshaft all the way to the rear and zero the dial indicator. Next, pry the crankshaft to the front as far as possible and check the reading on the dial indicator. The distance traveled is the endplay. A typical crankshaft endplay will fall between 0.003 to 0.010 inch (0.076 to 0.254 mm). If it is greater than that, check the crankshaft thrust surfaces for wear after it's removed. If no wear is evident, new main bearings should correct the endplay.

3 If a dial indicator isn't available, feeler gauges can be used. Gently pry the crankshaft all the way to the front of the engine. Slip feeler gauges between the crankshaft and the front face of the thrust bearing or washer to determine the clearance (see illustration).

4 Loosen the main bearing cap assembly bolts 1/4-turn at a time each, until they can be removed by hand.

5 Gently tap the main bearing cap assembly with a soft-face hammer around the perimeter of the assembly. Pull the main bearing cap assembly straight up and off the cylinder block. Try not to drop the bearing inserts if they come out with the assembly.

6 Carefully lift the crankshaft out of the engine. It may be a good idea to have an assistant available, since the crankshaft is quite heavy and awkward to handle. With the bearing inserts in place inside the engine block and main bearing caps, reinstall the main bearing cap assembly onto the engine block and tighten the bolts finger tight. Make sure you install the main bearing cap assembly with the arrow facing the front end of the engine.

10.1 Checking crankshaft endplay with a dial indicator

INSTALLATION

7 Crankshaft installation is the first step in engine reassembly. It's assumed at this point that the engine block and crankshaft have been cleaned, inspected and repaired or reconditioned.

8 Position the engine block with the bottom facing up.

9 Remove the mounting bolts and lift off the main bearing cap assembly.

10 If they're still in place, remove the original bearing inserts from the block and from the main bearing cap assembly. Wipe the bearing surfaces of the block and main bearing cap assembly with a clean, lint-free cloth. They must be kept spotlessly clean. This is critical for determining the correct bearing oil clearance.

MAIN BEARING OIL CLEARANCE CHECK

▶ **Refer to illustrations 10.17, 10.19 and 10.21**

11 Without mixing them up, clean the back sides of the new upper main bearing inserts (with grooves and oil holes) and lay one in each main bearing saddle in the block. Each upper bearing has an oil groove and oil hole in it.

✳✳ **CAUTION:**

The oil holes in the block must line up with the oil holes in the upper bearing inserts.

The thrust washer or thrust bearing insert must be installed in the number 3 crankshaft journal. Clean the back sides of the lower main bearing inserts and lay them in the corresponding location in the main bearing cap assembly. Make sure the tab on the bearing insert fits into the recess in the block or main bearing cap assembly. The upper bearings with the oil holes are installed into the engine block while the lower bearings without the oil holes are installed in the crankshaft bedplate (main bearing assembly).

✳✳ **CAUTION:**

Do not hammer the bearing insert into place and don't nick or gouge the bearing faces. DO NOT apply any lubrication at this time.

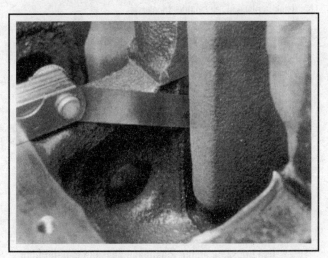

10.3 Checking crankshaft endplay with feeler gauges at the thrust bearing journal

10.17 Place the Plastigage onto the crankshaft bearing journal as shown

12 Clean the faces of the bearing inserts in the block and the crankshaft main bearing journals with a clean, lint-free cloth.

13 Check or clean the oil holes in the crankshaft, as any dirt here can go only one way - straight through the new bearings.

14 Once you're certain the crankshaft is clean, carefully lay it in position in the cylinder block.

15 Before the crankshaft can be permanently installed, the main bearing oil clearance must be checked.

16 Cut several strips of the appropriate size of Plastigage. They must be slightly shorter than the width of the main bearing journal.

17 Place one piece on each crankshaft main bearing journal, parallel with the journal axis as shown (see illustration).

18 Clean the faces of the bearing inserts in the main bearing cap assembly. Hold the bearing inserts in place and install the assembly onto the crankshaft and cylinder block. DO NOT disturb the Plastigage. Make sure you install the main bearing cap assembly with the arrow facing the front (timing belt end) of the engine.

19 Apply clean engine oil to all bolt threads prior to installation, then install all bolts finger-tight. Tighten main bearing cap assembly bolts (numbers 1 through 10) in the sequence shown (see illustration) progressing in steps, to the torque listed in this Chapter's Specifications. DO NOT rotate the crankshaft at any time during this operation.

➡Note: Be sure the three locating dowels are in place in the engine block. Their locations are indicated by the stars in the diagram.

20 Remove the bolts in the reverse order of the tightening sequence and carefully lift the main bearing cap assembly straight up and off the block. Do not disturb the Plastigage or rotate the crankshaft. If the main bearing cap assembly is difficult to remove, tap it gently from side-to-side with a soft-face hammer to loosen it.

21 Compare the width of the crushed Plastigage on each journal to the scale printed on the Plastigage envelope to determine the main bearing oil clearance (see illustration). Check with an automotive machine shop for the crankshaft endplay service limits.

22 If the clearance is not as specified, the bearing inserts may be the wrong size (which means different ones will be required). Before deciding if different inserts are needed, make sure that no dirt or oil was between the bearing inserts and the cap assembly or block when the clearance was measured. If the Plastigage was wider at one end than the other, the crankshaft journal may be tapered. If the clearance still exceeds the limit specified, the bearing insert(s) will have to be replaced with an undersize bearing insert(s).

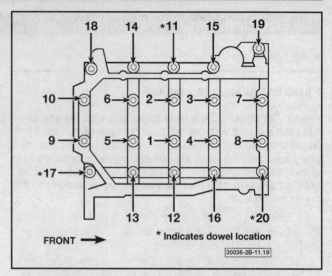

10.19 Main bearing cap bolt and bedplate bolt tightening sequence

⁂ CAUTION:

When installing a new crankshaft always install a standard bearing insert set.

23 Carefully scrape all traces of the Plastigage material off the main bearing journals and/or the bearing insert faces. Be sure to remove all residue from the oil holes. Use your fingernail or the edge of a plastic card - don't nick or scratch the bearing faces.

FINAL INSTALLATION

▶ **Refer to illustration 10.27**

24 Carefully lift the crankshaft out of the cylinder block.

25 Clean the bearing insert faces in the cylinder block, then apply a thin, uniform layer of moly-base grease or engine assembly lube to each of the bearing surfaces. Be sure to coat the thrust faces as well as the journal face of the thrust bearing.

26 Make sure the crankshaft journals are clean, then lay the crankshaft back in place in the cylinder block.

10.21 Use the scale on the Plastigage package to determine the bearing oil clearance - be sure to measure the widest part of the Plastigage and use the correct scale; it comes with both standard and metric scales

27 Clean the bearing insert faces and then apply the same lubricant to them.

➡Note: Make sure the three locating dowels are installed in the engine block (see Step 19).

Clean the engine block and the bedplate thoroughly. The surfaces must be free of oil residue. Apply a 1/16 inch bead of anaerobic sealant (Mopar Torque Cure Gasket Maker or equivalent) to the engine block (see illustration).

28 Hold the bedplate (main bearing assembly) and bearings in place and install the main bearing assembly onto the crankshaft and cylinder block. Install the main bearing assembly until the dowels lock together with the assembly.

29 Prior to installation, apply clean engine oil to all bolt threads wiping off any excess, then install all bolts finger-tight.

30 On 2000 models, tighten the main bearing cap assembly as follows (see illustration 10.19):

a) *Tighten bolts 11, 17 and 20 until the assembly contacts the engine block.*

b) *Tighten bolts 1 through 10 in the sequence shown, to the torque listed in this Chapter's Specifications.*

c) *Tighten bolts 11 through 20 in the sequence shown to the torque listed in this Chapter's Specifications. Be sure the baffle studs are located in bolts 12, 13 and 16.*

31 On 2001 through 2003 models, tighten the main bearing cap assembly as follows (see illustration 10.19):

a) *Tighten bolts 11, 17 and 20 until the assembly contacts the engine block.*

b) *To ensure correct thrust bearing alignment, rotate the crankshaft until the No. 4 piston is at TDC.*

c) *Carefully pry the crankshaft all the way towards the rear of the block and then towards the front of the block.*

d) *Wedge an appropriate tool such as a block of wood, between the engine block and the crankshaft counterweight to hold the crankshaft in the most forward position.* **DO NOT** *drive the wedge between the main bearing cap assembly and the crankshaft.*

e) *Tighten bolts 1 through 10 in the sequence shown, to the Step 1 torque listed in this Chapter's Specifications.*

f) *Tighten bolts 11 through 20 in the sequence shown to the torque listed in this Chapter's Specifications.*

g) *Loosen all bolts, remove the wedge, then tighten bolts 1 through 10 in three steps (see this Chapter's Specifications).*

h) *Tighten bolts 11 through 20 to the torque listed in this Chapter's Specifications.*

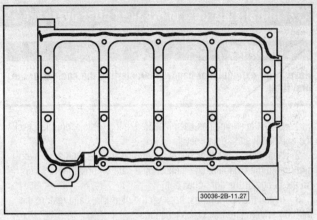

10.27 Apply a 1.5 to 2.0 mm bead of special bedplate sealant (or equivalent) to the darkened areas indicated

32 On 2004 and later models, tighten the main bearing cap assembly as follows (see illustration 10.19):

a) *Tighten bolts 11, 17 and 20 until the assembly contacts the engine block.*

b) *To ensure correct thrust bearing alignment, rotate the crankshaft until the No. 4 piston is at TDC.*

c) *Carefully pry the crankshaft all the way towards the rear of the block and then towards the front of the block.*

d) *Wedge an appropriate tool such as a block of wood between the engine block and the crankshaft counterweight to hold the crankshaft in the most forward position.* **DO NOT** *drive the wedge between the main bearing cap assembly and the crankshaft.*

e) *Tighten bolts 1 through 10 in the sequence shown, in 2 steps to the torque listed in this Chapter's Specifications.*

f) *Tighten bolts 11 through 20 (Step 3) in the sequence shown, to the torque listed in this Chapter's Specifications.*

g) *Tighten bolts 1 through 10 (Step 4) in the sequence shown, to the torque listed in this Chapter's Specifications.*

h) *Tighten bolts 11 through 20 (Step 5) in the sequence shown, to the torque listed in this Chapter's Specifications.*

33 Recheck crankshaft endplay with a feeler gauge or a dial indicator. The endplay should be correct if the crankshaft thrust faces aren't worn or damaged and if new bearings have been installed.

34 Rotate the crankshaft a number of times by hand to check for any obvious binding. It should rotate with a running torque of 50 in-lbs or less. If the running torque is too high, correct the problem at this time.

35 Install the new rear main oil seal (see Chapter 2A).

11 Engine overhaul - reassembly sequence

1 Before beginning engine reassembly, make sure you have all the necessary new parts, gaskets and seals as well as the following items on hand:

Common hand tools
A 1/2-inch drive torque wrench
New engine oil
Gasket sealant
Thread locking compound

2 If you obtained a short block it will be necessary to install the cylinder head, the oil pump and pick-up tube, the oil pan, the water pump, the timing belt and timing cover, and the valve cover (see Chapter 2A or 2B). In order to save time and avoid problems, the external

components must be installed in the following general order:

Thermostat and housing cover
Water pump
Intake and exhaust manifolds
Fuel injection components
Emission control components
Spark plug wires and spark plugs
Ignition coils
Oil filter
Engine mounts and mount brackets
Clutch and flywheel (manual transaxle)
Driveplate (automatic transaxle)

12 Initial start-up and break-in after overhaul

✳✳ WARNING:

Have a fire extinguisher handy when starting the engine for the first time.

1   Once the engine has been installed in the vehicle, double-check the engine oil and coolant levels.

2   With the spark plugs out of the engine and the ignition system and fuel pump disabled, crank the engine until oil pressure registers on the gauge or the light goes out.

3   Install the spark plugs, hook up the plug wires and restore the ignition system and fuel pump functions.

4   Start the engine. It may take a few moments for the fuel system to build up pressure, but the engine should start without a great deal of effort.

5   After the engine starts, it should be allowed to warm up to normal operating temperature. While the engine is warming up, make a thorough check for fuel, oil and coolant leaks.

6   Shut the engine off and recheck the engine oil and coolant levels.

7   Drive the vehicle to an area with minimum traffic, accelerate from 30 to 50 mph, then allow the vehicle to slow to 30 mph with the throttle closed. Repeat the procedure 10 or 12 times. This will load the piston rings and cause them to seat properly against the cylinder walls. Check again for oil and coolant leaks.

8   Drive the vehicle gently for the first 500 miles (no sustained high speeds) and keep a constant check on the oil level. It is not unusual for an engine to use oil during the break-in period.

9   At approximately 500 to 600 miles, change the oil and filter.

10   For the next few hundred miles, drive the vehicle normally. Do not pamper it or abuse it.

11   After 2000 miles, change the oil and filter again and consider the engine broken in.

Specifications

General

Displacement	122 cubic inches (2.0 liters)
Bore and Stroke	3.445 x 3.268 inches (87.5 x 83.0 mm)
Cylinder compression pressure	170 to 225 psi (1,172 to 1,551 kPa)
Oil pressure	
At curb idle (minimum)	4 psi (25 kPa)
At 3,000 rpm	25 to 80 psi (170 to 550 kPa)

Torque specifications

	Ft-lbs (unless otherwise indicated)	Nm
Connecting rod bearing cap bolts		
Step 1	20	27
Step 2	Tighten an additional 1/4 turn (90 degrees)	
Main bearing assembly (see illustration 10.19)		
2000 models		
Main bearing cap bolts (numbers 1 through 10)	60	81
Bedplate bolts (numbers 11 through 20)	25	33
2001 and later models		
Main bearing cap bolts (numbers 1 through 10)		
Step 1	30	41
Step 2	30	41
Step 3	Tighten an additional 1/4 turn (90 degrees)	
Bedplate bolts (numbers 11 through 20)	250 in-lbs	28
2004 and later models		
Step 1: Bolts 1 through 10	30	41
Step 2: Bolts 1 through 10	30	41
Step 3: Bolts 11 through 20	25	34
Step 4: Bolts 1 through 10	60	81
Step 5: Bolts 11 through 20	25	34

GLOSSARY

B

Backlash - The amount of play between two parts. Usually refers to how much one gear can be moved back and forth without moving gear with which it's meshed.

Bearing Caps - The caps held in place by nuts or bolts which, in turn, hold the bearing surface. This space is for lubricating oil to enter.

Bearing clearance - The amount of space left between shaft and bearing surface. This space is for lubricating oil to enter.

Bearing crush - The additional height which is purposely manufactured into each bearing half to ensure complete contact of the bearing back with the housing bore when the engine is assembled.

Bearing knock - The noise created by movement of a part in a loose or worn bearing.

Blueprinting - Dismantling an engine and reassembling it to EXACT specifications.

Bore - An engine cylinder, or any cylindrical hole; also used to describe the process of enlarging or accurately refinishing a hole with a cutting tool, as to bore an engine cylinder. The bore size is the diameter of the hole.

Boring - Renewing the cylinders by cutting them out to a specified size. A boring bar is used to make the cut.

Bottom end - A term which refers collectively to the engine block, crankshaft, main bearings and the big ends of the connecting rods.

Break-in - The period of operation between installation of new or rebuilt parts and time in which parts are worn to the correct fit. Driving at reduced and varying speed for a specified mileage to permit parts to wear to the correct fit.

Bushing - A one-piece sleeve placed in a bore to serve as a bearing surface for shaft, piston pin, etc. Usually replaceable.

C

Camshaft - The shaft in the engine, on which a series of lobes are located for operating the valve mechanisms. The camshaft is driven by gears or sprockets and a timing chain. Usually referred to simply as the cam.

Carbon - Hard, or soft, black deposits found in combustion chamber, on plugs, under rings, on and under valve heads.

Cast iron - An alloy of iron and more than two percent carbon, used for engine blocks and heads because it's relatively inexpensive and easy to mold into complex shapes.

Chamfer - To bevel across (or a bevel on) the sharp edge of an object.

Chase - To repair damaged threads with a tap or die.

Combustion chamber - The space between the piston and the cylinder head, with the piston at top dead center, in which air-fuel mixture is burned.

Compression ratio - The relationship between cylinder volume (clearance volume) when the piston is at top dead center and cylinder volume when the piston is at bottom dead center.

Connecting rod - The rod that connects the crank on the crankshaft with the piston. Sometimes called a con rod.

Connecting rod cap - The part of the connecting rod assembly that attaches the rod to the crankpin.

Core plug - Soft metal plug used to plug the casting holes for the coolant passages in the block.

Crankcase - The lower part of the engine in which the crankshaft rotates; includes the lower section of the cylinder block and the oil pan.

Crank kit - A reground or reconditioned crankshaft and new main and connecting rod bearings.

Crankpin - The part of a crankshaft to which a connecting rod is attached.

Crankshaft - The main rotating member, or shaft, running the length of the crankcase, with offset throws to which the connecting rods are attached; changes the reciprocating motion of the pistons into rotating motion.

Cylinder sleeve - A replaceable sleeve, or liner, pressed into the cylinder block to form the cylinder bore.

D

Deburring - Removing the burrs (rough edges or areas) from a bearing.

Deglazer - A tool, rotated by an electric motor, used to remove glaze from cylinder walls so a new set of rings will seat.

E

Endplay - The amount of lengthwise movement between two parts. As applied to a crankshaft, the distance that the crankshaft can move forward and back in the cylinder block.

F

Face - A machinist's term that refers to removing metal from the end of a shaft or the face of a larger part, such as a flywheel.

Fatigue - A breakdown of material through a large number of loading and unloading cycles. The first signs are cracks followed shortly by breaks.

Feeler gauge - A thin strip of hardened steel, ground to an exact thickness, used to check clearances between parts.

Free height - The unloaded length or height of a spring.

Freeplay - The looseness in a linkage, or an assembly of parts, between the initial application of force and actual movement. Usually perceived as slop or slight delay.

Freeze plug - See Core plug.

G

Gallery - A large passage in the block that forms a reservoir for engine oil pressure.

Glaze - The very smooth, glassy finish that develops on cylinder walls while an engine is in service.

H

Heli-Coil - A rethreading device used when threads are worn or damaged. The device is installed in a retapped hole to reduce the thread size to the original size.

I

Installed height - The spring's measured length or height, as installed on the cylinder head. Installed height is measured from the spring seat to the underside of the spring retainer.

J

Journal - The surface of a rotating shaft which turns in a bearing.

K

Keeper - The split lock that holds the valve spring retainer in position on the valve stem.

Key - A small piece of metal inserted into matching grooves machined into two parts fitted together - such as a gear pressed onto a shaft - which prevents slippage between the two parts.

Knock - The heavy metallic engine sound, produced in the combustion chamber as a result of abnormal combustion - usually detonation. Knock is usually caused by a loose or worn bearing. Also referred to as detonation, pinging and spark knock. Connecting rod or main bearing knocks are created by too much oil clearance or insufficient lubrication.

L

Lands - The portions of metal between the piston ring grooves.

Lapping the valves - Grinding a valve face and its seat together with lapping compound.

Lash - The amount of free motion in a gear train, between gears, or in a mechanical assembly, that occurs before movement can begin. Usually refers to the lash in a valve train.

Lifter - The part that rides against the cam to transfer motion to the rest of the valve train.

M

Machining - The process of using a machine to remove metal from a metal part.

Main bearings - The plain, or babbitt, bearings that support the crankshaft.

Main bearing caps - The cast iron caps, bolted to the bottom of the block, that support the main bearings.

O

O.D. - Outside diameter.

Oil gallery - A pipe or drilled passageway in the engine used to carry engine oil from one area to another.

Oil ring - The lower ring, or rings, of a piston; designed to prevent excessive amounts of oil from working up the cylinder walls and into the combustion chamber. Also called an oil-control ring.

Oil seal - A seal which keeps oil from leaking out of a compartment. Usually refers to a dynamic seal around a rotating shaft or other moving part.

O-ring - A type of sealing ring made of a special rubberlike material; in use, the O-ring is compressed into a groove to provide the sealing action.

Overhaul - To completely disassemble a unit, clean and inspect all parts, reassemble it with the original or new parts and make all adjustments necessary for proper operation.

P

Pilot bearing - A small bearing installed in the center of the flywheel (or the rear end of the crankshaft) to support the front end of the input shaft of the transmission.

Pip mark - A little dot or indentation which indicates the top side of a compression ring.

Piston - The cylindrical part, attached to the connecting rod, that moves up and down in the cylinder as the crankshaft rotates. When the fuel charge is fired, the piston transfers the force of the explosion to the connecting rod, then to the crankshaft.

Piston pin (or wrist pin) - The cylindrical and usually hollow steel pin that passes through the piston. The piston pin fastens the piston to the upper end of the connecting rod.

Piston ring - The split ring fitted to the groove in a piston. The ring contacts the sides of the ring groove and also rubs against the cylinder wall, thus sealing space between piston and wall. There are two types of rings: Compression rings seal the compression pressure in the combustion chamber; oil rings scrape excessive oil off the cylinder wall.

Piston ring groove - The slots or grooves cut in piston heads to hold piston rings in position.

Piston skirt - The portion of the piston below the rings and the piston pin hole.

Plastigage - A thin strip of plastic thread, available in different sizes, used for measuring clearances. For example, a strip of plastigage is laid across a bearing journal and mashed as parts are assembled. Then parts are disassembled and the width of the strip is measured to determine clearance between journal and bearing. Commonly used to measure crankshaft main-bearing and connecting rod bearing clearances.

Press-fit - A tight fit between two parts that requires pressure to force the parts together. Also referred to as drive, or force, fit.

Prussian blue - A blue pigment; in solution, useful in determining the area of contact between two surfaces. Prussian blue is commonly used to determine the width and location of the contact area between the valve face and the valve seat.

R

Race (bearing) - The inner or outer ring that provides a contact surface for balls or rollers in bearing.

Ream - To size, enlarge or smooth a hole by using a round cutting tool with fluted edges.

Ring job - The process of reconditioning the cylinders and installing new rings.

Runout - Wobble. The amount a shaft rotates out-of-true.

S

Saddle - The upper main bearing seat.

Scored - Scratched or grooved, as a cylinder wall may be scored by abrasive particles moved up and down by the piston rings.

Scuffing - A type of wear in which there's a transfer of material between parts moving against each other; shows up as pits or grooves in the mating surfaces.

Seat - The surface upon which another part rests or seats. For example, the valve seat is the matched surface upon which the valve face rests. Also used to refer to wearing into a good fit; for example, piston rings seat after a few miles of driving.

Short block - An engine block complete with crankshaft and piston and, usually, camshaft assemblies.

Static balance - The balance of an object while it's stationary.

Step - The wear on the lower portion of a ring land caused by excessive side and back-clearance. The height of the step indicates the ring's extra side clearance and the length of the step projecting from the back wall of the groove represents the ring's back clearance.

Stroke - The distance the piston moves when traveling from top dead center to bottom dead center, or from bottom dead center to top dead center.

Stud - A metal rod with threads on both ends.

T

Tang - A lip on the end of a plain bearing used to align the bearing during assembly.

Tap - To cut threads in a hole. Also refers to the fluted tool used to cut threads.

Taper - A gradual reduction in the width of a shaft or hole; in an engine cylinder, taper usually takes the form of uneven wear, more pronounced at the top than at the bottom.

Throws - The offset portions of the crankshaft to which the connecting rods are affixed.

Thrust bearing - The main bearing that has thrust faces to prevent excessive end-play, or forward and backward movement of the crankshaft.

Thrust washer - A bronze or hardened steel washer placed between two moving parts. The washer prevents longitudinal movement and provides a bearing surface for thrust surfaces of parts.

Tolerance - The amount of variation permitted from an exact size of measurement. Actual amount from smallest acceptable dimension to largest acceptable dimension.

U

Umbrella - An oil deflector placed near the valve tip to throw oil from the valve stem area.

Undercut - A machined groove below the normal surface.

Undersize bearings - Smaller diameter bearings used with re-ground crankshaft journals.

V

Valve grinding - Refacing a valve in a valve-refacing machine.

Valve train - The valve-operating mechanism of an engine; includes all components from the camshaft to the valve.

Vibration damper - A cylindrical weight attached to the front of the crankshaft to minimize torsional vibration (the twist-untwist actions of the crankshaft caused by the cylinder firing impulses). Also called a harmonic balancer.

W

Water jacket - The spaces around the cylinders, between the inner and outer shells of the cylinder block or head, through which coolant circulates.

Web - A supporting structure across a cavity.

Woodruff key - A key with a radiused backside (viewed from the side).

Section

Reference to other Chapters

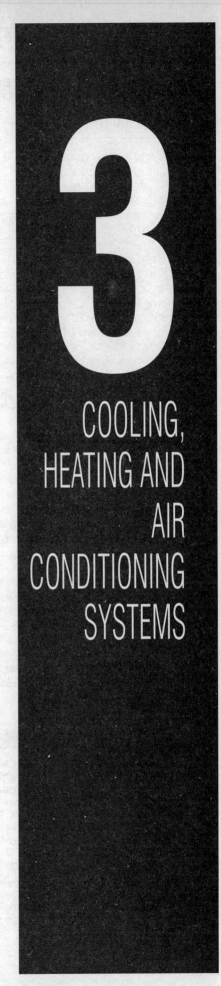

3

COOLING, HEATING AND AIR CONDITIONING SYSTEMS

1 General information

ENGINE COOLING SYSTEM

All vehicles covered by this manual employ a pressurized engine cooling system with thermostatically controlled coolant circulation. An impeller-type water pump mounted on the front of the engine pumps coolant through the engine. The pump mounts directly on the engine block. The coolant flows around the combustion chambers and toward the rear of the engine. Cast-in coolant passages direct coolant near the intake ports, exhaust ports, and spark plug areas. Remove the timing belt to access the water pump.

A wax pellet-type thermostat is located in a housing near the side of the engine. During warm-up, the closed thermostat prevents coolant from circulating through the radiator. As the engine nears normal operating temperature, the thermostat opens and allows hot coolant to travel through the radiator, where it's cooled before returning to the engine.

The cooling system is sealed by a pressure-type cap, which raises the boiling point of the coolant and increases the cooling efficiency of the system. If the system pressure exceeds the cap pressure relief value, the excess pressure in the system forces the spring-loaded valve inside the cap off its seat and allows the coolant to escape through the overflow tube into a coolant reservoir. When the system cools the excess coolant is automatically drawn from the reservoir back into the radiator.

The coolant reservoir does double duty as both the point at which fresh coolant is added to the cooling system to maintain the proper fluid level and as a holding tank for overheated coolant. This type of cooling system is known as a closed design because coolant that escapes past the pressure cap is saved and reused.

These models are equipped with a single speed electric cooling fan. Depending on the engine/transaxle combination installed in the vehicle, some models are equipped with two cooling fans. The cooling fan operation is controlled by two different conditions. Whenever the air conditioner is running, the cooling fan(s) will always be operating. The fan(s) will also run when the coolant reaches a specified temperature determined by the information relayed by the engine coolant temperature sensor (ECT) to the PCM. The PCM turns the fan on depending upon the driving conditions and the temperature of the engine. Power to the fan is switched using relays.

HEATING SYSTEM

The heating system consists of a blower fan and heater core located in the heater/ventilation air conditioning (HVAC) housing, with hoses connecting the heater core to the engine cooling system. Hot engine coolant is circulated through the heater core. When the heater mode on the heater/air conditioning control panel on the instrument panel is activated, a flap door opens to expose the heater core to the passenger compartment. A fan switch on the control panel activates the blower motor, which forces air through the core, heating the air.

AIR CONDITIONING SYSTEM

The air conditioning system consists of a condenser mounted in front of the radiator, an evaporator mounted adjacent to the heater core, a compressor mounted on the engine, a receiver/drier (2000 through 2002 models) or accumulator (2003 models) and the plumbing connecting all of the above components.

2000 through 2002 models are equipped with a thermostatic switch, a receiver/drier and expansion valve to sense evaporator temperatures and to control refrigerant flow. 2003 models are equipped with an accumulator and a fixed orifice tube system instead.

A blower fan forces the warmer air of the passenger compartment through the evaporator core, transferring the heat from the air to the refrigerant (sort of a "radiator in reverse"). The liquid refrigerant boils off into low pressure vapor, taking the heat with it when it leaves the evaporator.

2 Antifreeze - general information

▶ Refer to illustration 2.5

✳ WARNING:

Do not allow antifreeze to come in contact with your skin or painted surfaces of the vehicle. Rinse off spills immediately with plenty of water. Antifreeze is highly toxic if ingested. Never leave antifreeze lying around in an open container or in puddles on the floor; children and pets are attracted by it's sweet smell and may drink it. Check with local authorities about disposing of used antifreeze. Many communities have collection centers which will see that antifreeze is disposed of safely. Never dump used antifreeze on the ground or pour it into drains.

➡Note: Non-toxic antifreeze is now manufactured and available at local auto parts stores, but even this type must be disposed of properly.

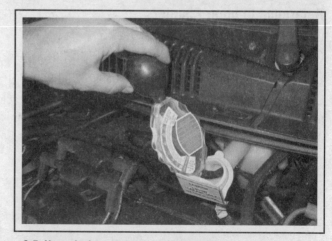

2.5 Use a hydrometer to test the coolant

The cooling system should be filled with the recommended coolant (see Chapter 1), which will prevent freezing down to at least -20-degrees F, or lower if local climate requires it. It also provides protection against corrosion and increases the antifreeze (coolant) boiling point. Mixing incorrect coolant types in this system will severely reduce the longevity and efficiency of this system. The manufacturer recommends that the correct type of coolant be used and strongly urges that coolant types not be mixed.

Drain, flush and refill the cooling system at least every other year (see Chapter 1). The use of antifreeze solutions for periods of longer than two years is likely to cause damage and encourage the formation of rust and scale in the system.

Before adding antifreeze to the system, inspect all hose connections. Antifreeze can leak through very minute openings.

The exact mixture of antifreeze to water, which you should use, depends on the relative weather conditions. The mixture should contain at least 50-percent antifreeze, but should never contain more than 70-percent anti-freeze. Consult the mixture ratio chart on the container before adding coolant.

Hydrometers are available at most auto parts stores to test the coolant (see illustration). Use antifreeze that meets specifications listed in this Chapter.

3 Thermostat - check and replacement

✳✳ WARNING:

Do not remove the radiator cap, drain the coolant or replace the thermostat until the engine has cooled completely.

CHECK

1   Before assuming the thermostat is to blame for a cooling system problem, check the coolant level, drivebelt tension (see Chapter 1) and temperature gauge operation.

2   If the engine seems to be taking a long time to warm up (based on heater output or temperature gauge operation), the thermostat is probably stuck open. Replace the thermostat with a new one.

3   If the engine runs hot, use your hand to check the temperature of the upper radiator hose. If the hose isn't hot, but the engine is, the thermostat is probably stuck closed, preventing the coolant inside the engine from escaping to the radiator. Replace the thermostat.

✳✳ CAUTION:

Don't drive the vehicle without a thermostat. The computer may stay in open loop and emissions and fuel economy will suffer.

4   If the upper radiator hose is hot, it means that the coolant is flowing and the thermostat is open. Consult the *Troubleshooting* Section at the front of this manual for cooling system diagnosis.

REPLACEMENT

▶ **Refer to illustrations 3.8, 3.9, 3.13a and 3.13b**

5   Disconnect the cable from the negative battery terminal (see Chapter 5).

6   Drain the cooling system (see Chapter 1). If the coolant is relatively new and still in good condition, save it and reuse it.

7   Disconnect the coolant reservoir hose from the thermostat housing cover (see Section 5).

8   Remove the upper radiator hose from the thermostat housing cover (see illustration). Detach the hose from the fitting. If it's stuck, grasp it near the end with a pair of adjustable pliers and twist it to break the seal, then pull it off. If the hose is old or if it has deteriorated, cut it off and install a new one.

9   Remove the thermostat housing cover mounting bolts (see illustration). If the housing cover is stuck, tap it with a soft-face hammer to

3.8 The thermostat housing cover is located at the right (timing belt) end of the cylinder head; the radiator cap, radiator outlet connector and the thermostat are also integral components of the thermostat housing cover. Remove the upper radiator hose clamp

3.9 Remove the thermostat housing cover mounting bolts

jar it loose. Be prepared for some coolant to spill as the gasket seal is broken.

10 If the outer surface of the thermostat housing cover, which mates with the hose, is already corroded, pitted, or otherwise deteriorated, it might be damaged even more by hose removal. If it is, replace the thermostat housing cover.

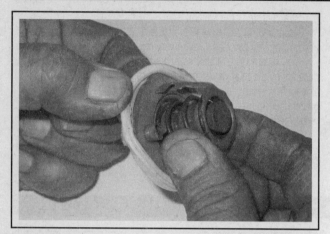

3.13a Be sure to install a new rubber O-ring type gasket on the thermostat

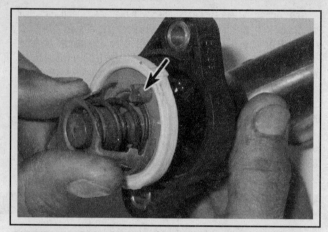

3.13b When installing the thermostat, make sure that the jiggle valve is positioned at the 1 o'clock position

11 Note how it's installed (which end is facing up, or out) and then remove the thermostat.

12 Remove all traces of old gasket material and sealant from the housing and cover with a gasket scraper.

13 Install a new rubber gasket on the thermostat (see illustration) and then install the thermostat into the housing cover, spring-end out. Make sure that you install the thermostat with the "jiggle valve" at 1 o'clock (see illustration).

14 Install the thermostat housing cover and bolts and then tighten the bolts to the torque listed in this Chapter's Specifications.

15 Reattach the hose to the thermostat housing cover. Make sure that the hose clamp is still tight. If it isn't, replace it.

➡**Note: These models are equipped with a spring hose clamp. If the clamp is replaced, be sure to use only a factory replacement part that matches the original.**

16 Refill the cooling system (see Chapter 1).

17 Start the engine and allow it to reach normal operating temperature, then check for leaks and proper thermostat operation (as described in Steps 2 through 4). See Chapter 1 for the cooling system air-bleeding procedure.

4   Engine cooling fans - check and replacement

✳✳ **WARNING:**

To avoid possible injury or damage, DO NOT operate the engine with a damaged fan. Do not attempt to repair fan blades - replace a damaged fan with a new one.

➡**Note: Always be sure to check for blown fuses before attempting to diagnose an electrical circuit problem.**

CHECK

1   If the engine is overheating and the cooling fan is not coming on when the engine temperature rises to an excessive level, unplug the fan motor electrical connector (see illustration 4.10) and then connect the motor directly to the battery with fused jumper wires. If the fan motor doesn't come on, replace the motor.

2   If the radiator fan motor is okay, but it isn't coming on when the engine gets hot, the fan relay might be defective, The cooling fan operation is controlled by two different conditions. Whenever the air conditioner is running, the cooling fan(s) will always be operating. The fan(s) will also run when the coolant reaches a specified temperature determined by the information relayed by the engine coolant temperature sensor (ECT) to the PCM. The PCM turns the fan on depending upon the driving conditions and the temperature of the engine. These control circuits are very complex and should be checked by a dealer service department. Sometimes, the cooling fan system can be fixed by simply identifying and replacing a bad fuse and/or relay (see Chapter 12).

3   Locate the fan relay in the engine compartment fuse/relay box (see Chapter 12).

4   Test the relay (see Chapter 12).

5   If the relay is okay, check all wiring and connections to the fan motor. Refer to the wiring diagrams at the end of Chapter 12. If no obvious problems are found, the problem could be the Engine Coolant Temperature (ECT) sensor or the Powertrain Control Module (PCM). Have the cooling fan system and circuit diagnosed by a dealer service department or repair shop with the proper diagnostic equipment.

REPLACEMENT

⬧ **Refer to illustrations 4.10, 4.12a, 4.12b and 4.13**

6   Disconnect the cable from the negative battery terminal (see Chapter 5).

7   Remove the air filter housing (see Chapter 4).

8   Drain the cooling system (see Chapter 1). If the coolant is relatively new and still in good condition, save it and reuse it.

9   Raise the vehicle and secure it on jackstands.

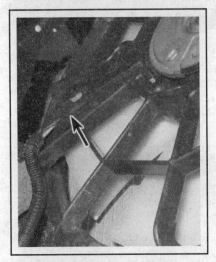

4.10 Disconnect the cooling fan electrical connector

4.12a Remove the bolts from the left side radiator support bracket . . .

4.12b . . . and the right side radiator support bracket

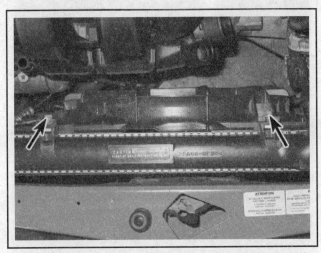

4.13 Remove the lower cooling fan mounting bolts from the bottom of the radiator

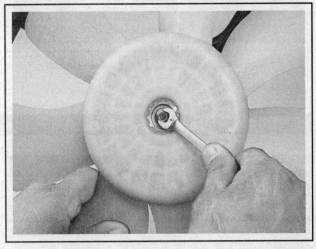

4.17 The fan blade is retained to the motor with either a nut or a clip (radiator fan shown, condenser fan similar)

10 Disconnect the fan motor electrical connector (see illustration).

11 Remove the upper radiator hose. Loosen the hose clamp by squeezing the ends together. Hose clamp pliers work best, but regular pliers will work also. If the radiator hose is stuck, grasp it near the end with a pair of adjustable pliers and twist it to break the seal, then pull it off. If the hose is old or if it has deteriorated, cut it off and install a new one.

12 Remove the upper radiator support mounts (see illustrations).

13 Remove the fan mounting screws (see illustration).

2001 and later models

14 Remove the PCM from the engine compartment (see Chapter 6).

15 Remove the air conditioning line support bracket and position the line off to the side.

All models

▶ **Refer to illustration 4.17**

16 Remove the fan by lifting the fan assembly from above the engine compartment.

17 To detach the fan blade from the motor, remove the nut or retaining clip from the motor shaft (see illustration). Remove the fan blade from the motor.

18 To detach the motor from the shroud, remove the retaining nuts or screws. Remove the motor from the shroud.

19 Installation is the reverse of removal.

➡**Note: When reinstalling the fan assembly, make sure the rubber air shields around the assembly are still in place - without them, the cooling system may not work efficiently.**

5 Coolant reservoir - removal and installation

▶ **Refer to illustrations 5.2 and 5.3**

1 Drain the cooling system (see Chapter 1).
2 Disconnect the coolant reservoir hose (see illustration). Plug the hose to prevent leakage.
3 Remove the coolant reservoir mounting bolts (see illustration)

and lift the reservoir from the engine compartment.
4 Clean out the reservoir with soapy water and a brush to remove any deposits inside. Inspect the reservoir carefully for cracks. If you find a crack, replace the reservoir.
5 Installation is the reverse of removal.

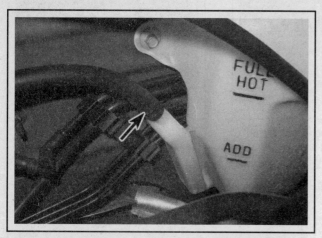

5.2 Remove the hose clamp from the coolant reservoir

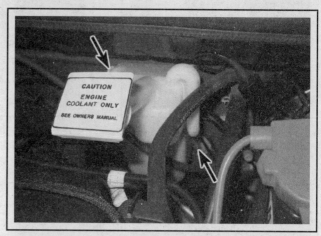

5.3 Coolant reservoir mounting bolt locations

6 Radiator - removal and installation

❊❊ WARNING:

Wait until the engine is completely cool before beginning this procedure.

REMOVAL

▶ **Refer to illustrations 6.5, 6.6, 6.8 and 6.9**

1 Disconnect the cable from the negative battery terminal (see Chapter 5).

2 Drain the cooling system (see Chapter 1). If the coolant is relatively new and in good condition, save it and reuse it.
3 Remove the cooling fan(s) (see Section 4).
4 Raise the vehicle and place it securely on jackstands.
5 On automatic transaxle models, disconnect the transaxle oil cooler lines (see illustration) from the radiator (see Chapter 7B).
➡**Note: It will be necessary to replace the transaxle lines once they are removed from the radiator. The inner wall of the transaxle lines will become damaged and prone to leaking.**

6 Disconnect the upper (see illustration) and the lower radiator hose. Loosen the hose clamp by squeezing the ends together. Hose clamp pliers work best, but regular pliers will work also. If the radiator

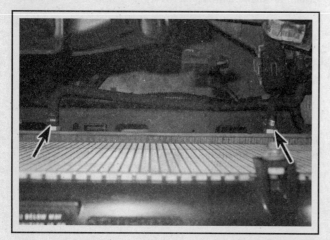

6.5 Location of the transaxle oil cooler lines at the radiator

6.6 Remove the upper radiator hose clamp

6.8 Remove the condenser bracket mounting bolt (right side shown, left side similar)

6.9 Lift the radiator from the engine compartment

hose is stuck, grasp it near the end with a pair of adjustable pliers and twist it to break the seal, then pull it off. If the hose is old or if it has deteriorated, cut it off and install a new one.

7   Remove the radiator mounting brackets from the radiator support and the radiator (see illustrations 4.12a and 4.12b).

8   Remove the air conditioning condenser mounting screws (see illustration) and lean the condenser forward, making room for the radiator.

9   Carefully lift out the radiator as a single assembly (see illustration). Don't spill coolant on the vehicle or scratch the paint. Make sure the rubber radiator seals or insulators that fit on the bottom of the radiator and into the sockets in the body remain in place in the body for proper reinstallation of the radiator.

10 Remove bugs and dirt from the radiator with compressed air and a soft brush. Don't bend the cooling fins. Inspect the radiator for leaks and damage. If it needs repair, have a radiator shop or a dealer service department do the work.

INSTALLATION

11 Inspect the rubber seals or insulators in the lower crossmember and upper crossmember for cracks and deterioration. Make sure that they're free of dirt and gravel. When installing the radiator, make sure that it's correctly seated on the insulators before fastening the top brackets.

12 Installation is otherwise the reverse of the removal procedure. After installation, fill the cooling system with the correct mixture of antifreeze and water (see Chapter 1).

13 Start the engine and check for leaks. Allow the engine to reach normal operating temperature, indicated by the upper radiator hose becoming hot. Recheck the coolant level and add more if required.

14 If you're working on an automatic transaxle equipped vehicle, check and add fluid as needed.

7   Water pump - check

1   A failure in the water pump can cause serious engine damage due to overheating.

2   There are three ways to check the operation of the water pump while it's installed on the engine. If the pump is defective, it should be replaced with a new or rebuilt unit.

3   Water pumps are equipped with weep or vent holes. If a failure occurs in the pump seal, coolant will leak from the hole. In most cases you'll need a flashlight to find the hole on the water pump from underneath to check for leaks.

➡**Note: Some small black staining around the weep hole is nor-**

mal. If the stain is heavy brown or actual coolant is evident, replace the pump.

4   If the water pump shaft bearings fail there may be a howling sound at the front of the engine while it's running. With the engine off, shaft wear can be felt if the water pump pulley is rocked up-and-down. Don't mistake drivebelt slippage, which causes a squealing sound, for water pump bearing failure.

5   A quick water pump performance check is to put the heater on. If the pump is failing, it won't be able to efficiently circulate hot water all the way to the heater core as it should.

8   Water pump - replacement

▶ **Refer to illustrations 8.5a, 8.5b and 8.8**

1   Disconnect the cable from the negative battery terminal (see Chapter 5).

2   Drain the cooling system (see Chapter 1).

3   Remove the accessory drivebelts (see Chapter 1).

4   Remove the timing belt cover, the timing belt, the timing belt tensioner and, if applicable, the rear timing belt cover (see Chapter 2A).

5   Remove the bolts attaching the water pump to the engine block

8.5a Using a criss-cross pattern, remove the bolts securing the water pump . . .

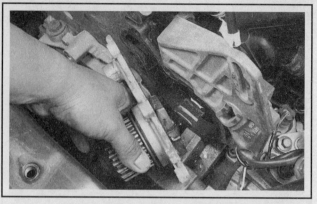

8.5b . . . and then remove the water pump from the engine; if necessary, tap it gently with a soft-faced hammer to break the seal

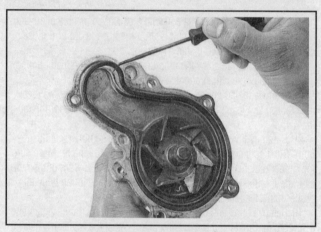

8.8 Remove the old O-ring seal, install a new one, then apply a thin film of RTV sealant to hold the seal in place during installation; make sure that the O-ring seal is correctly seated in the water pump and that it stays in place during pump installation

and remove the pump from the engine (see illustrations). If the water pump is stuck, gently tap it with a soft-faced hammer to break the seal.

6   Clean the bolt threads and the threaded holes in the engine to remove the corrosion and sealant. Remove all traces of old gasket material from the sealing surfaces.

7   Compare the new pump to the old one to make sure that they're identical.

8   Install a new O-ring seal in the groove that lines the water pump body (see illustration) and then apply a thin film of RTV sealant to hold the seal in place during installation.

❄❄ CAUTION:

Make sure that the O-ring seal is correctly seated in the water pump groove to avoid a coolant leak. Carefully mate the pump to the engine.

9   Install the water pump bolts and then tighten them to the torque listed in this Chapter's Specifications. Don't overtighten the water pump bolts; doing so will damage the pump.

10   The remainder of installation is the reverse of removal. Refill the cooling system (see Chapter 1) when you're done. Then run the engine and check for leaks.

9   Coolant temperature sending unit - check and replacement

❄❄ WARNING:

Wait until the engine is completely cool before beginning this procedure.

CHECK

1   The coolant temperature indicator system consists of a warning light or a temperature gauge on the dash and a coolant temperature sending unit mounted on the engine. On the models covered by this manual, the Engine Coolant Temperature (ECT) sensor, which is an information sensor for the Powertrain Control Module (PCM), also functions as the coolant temperature sending unit.

2   If an overheating indication occurs, check the coolant level in the system and then make sure all connectors in the wiring harness between the sending unit and the indicator light or gauge are tight.

3   When the ignition switch is turned to START and the starter motor is turning, the indicator light (if equipped) should come on. This doesn't mean the engine is overheated; it just means that the bulb is good.

4   If the light doesn't come on when the ignition key is turned to START, the bulb might be burned out, the ignition switch might be faulty or the circuit might be open.

5   As soon as the engine starts, the indicator light should go out and remain off, unless the engine overheats. If the light doesn't go out, the wire between the sending unit and the light could be grounded (see the Wiring Diagrams at the end of Chapter 12); the sending unit might be defective (have it checked by a dealer service department); or the ignition switch might be faulty (see Chapter 12). Check the coolant to make sure it's correctly mixed; plain water, with no antifreeze, or coolant that's mainly water, might have too low a boiling point to activate the sending unit (see Chapter 1).

REPLACEMENT

6   See Chapter 6.

10 Blower motor resistor and blower motor - replacement

> ✳✳ **WARNING:**
>
> **These models are equipped with airbags. Always disable the airbag system before working in the vicinity of the impact sensors, steering column or instrument panel to avoid the possibility of accidental deployment of the airbag, which could cause personal injury (see Chapter 12).**

BLOWER MOTOR RESISTOR

▸ **Refer to illustration 10.4**

1 Disconnect the cable from the negative battery terminal (see Chapter 5).

2 Remove the windshield wiper arms (see Chapter 12).

3 Remove the cowl top screen at the base of the windshield (see Chapter 11).

4 Working on the right side of the cowl area, disconnect the electrical connector from the blower motor resistor (see illustration).

5 Remove the blower motor resistor by carefully prying the resistor with a flat bladed screwdriver to release the snap fit. There are two different types of resistors: an air conditioning type or non-air conditioning type. Do not interchange these two different resistors.

6 Installation is the reverse of removal.

BLOWER MOTOR

▸ **Refer to illustration 10.9**

7 Disconnect the cable from the negative battery terminal (see Chapter 5).

8 Remove the lower dash panel from the right side of the dashboard below the glovebox (see Chapter 11).

9 Pull back the carpet and disconnect the blower motor electrical connector (see illustration).

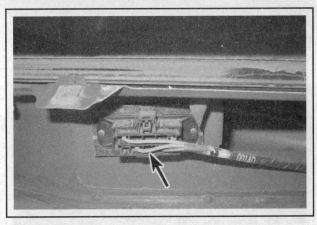

10.4 The blower motor resistor and electrical connector snap fit into the engine compartment firewall

Air conditioned models

▸ **Refer to illustration 10.10**

10 Remove the blower motor mounting screws (see illustration).

11 Remove the blower motor from the blower housing.

Non-air conditioned models

12 Pull the release tab down and simultaneously rotate the blower motor 1/8 turn counterclockwise.

13 Separate the blower motor from the blower housing.

All models

▸ **Refer to illustration 10.14**

14 Remove the circlip from the fan (see illustration) and separate the fan from the blower motor frame.

15 Installation is the reverse of removal.

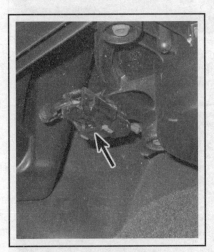

10.9 Disconnect the blower fan connector located on the blower fan assembly

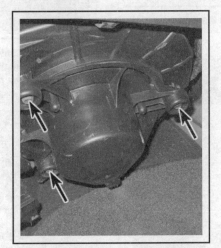

10.10 Remove the blower motor mounting bolts

10.14 Remove the clip and separate the blower motor from the fan

11 Heater/air conditioner control assembly - removal and installation

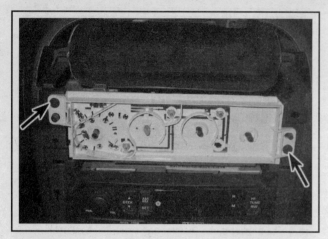

11.4 Remove the mounting bolts from the heater/air conditioner control assembly

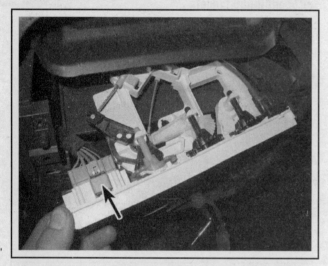

11.5 Remove the electrical connector from the heater/air conditioner control assembly

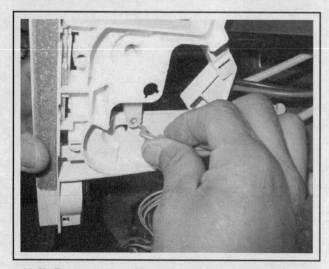

11.6b To remove the cable end, angle the cable from the mounting hole

▶ Refer to illustrations 11.4, 11.5, 11.6a, 11.6b and 11.6c

✳✳ WARNING:

These models are equipped with airbags. Always disable the airbag system before working in the vicinity of the impact sensors, steering column or instrument panel to avoid the possibility of accidental deployment of the airbag, which could cause personal injury (see Chapter 12).

1   Disconnect the cable from the negative battery terminal (see Chapter 5).

2   Remove the center air duct.

3   Remove the instrument panel center bezel (see Chapter 11).

4   Remove the heater/air conditioner control assembly retaining screws (see illustration).

5   Disconnect the electrical connections and the vacuum harness connection from the heater/air conditioner control unit (see illustration) and then partially remove the assembly from the dash to access the cables.

6   Disconnect the control cables from the mode selector lever, from the temperature control lever and from the air selector lever (see illustrations).

7   Installation is the reverse of removal.

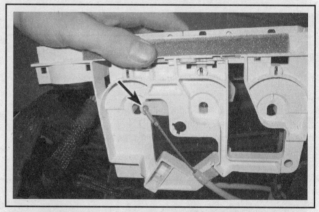

11.6 Pull the control assembly out to access the cables. Remove the temperature control cable from the mounting bracket first then remove the cable end from the selector arm

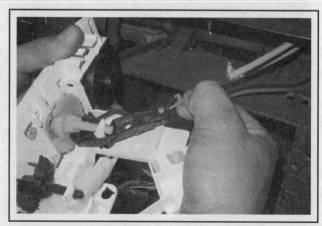

11.6c To remove the mode select cable, align the cable end with the slot in the control arm, then lift the cable up

12 Heater core - replacement

▶ Refer to illustrations 12.2, 12.5a, 12.5b, 12.8, 12.9, 12.10, 12.11, 12.13a, 12.13b, 12.14a, 12.14b and 12.15

❈❈ WARNING 1:

All models are equipped with airbags, so to avoid the possibility of accidental deployment of the airbag, which could cause personal injury, always make sure that you disable the airbag system before working in the vicinity of the impact sensors, steering column or instrument panel (see Chapter 12).

❈❈ WARNING 2:

The heating/ventilation/air conditioning (HVAC) housing must be removed before the heater core or the air conditioning evaporator can be removed. So if you are going to remove the heater core on any of these models, have the air conditioning system discharged by an automotive air conditioning shop BEFORE beginning this procedure. Do NOT loosen any refrigerant line fittings until the system has been discharged. The system is under high pressure and could cause a serious injury if opened while still pressurized. You will not be able to drive the vehicle to an automotive air conditioning shop to have the system discharged once you have begun this procedure because the instrument panel must be removed and the vehicle will not be driveable.

1   Have the air conditioning system discharged by a dealer service department or by an automotive air conditioning shop before proceeding (see **Warning 2** above).

2   Drain the cooling system (see Chapter 1) and then disconnect the heater hoses from the heater core inlet and outlet pipes at the firewall (see illustration).

3   Remove the coolant reservoir (see Section 5).

4   Remove the instrument panel (see Chapter 11).

5   Working in the engine compartment, disconnect the refrigerant line connector block, if equipped (see illustrations).

➡Note: 2000 through 2002 models are equipped with an expansion valve. 2003 and later models are equipped with only the refrigerant line connector block.

6   Working in the passenger compartment, disconnect the expansion valve from the evaporator.

7   Disconnect the vacuum lines from the brake power booster (see Chapter 9).

8   Working in the engine compartment, remove the drain tube from the lower section of the heater/ventilation/air conditioning (HVAC) housing (see illustration).

9   Working in the passenger compartment, remove the defroster duct from the HVAC housing (see illustration).

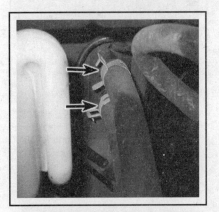

12.2 Trace the heater hoses to the firewall, remove the hose clamps and then disconnect the hoses from the heater core pipes

12.5a First, remove the mounting bolt from the air conditioning line connector and . . .

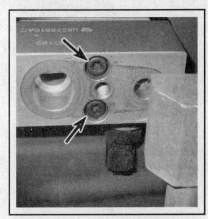

12.5b . . . then remove the Torx bolts from the expansion valve (2000 model shown)

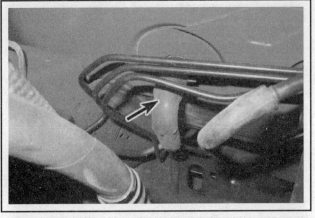

12.8 Look for the evaporator drain hose on the firewall in the engine compartment; remove the hose clamp and simply pull it off

12.9 Remove the defroster duct from the HVAC assembly

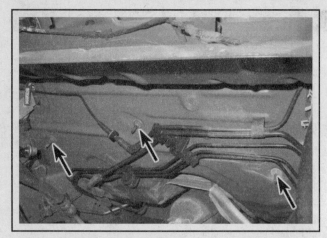

12.10 Remove the HVAC housing mounting bolts from the engine compartment

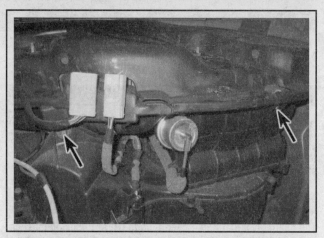

12.11 Remove the bolts that mount the HVAC assembly to the upper dash support structure

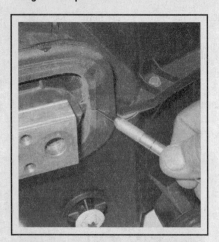

12.13a Cut the foam seal around the evaporator lines parallel with the case halves

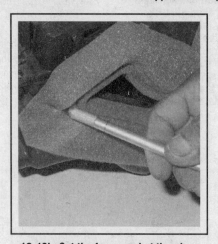

12.13b Cut the foam seal at the air distribution outlet in line with the HVAC housing seams

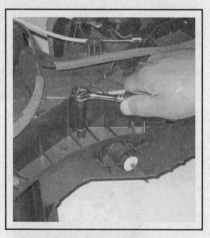

12.14a Remove the upper and lower heater case mounting screws

10 Working in the engine compartment, remove the HVAC mounting nuts (see illustration).

11 Working in the passenger compartment, first remove the right side mounting screw from the HVAC housing then remove the remain-ing fasteners (see illustration).

12 Lift the HVAC housing from the passenger compartment.

13 Remove the foam seal from the air distribution outlet, the evapo-rator lines and the heater core tubes (see illustration).

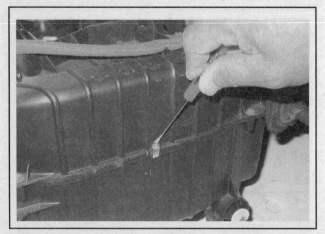

12.14b Remove the clips that secure the case halves together

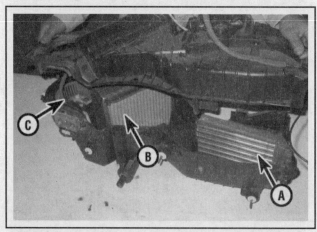

12.15 Lift the top half of the HVAC housing to expose the heater core (A), the evaporator (B) and the blower fan (C)

14 Remove the upper and lower heater case mounting screws and clips (see illustrations).

15 Pull the heater core out of the housing (see illustration).

16 If the old heater core has been leaking, clean the coolant from the housing.

17 Installation is the reverse of removal. Don't forget to reconnect the heater core inlet and outlet hoses at the firewall.

18 Refill the cooling system (see Chapter 1) when you're done.

13 Air conditioning and heating system - check and maintenance

❄❄ WARNING:

The air conditioning system is under high pressure. Do not loosen any hose fittings or remove any components until after the system has been discharged by a dealer service department or service station. Always wear eye protection when disconnecting air conditioning system fittings.

1   The following maintenance checks should be performed on a regular basis to ensure the air conditioner continues to operate at peak efficiency.

a) *Check the compressor drivebelt. If it's worn or deteriorated, replace it* (see Chapter 1).

b) *Check the drivebelt tension and, if necessary, adjust it* (see Chapter 1).

c) *Check the system hoses. Look for cracks, bubbles, hard spots and deterioration. Inspect the hoses and all fittings for oil bubbles and seepage. If there's any evidence of wear, damage or leaks, replace the hose(s).*

d) *Inspect the condenser fins for leaves, bugs and other debris. Use a "fin comb" or compressed air to clean the condenser.*

e) *Make sure the system has the correct refrigerant charge.*

f) *Check the evaporator housing drain tube* (see illustration 12.8) *for blockage.*

2   It's a good idea to operate the system for about 10 minutes at least once a month, particularly during the winter. Long term non-use can cause hardening, and subsequent failure, of the seals.

3   Because of the complexity of the air conditioning system and the special equipment necessary to service it, in-depth troubleshooting and repairs are not included in this manual. However, simple checks and component replacement procedures are provided in this Chapter.

4   The most common cause of poor cooling is simply a low system refrigerant charge. If a noticeable drop in cool air output occurs, the following quick check will help you determine if the refrigerant level is low.

CHECKING THE REFRIGERANT CHARGE

5   Warm the engine up to normal operating temperature.

6   Place the air conditioning temperature selector at the coldest setting and the blower at the highest setting. Open the doors (to make sure the air conditioning system doesn't cycle off as soon as it cools the passenger compartment).

7   With the compressor engaged - the clutch will make an audible click and the center of the clutch will rotate. If the compressor discharge line feels warm and the compressor inlet pipe feels cool, the system is properly charged.

8   Place a thermometer in the dashboard vent nearest the evapora-

tor and operate the system until the indicated temperature is around 40 to 45 degrees F. If the ambient (outside) air temperature is very high, say 110 degrees F, the duct air temperature may be as high as 60 degrees F, but generally the air conditioning is 30-50 degrees F cooler than the ambient air.

➡**Note: Humidity of the ambient air also affects the cooling capacity of the system. Higher ambient humidity lowers the effectiveness of the air conditioning system.**

ADDING REFRIGERANT

▶ **Refer to illustrations 13.12 and 13.15**

9   Buy an automotive charging kit at an auto parts store. A charging kit includes a 14-ounce can of refrigerant, a tap valve and a short section of hose that can be attached between the tap valve and the system low side service valve. Because one can of refrigerant may not be sufficient to bring the system charge up to the proper level, it's a good idea to buy an additional can. Make sure that one of the cans contains red refrigerant dye. If the system is leaking, the red dye will leak out with the refrigerant and help you pinpoint the location of the leak.

❄❄ CAUTION:

There are two types of refrigerant used in automotive systems; R-12 - which has been widely used on earlier models and the more environmentally-friendly R-134a used in all models covered by this manual. These two refrigerants (and their appropriate refrigerant oils) are not compatible and must never be mixed or components will be damaged. Use only R-134a refrigerant in the models covered by this manual.

10 Hook up the charging kit by following the manufacturer's instructions.

❄❄ WARNING:

DO NOT hook the charging kit hose to the system high side! The fittings on the charging kit are designed to fit only on the low side of the system.

11 Back off the valve handle on the charging kit and screw the kit onto the refrigerant can, making sure first that the O-ring or rubber seal inside the threaded portion of the kit is in place.

❄❄ WARNING:

Wear protective eyewear when dealing with pressurized refrigerant cans.

13.12 Cans of R-134A refrigerant available at auto parts stores can be added to the low side of the air conditioning system with a simple recharging kit

13.15 Insert a thermometer in the center vent, turn on the air conditioning system and wait for it to cool down; depending on the humidity, the output air should be 30 to 40 degrees cooler than the ambient air temperature

12 Remove the dust cap from the low-side charging connection and attach the quick-connect fitting on the kit hose (see illustration).

13 Warm up the engine and turn on the air conditioner. Keep the charging kit hose away from the fan and other moving parts.

➥**Note: The charging process requires the compressor to be running. Your compressor may cycle off if the pressure is low due to a low charge. If the clutch cycles off, you can pull the low-pressure cycling switch plug and attach a jumper wire. This will keep the compressor ON.**

14 Turn the valve handle on the kit until the stem pierces the can, then back the handle out to release the refrigerant. You should be able to hear the rush of gas. Add refrigerant to the low side of the system until both the receiver-drier surface and the evaporator inlet pipe feel about the same temperature. Allow stabilization time between each addition.

15 If you have an accurate thermometer, place it in the center air conditioning vent (see illustration) and note the temperature of the air coming out of the vent. A fully-charged system which is working correctly should cool down to about 40 degrees F. Generally, an air conditioning system will put out air that is 30 to 40 degrees F cooler than the ambient air. For example, if the ambient (outside) air temperature is very high (over 100 degrees F), the temperature of air coming out of the registers should be 60 to 70 degrees F.

16 When the can is empty, turn the valve handle to the closed position and release the connection from the low-side port. Replace the dust cap.

17 Remove the charging kit from the can and store the kit for future use with the piercing valve in the UP position, to prevent inadvertently piercing the can on the next use.

HEATING SYSTEMS

18 If the carpet under the heater core is damp, or if antifreeze vapor or steam is coming through the vents, the heater core is leaking. Remove it (see Section 11) and install a new unit (most radiator shops will not repair a leaking heater core).

19 If the air coming out of the heater vents isn't hot, the problem could stem from any of the following causes:

a) *The thermostat is stuck open, preventing the engine coolant from warming up enough to carry heat to the heater core. Replace the thermostat* (see Section 3).

b) *There is a blockage in the system, preventing the flow of coolant through the heater core. Feel both heater hoses at the firewall. They should be hot. If one of them is cold, there is an obstruction in one of the hoses or in the heater core, or the heater control valve is shut. Detach the hoses and back flush the heater core with a water hose. If the heater core is clear but circulation is impeded, remove the two hoses and flush them out with a water hose.*

c) *If flushing fails to remove the blockage from the heater core, the core must be replaced* (see Section 11).

ELIMINATING AIR CONDITIONING ODORS

20 Unpleasant odors that often develop in air conditioning systems are caused by the growth of a fungus, usually on the surface of the evaporator core. The warm, humid environment there is a perfect breeding ground for mildew to develop.

21 The evaporator core on most vehicles is difficult to access, and factory dealerships have a lengthy, expensive process for eliminating the fungus by opening up the evaporator case and using a powerful disinfectant and rinse on the core until the fungus is gone. You can service your own system at home, but it takes something much stronger than basic household germ-killers or deodorizers.

22 Aerosol disinfectants for automotive air conditioning systems are available in most auto parts stores, but remember when shopping for them that the most effective treatments are also the most expensive. The basic procedure for using these sprays is to start by running the system in the RECIRCULATION mode for ten minutes with the blower on its highest speed. Use the highest heat mode to dry out the system and keep the compressor from engaging by disconnecting the wiring connector at the compressor (see Section 14).

23 Make sure that the disinfectant can comes with a long spray hose. Remove the evaporator drain tube and point the nozzle so that it protrudes inside the duct that leads to the evaporator housing (see illustration 12.8), and then spray according to the manufacturer's recommendations. Try to cover the whole surface of the evaporator core, by aiming the spray up, down and sideways. Follow the manufacturer's recommendations for the length of spray and waiting time between applications.

24 Once the evaporator has been cleaned, the best way to prevent the mildew from coming back again is to make sure your evaporator housing drain tube is clear (see illustration 12.8).

14 Air conditioning compressor - removal and installation

✳✳ WARNING:

The air conditioning system is under high pressure. DO NOT disassemble any part of the system (hoses, compressor, line fittings, etc.) until after the system has been discharged by a dealer service department or by an automotive air conditioning shop.

➡Note: If you're replacing the compressor, also replace the accumulator (see Section 15).

REMOVAL

♦ Refer to illustrations 14.5, 14.7 and 14.8

1   Have the air conditioning system discharged (see the **Warning** above).

2   Disconnect the cable from the negative battery terminal (see Chapter 5).

3   Raise the vehicle and secure it on jackstands.

4   Remove the accessory drivebelts (see Chapter 1).

5   Unplug the electrical connector (see illustration) from the compressor clutch.

6   Disconnect the refrigerant lines from the compressor. Plug the open fittings to prevent entry of dirt and moisture. Be sure to remove and discard the old refrigerant line fitting O-rings.

7   Disconnect the high pressure cut-out switch (see illustration) from the back of the compressor.

8   Remove the compressor mounting bolts (see illustration) and then remove it from the vehicle.

INSTALLATION

9   If a new compressor is being installed, pour out the oil from the old compressor into a graduated container and add that amount of new refrigerant oil to the new compressor. Also follow any directions included with the new compressor.

10   The clutch might have to be transferred from the original compressor to the new unit.

11   Installation is the reverse of removal. Replace all O-rings with new ones specifically made for use with R-134a refrigerant and lubricate them with R-134a-compatible refrigerant oil.

12   Have the system evacuated, recharged and leak tested by the shop that discharged it.

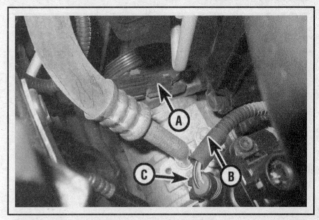

14.5 Details of the compressor clutch electrical connector and refrigerant lines

A	Compressor clutch electrical connector	B	Discharge port (high side)
		C	Suction port (low side)

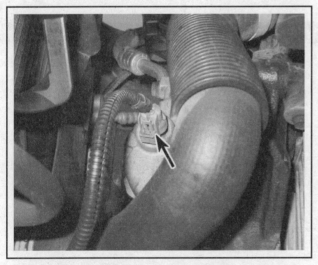

14.7 Location of the high pressure cut-out switch on the air conditioning compressor

14.8 Location of the compressor mounting bolts (other two bolts not visible)

15 Receiver/drier (2000 through 2002 models) - removal and installation

✳✳ WARNING:

The air conditioning system is under high pressure. DO NOT disassemble any part of the system (hose, compressor, line fittings, etc.) until after the system has been evacuated and the refrigerant recovered by a dealer service department or service station.

✳✳ CAUTION:

Replacement receiver/driers are so effective at absorbing moisture that they can quickly saturate upon exposure to the atmosphere. When installing a new unit, have all tools and supplies ready for quick reassembly to avoid having the system open any longer than necessary.

REMOVAL

▶ **Refer to illustration 15.3**

1 The receiver/drier is mounted in a rubber grommet on the right side of the engine compartment.

2 Have the system discharged (see the **Warning** at the beginning of this Section).

3 Disconnect the refrigerant lines from the receiver/drier (see illustration). Be sure to discard and replace the old O-rings in the fittings and then plug the open fittings to prevent entry of dirt and moisture.

4 Lift the receiver/drier from the rubber grommet.

INSTALLATION

5 Installation is the reverse of removal. If a new receiver/drier is being installed, add one ounce of refrigerant oil to it before installation.

6 Replace all the O-rings with new parts.

7 Take the vehicle back to the shop that discharged it. Have the system evacuated, recharged and leak tested.

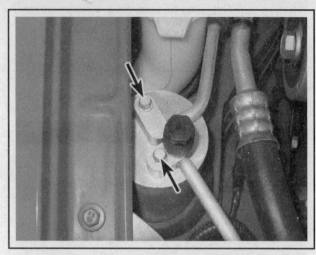

15.3 Location of the air conditioning refrigerant lines at the receiver/drier

16 Air conditioning accumulator (2003 and later models) - removal and installation

✳✳ WARNING:

The air conditioning system is under high pressure. DO NOT disassemble any part of the system (hose, compressor, line fittings, etc.) until after the system has been evacuated and the refrigerant recovered by a dealer service department or service station.

✳✳ CAUTION:

Replacement accumulator units are so effective at absorbing moisture that they can quickly saturate upon exposure to the atmosphere. When installing a new unit, have all tools and supplies ready for quick reassembly to avoid having the system open any longer than necessary.

REMOVAL

1 The accumulator is located between the evaporator outlet and the compressor suction (low side) port.

2 Have the system discharged (see the **Warning** at the beginning of this Section).

3 Remove the cruise control servo mounting bolts and position the assembly off to the side. Do not disconnect the servo components.

4 Remove the low pressure cycling switch from the accumulator (see Section 18).

5 Disconnect the refrigerant lines from the accumulator. Be sure to discard and replace the old O-rings in the fittings and then plug the open fittings to prevent entry of dirt and moisture.

6 Remove the bracket screw at the base of the accumulator.

7 Lift the accumulator from the engine compartment.

INSTALLATION

8 Installation is the reverse of removal. If a new accumulator is being installed, add one ounce of refrigerant oil to it before installation.

9 Replace all the O-rings with new parts.

10 Take the vehicle back to the shop that discharged it. Have the system evacuated, recharged and leak tested.

17 Air conditioning condenser - removal and installation

❋❋ WARNING:

The air conditioning system is under high pressure. Do not loosen any hose fittings or remove any components until after the system has been discharged by a dealer service department or service station. Always wear eye protection when disconnecting air conditioning system components.

➡Note: The accumulator should be replaced whenever the condenser is replaced (see Section 14).

REMOVAL

◗ Refer to illustrations 17.6 and 17.7

1   Have the system discharged (see the **Warning** at the beginning of this Section).
2   Disconnect the cable from the negative battery terminal (see Chapter 5).
3   Remove the battery, the battery tray and the support strut (see Chapter 5).
4   Remove the engine cooling fan(s) (see Section 4).
5   Remove the radiator support brackets (see Section 5).
6   Disconnect the refrigerant lines (see illustration). Be sure to remove and discard the old O-rings.
7   Remove the radiator to condenser mounting bolts (see illustration). The radiator is supported by the condenser mounting brackets.
8   Remove the condenser from the vehicle.

INSTALLATION

9   Be sure to inspect the rubber mounting bolt insulators for cracks and deterioration. Replace them if they're worn or damaged.
10 Installation is the reverse of removal. If a new condenser is being installed add one ounce of refrigerant oil to it before installation.
11 Take the vehicle back to the shop that discharged it. Have the system evacuated, recharged and leak tested.

17.6 Location of the refrigerant lines at the condenser

17.7 Remove the condenser mounting bracket bolt (right side shown, left side similar)

18 Air conditioning low pressure cycling switch - replacement

◗ Refer to illustration 18.2

❋❋ WARNING:

The air conditioning system is under high pressure. Do not loosen any hose fittings or remove any components until the system has been discharged. Air conditioning refrigerant should be properly discharged into an EPA-approved recovery/recycling unit by a dealer service department or an automotive air conditioning repair facility. Always wear eye protection when disconnecting air conditioning system fittings.

➡Note: The high pressure cycling switch is located in the air conditioning compressor on all models. The low pressure cycling switch is located in the expansion valve on 2000 through 2002 models or on the accumulator on 2003 and later models.

1   Have the refrigerant discharged by an air conditioning technician.
2   Unplug the electrical connector from the pressure cycling switch (see illustration).

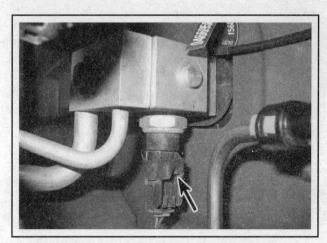

18.2 Location of the low pressure cycling switch on a 2000 model

➡Note: The low pressure cycling switch is located under the expansion valve on 2000 through 2002 models or on the accumulator on 2003 and later models.

3 Unscrew the pressure cycling switch.
4 Lubricate the switch O-ring with clean refrigerant oil of the correct type.

5 Screw the new switch onto the threads until hand tight, and then tighten it securely.
6 Reconnect the electrical connector.
7 Have the system evacuated, charged and leak tested by the shop that discharged it.

19 Orifice tube (2003 and later models) - removal and installation

❄❄ WARNING:

The air conditioning system is under high pressure. Do not loosen any hose fittings or remove any components until the system has been discharged. Air conditioning refrigerant should be properly discharged into an EPA-approved recovery/recycling unit by a dealer service department or an automotive air conditioning repair facility. Always wear eye protection when disconnecting air conditioning system fittings.

➡Note: The orifice tube is an integral part of the liquid line. The liquid line joins the outlet of the condenser to the inlet of the evaporator. The liquid line must be replaced as a complete unit when servicing the orifice tube. Consult a dealer parts department.

1 Have the refrigerant discharged by an air conditioning technician.
2 Disconnect the outlet refrigerant line (to evaporator) at the condenser.
3 Disconnect the liquid line at the connector block.
4 Separate the liquid line (orifice tube) from the condenser and the connector block.
5 Installation is the reverse of removal. Be sure to remove and discard the old O-rings.
6 Take the vehicle back to the shop that discharged it. Have the system evacuated, recharged and leak tested.

Specifications

General

Radiator cap pressure rating	14 to 18 psi (96 to 124 kPa)
Thermostat rating (opening temperature)	192 to 199 degrees F (88 to 93 degrees C)
Cooling system capacity	See Chapter 1
Refrigerant capacity	Refer to HVAC specification tag under the hood

Torque specifications

	Ft-lbs (unless otherwise indicated)	Nm
Thermostat housing bolts	105 in-lbs	12
Transaxle cooler lines	18 in-lbs	2
Condenser-to-radiator mounting bolts	65 in-lbs	7
Water pump mounting bolts	105 in-lbs	12

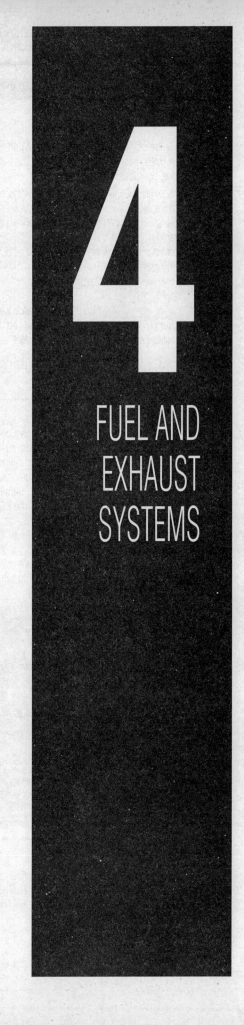

4
FUEL AND EXHAUST SYSTEMS

Section

Reference to other Chapters

1   General information

▶ **Refer to illustration 1.1**

All models are equipped with a sequential Multi-Port Electronic Fuel Injection (MPI) system (see illustration). The fuel system consists of the following components:

Air filter housing and air intake duct between throttle body and air intake plenum
Electric fuel pump/fuel level sending unit module (located in the fuel tank)
Fuel pressure regulator (integrated into the fuel pump/fuel level sending unit module)
Fuel rail and fuel injector assembly
Fuel tank
Throttle body assembly, which includes throttle valve, Throttle Position (TP) sensor and Idle Air Control (IAC) valve (for information on the TP sensor and IAC valve, see Chapter 6)

SEQUENTIAL MULTI-PORT ELECTRONIC FUEL INJECTION (MPI) SYSTEM

Intake air is drawn through the air filter housing and into the throttle body. Inside the throttle body, the throttle valve controls the amount of air passing through the air intake duct into the air intake plenum and then into the cylinders. The intake port for each cylinder is equipped with its own injector. The MPI system injects fuel into the intake ports in the same sequence as the firing order of the engine. The Powertrain Control Module (PCM) controls the injectors. The PCM constantly monitors the operating conditions of the engine - temperature, speed, load, etc. - and then delivers the optimal amount of fuel for these conditions. The amount of fuel delivered by each injector is determined by its "on time," which the PCM can vary so quickly that two successive injectors can deliver a different amount of fuel. In other words, the

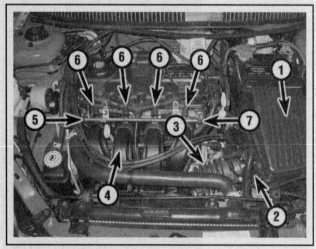

1.1 Fuel system components

1	*Air filter housing*	*5*	*Fuel rail*
2	*Throttle body*	*6*	*Fuel injectors*
3	*Air intake duct*	*7*	*Fuel rail test port*
4	*Air intake manifold*		

PCM responds instantly to any changes in the operating conditions of the engine.

FUEL PUMP

The electric fuel pump/fuel level sending unit module is located inside the fuel tank. To remove the fuel pump/fuel level sending unit module, lower the tank, unscrew the pump locknut and then remove the fuel pump through the hole in the top of the tank.

Voltage to the fuel pump circuit is supplied through the fuel pump relay, which is located inside the Power Distribution Center (PDC) which is located in the engine compartment. The Powertrain Control Module (PCM) controls the fuel pump relay. When the ignition key is turned to ON (but not cranked), the PCM energizes the fuel pump relay for one second, which closes the fuel pump circuit long enough to pressurize the fuel system. Then the PCM turns off the fuel pump relay.

When the ignition key is turned to START, current flows from the ignition switch to the starter relay coil and to the PCM, which also receives a signal from the Crankshaft Position (CKP) sensor. When the PCM receives these two signals (from the ignition switch and from the CKP sensor), a transistor inside the PCM that controls the fuel pump relay allows current to flow to the fuel pump relay, which allows current to flow to the fuel pump. As long as the transistor inside the PCM that controls the fuel pump relay continues to receive the CKP sensor signal, it continues to supply current to the fuel pump relay, which continues to supply current to the fuel pump.

When the engine is running, fuel is pumped by the fuel pump inside the fuel tank to the fuel rail and injectors through a metal line running along the underside of the vehicle. There is no fuel return line between the fuel rail and the fuel tank. The fuel pressure regulator, which is an integral part of the fuel pump/fuel level sending unit module, maintains fuel pressure between 43 and 53 psi (2000 models) or 54 and 64 psi (2001 and later models). If the fuel pressure exceeds this range, a fuel-return valve routes the excess fuel back to the fuel tank. The system fuel filter is also an integral part of the fuel pressure regulator. There is no external fuel filter.

FUEL LINE, FUEL HOSES AND QUICK-CONNECT FITTINGS

The metal fuel line underneath the vehicle, which carries pressurized fuel from the fuel tank to the fuel rail, is connected to the fuel pump module and to the fuel rail by hoses that use special quick-connect fittings. For more information on these fittings, refer to Section 4.

EXHAUST SYSTEM

The exhaust system consists of the exhaust manifold, the catalytic converter, the exhaust pipe, the muffler and the tail pipe. For information on servicing the exhaust system, refer to Section 16. The catalytic converter is an emission-control device in the exhaust system that reduces hydrocarbon (HC), carbon monoxide (CO) and oxides of nitrogen (NOx) pollutants. For more information regarding the catalytic converter, refer to Chapter 6.

2   Fuel pressure relief

▶ Refer to illustration 2.3

✳✳ WARNING:

Gasoline is extremely flammable, so take extra precautions when you work on any part of the fuel system. Don't smoke or allow open flames or bare light bulbs near the work area, and don't work in a garage where a gas-type appliance (such as a water heater or a clothes dryer) is present. Since gasoline is carcinogenic, wear latex gloves when there's a possibility of being exposed to fuel, and, if you spill any fuel on your skin, rinse it off immediately with soap and water. Mop up any spills immediately and do not store fuel-soaked rags where they could ignite. The fuel system is under constant pressure, so, if any fuel lines are to be disconnected, the fuel pressure in the system must be relieved first. When you perform any kind of work on the fuel system, wear safety glasses and have a Class B type fire extinguisher on hand.

1   Remove the fuel filler cap to relieve any pressure inside the fuel tank.

2   Remove the lid from the Power Distribution Center (PDC), the fuse and relay box which is located in the engine compartment, adjacent to the battery and the air filter housing.

3   Remove the fuel pump relay (see illustration) from the PDC.

4   Start the engine.

2.3 To relieve the pressure in the fuel system, disable the electric fuel pump by removing the fuel pump relay (1); if you're checking the fuel pump electrical circuit, always check the fuel pump relay and the fuel pump fuse (2)

5   Allow the engine to run until it stops. Turn the ignition switch to OFF and then disconnect the cable from the negative terminal of the battery before working on the fuel system.

6   The fuel system pressure is now relieved. When you're finished working on the fuel system, install the fuel pump relay and then connect the negative cable to the battery.

3   Fuel pump/fuel pressure - check

✳✳ WARNING:

Gasoline is extremely flammable, so take extra precautions when you work on any part of the fuel system. See the *Warning* in Section 2.

GENERAL CHECKS

1   If you suspect insufficient fuel delivery check the following items first:

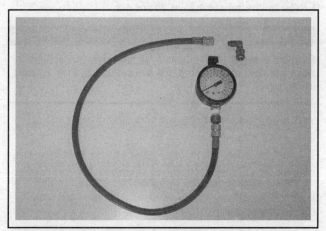

3.4 This typical aftermarket fuel pressure testing set-up includes a fuel pressure gauge capable of measuring the operating pressure range of fuel injection systems and a special Schrader-valve-type adapter to allow you to easily connect the pressure gauge hose to the fuel rail test port

a) *Check the battery and make sure that it's fully charged (see Chapter 5).*

b) *Check the fuel filter (see Section 5) for obstructions (the fuel filter is an integral part of the fuel pressure regulator, which is located on the fuel pump/fuel level sending unit module).*

c) *Inspect the fuel line, hoses and quick-connect fittings (see Section 4). Verify that the problem is not simply a leak in a line.*

2   Verify that the fuel pump actually runs. Remove the fuel filler cap and then have an assistant turn the ignition switch to ON while you listen carefully for the sound of the fuel pump operating. You should hear a brief whirring noise (for about one second) as the pump comes on and pressurizes the system. If the fuel pump makes no sound, check the fuel pump electrical circuit (see the wiring diagrams at the end of Chapter 12). If the fuel pump runs, but a fuel system problem persists, proceed to the fuel pump pressure check.

FUEL PUMP PRESSURE CHECK

▶ Refer to illustrations 3.4, 3.5a and 3.5b

➡Note: Before proceeding, make sure that you have a fuel pressure gauge capable of measuring fuel pressure in the range of 44 to 54 psi (2000 models) or 53 to 63 psi (2001 and later models) and an adapter suitable for connecting the fuel pressure gauge to the Schrader-valve-type test port on the fuel rail.

3   Relieve the fuel system pressure (see Section 2).

4   Besides a fuel pressure gauge capable of reading fuel pressure in a range of 53 to 63 psi, you'll need a hose and an adapter suitable for hooking up to the Schrader-valve-type test port on the fuel rail (see illustration).

3.5a To check the fuel pressure, unscrew this cap from the test port on the left end of the fuel rail . . .

3.5b . . . and then hook up the fuel pressure gauge to the test port with the Schrader-valve-type adapter

5   Unscrew the test port cap (see illustration) located on the left end of the fuel rail and then connect the fuel pressure gauge and hose to the test port (see illustration).

6   Turn off all the accessories and then start the engine and let it idle. The fuel pressure should be within the operating range listed in this Chapter's Specifications. If the pressure reading is within the specified range, the system is operating correctly.

7   If the fuel pressure is higher than specified, replace the fuel pressure regulator (see Section 5).

8   If the fuel pressure is lower than specified, check the fuel line, the fuel hoses, the connectors and the fuel injectors for leaks. Remove the fuel pump module and inspect the fuel strainer for restrictions. If all these components are okay, the fuel filter could be clogged or the fuel pressure regulator or the fuel pump could be defective. Replace the fuel pump module to prevent any future fuel pressure problems.

➡Note: Since it is difficult to identify a restricted fuel filter we recommend the home mechanic simply replace the filter if it is suspect.

9   After the testing is complete, relieve the fuel pressure (see Section 2), remove the fuel pressure gauge and then install the cap on the test port.

FUEL PUMP ELECTRICAL CIRCUIT CHECK

10   If the pump does not turn on, check the fuel pump fuse and the fuel pump relay, both of which are located in the Power Distribution Center (PDC) (see illustration 2.3).

11   If the fuse and relay are good and the fuel pump does not operate, have the circuit checked by a dealer service department or other qualified shop.

4   Fuel lines and fittings - general information

4.2a The metal fuel supply line (and plastic EVAP lines) running under the vehicle are attached to the pan by a series of clips such as this one; the clips are secured to the pan by Phillips head nuts screwed onto studs welded to the pan

▶ Refer to illustrations 4.2a, 4.2b and 4.3

✳✳ WARNING:

Gasoline is extremely flammable, so take extra precautions when you work on any part of the fuel system. See the *Warning* in Section 2.

1   Always relieve the system fuel pressure before servicing fuel lines or fittings (see Section 2).

2   The fuel supply line extends from the fuel tank to the engine compartment. The lines are secured to the underbody with clips (see illustration). Anytime you raise the vehicle for underbody service, inspect the lines underneath the vehicle for leaks, kinks and dents. If a clip is damaged, replace it (see illustration).

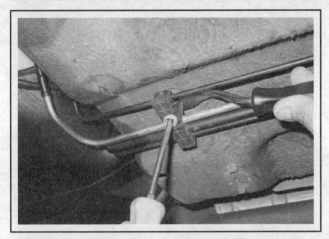

4.2b To detach a fuel line clip from the pan, loosen the Phillips head nut and then carefully pry the clip off the stud (even after a clip fastener is unscrewed, the clip won't just fall down because it's held in place by the tension of the metal fuel line and the plastic EVAP lines)

3   The fuel supply line is attached to the fuel supply hoses at the fuel tank and in the engine compartment by quick-connect fittings (see illustration).

4   If evidence of dirt is found in the system during disassembly, disconnect the fuel supply line and then blow it out with compressed air. And be sure to inspect the fuel filter on the fuel pump module (see Section 5) for damage and deterioration.

STEEL TUBING

5   If replacement of a fuel line or emission line is called for, use welded steel tubing meeting the manufacturer's specifications or its equivalent.

6   Don't use copper or aluminum tubing to replace steel tubing. These materials cannot withstand normal vehicle vibration.

7   Because fuel lines used on fuel-injected vehicles are under high pressure, they require special consideration.

8   Some fuel lines have threaded fittings with O-rings. Any time the fittings are loosened to service or replace components:

   a) *Use a backup wrench while loosening and tightening the fittings.*

   b) *Check all O-rings for cuts, cracks and deterioration. Replace any that appear hardened, worn or damaged.*

   c) *If the lines are replaced, always use original equipment parts, or parts that meet the original equipment standards specified in this Section.*

FLEXIBLE HOSE

✳✳ CAUTION:

Use only original equipment replacement hoses or their equivalent. Others may fail from the high pressures of this system.

9   Don't route fuel hose within four inches of any part of the exhaust system or within ten inches of the catalytic converter. Metal lines and rubber hoses must never be allowed to chafe against the frame. A minimum of 1/4-inch clearance must be maintained around a line or hose to prevent contact with the frame.

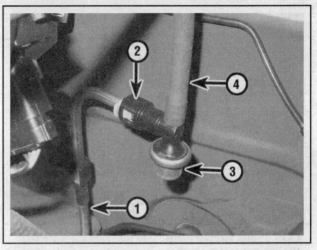

4.3 The metal fuel supply line is connected to the fuel supply hoses back at the fuel tank and in the engine compartment by quick-connect fittings; this is a typical two-tab type fitting at the connection between the underbody fuel line and the engine compartment fuel hose that goes to the fuel rail

   1   Metal fuel supply line from fuel tank
   2   Quick-connect fitting (two-tab type shown)
   3   Fuel pulsation damper
   4   Fuel supply hose (delivers fuel to fuel rail)

10   Some models may be equipped with nylon fuel line and quick-connect fittings at the fuel filter and/or fuel pump. The quick-connect fittings cannot be serviced separately. Do not attempt to service these types of fuel lines in the event the retainer tabs or the line becomes damaged. Replace the entire fuel line as an assembly.

REPLACEMENT

11   If a fuel line or fuel hose is damaged, replace it with factory replacement parts. Do not substitute line or hose of inferior quality; it might not be suitable for, and it might fail from, the operating pressure of this system.

12   Relieve the fuel system pressure (see Section 2).

13   Disconnect the negative battery cable.

14   Remove all clamps and/or clips attaching the line to the vehicle body. Pay close attention to all clips; they not only secure the fuel line and hoses, they also route them correctly. The hoses and line must be reattached to their respective clips when reassembled.

Quick-connect fittings

15   Various types of quick-connect fittings are used on the vehicles covered by this manual. After you locate the fitting, determine which type it is and then use whichever of the following procedures that applies to that fitting to disconnect and reconnect it.

Single-tab type

➡**Note: The single-tab type quick-connect fitting looks just like a two-tab type and works the same way too. It just has one tab instead of two.**

16   Squeeze the retainer tab and then pull the quick-connect fitting and the fuel line apart. The retainer remains on the fuel line.

17   Wipe off the fuel line and the retainer with a clean shop rag and then coat the fuel line with clean engine oil.

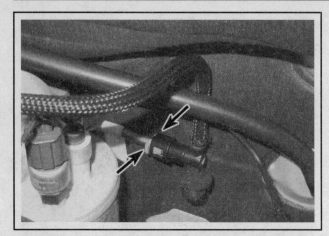

4.21a   To disconnect a two-tab type fitting, squeeze the two plastic retainer tabs together . . .

4.21b   . . . and then pull the metal line out of the quick-connect fitting

18 Inspect the condition of the fitting. If any part is damaged, worn or broken, replace the fuel line and the fitting as an assembly.

19 Push the quick-connect fitting onto the fuel line until you hear a click, but do NOT rely upon this audible click to verify that the retainer is fully seated. The fitting has "windows" through which the retainer locking ears protrude. If the fitting is correctly attached to the fuel line, the retainer locking ears and the fuel line shoulder will be visible in this window. If they're not visible, the retainer is not fully seated.

20 Reconnect the negative battery cable and then start the engine and check the fitting for leaks.

Two-tab type

▶ **Refer to illustrations 4.21a and 4.21b**

➡**Note: The retainers for this type of quick-connect fitting can be replaced separately, but not the O-rings and spacers. So this type of fitting, if damaged, cannot be repaired. Instead, replace the complete fuel line assembly.**

21 Squeeze the plastic retainer tabs against the fuel line with your finger and then pull the fuel line and the fitting apart (see illustrations). The plastic retainer remains with the fuel line. The O-rings and spacer remain inside the quick-connect fitting connector body.

22 Wipe off both halves of the quick-connect fitting with a clean shop rag and then lubricate the fuel line with clean engine oil.

23 Push the quick-connect fitting onto the fuel line until you hear a click, which indicates that the retainer has seated. Verify that the two

halves are locked together by trying to pull them apart.

24 Reconnect the negative battery cable and start the engine and check the fitting for leaks.

Plastic retainer type

25 This type of fitting is distinguished by its one-piece retainer ring, which is usually black in color and encircles the fuel line.

26 Firmly push the fitting toward the fuel line while simultaneously pushing the plastic retainer ring into the fitting and then, with the plastic ring depressed, pull the fuel line and the fitting apart.

➡**Note: Press the retainer ring squarely into the fitting. If the retainer is cocked, disconnection of the fuel line and the fitting is difficult. If necessary, use a tool such as a small adjustable wrench or a small pair of pliers to push the retainer into the fitting.**

After disconnection, the plastic retainer ring remains with the quick-connect fitting.

27 Wipe off the fitting and the fuel line with a clean shop rag and inspect the fitting parts for damage and wear. Replace as necessary.

28 Lubricate the fuel line with clean engine oil and insert it into the quick-connect fitting until you hear a click. Verify that the fitting is locked by trying to pull apart the line and the fitting.

29 Reconnect the negative battery cable and start the engine and check the fitting for leaks.

5   Fuel filter/fuel pressure regulator - removal and installation

▶ **Refer to illustrations 5.4a, 5.4b, 5.5a, 5.5b, 5.6 and 5.8**

1   The fuel filter/fuel pressure regulator is located on top of the fuel pump/fuel level sending unit module. It can be removed without disturbing the fuel tank or the fuel pump/fuel level sending unit module.

2   Relieve system fuel pressure (see Section 2) and disconnect the cable from the negative battery terminal.

3   Raise the rear of the vehicle and place it securely on jackstands.

4   Disconnect the fuel supply line connector from the fuel filter/fuel pressure regulator (see illustrations). (Refer to Section 4 if you need help with this fitting.)

5   Disconnect the ground wire from the fuel filter/fuel pressure regulator, depress the locking spring tab on the side of the filter/regulator, rotate the filter/regulator counterclockwise 90 degrees until the mounting flanges are free of the hold-down tabs and pull it off the fuel pump/fuel level sending unit module (see illustrations).

6   Remove the old rubber grommet (see illustration) and inspect it for cracks, tears and other deterioration. If the grommet is in good condition, it's okay to reuse it (but we recommend replacing it anytime you remove the fuel filter/fuel pressure regulator). Install the grommet.

7   Remove the old O-rings from the fuel filter/fuel pressure regulator (see illustration 5.6) and inspect them for cracks, tears and other

5.4a To disconnect the fuel supply line connector from the fuel filter/fuel pressure regulator . . .

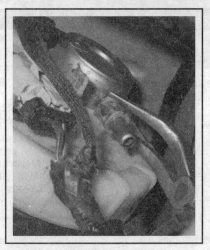

5.4b . . . squeeze these tangs together and pull on the female side of the fitting

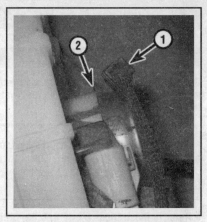

5.5a Disconnect the ground wire (1) from the spade terminal on the fuel filter/fuel pressure regulator, then depress the locking spring tab (2) on the side of the fuel filter/fuel pressure regulator . . .

5.5b . . . rotate the filter/regulator counterclockwise 90 degrees and pull it off the fuel pump/fuel level sending unit module

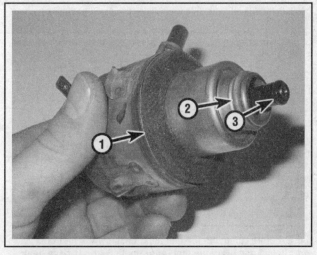

5.6 Remove the old rubber grommet (1) from the fuel filter/fuel pressure regulator and inspect it for cracks, tears and other deterioration. Remove the old O-rings (2 and 3) and inspect them for cracks, tears and other deterioration; when installing the O-rings on the filter/regulator, be sure to lightly oil them

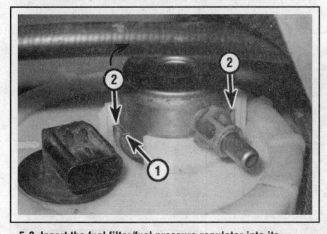

5.8 Insert the fuel filter/fuel pressure regulator into its receptacle on top of the fuel pump/fuel level sending unit module, align the mounting flanges (1) (other flange not visible in this photo) with the two hold-down tabs (2), then push down on the fuel filter/fuel pressure regulator and rotate it clockwise until the locking spring tab snaps into the locating slot

deterioration. If the old O-rings are in good condition, you can use them again (but we recommend replacing them anytime you remove the fuel filter/fuel pressure regulator). Lightly oil the O-rings and install them on the fuel filter/fuel pressure regulator.

8   Insert the fuel filter/fuel pressure regulator into its receptacle on top of the fuel pump/fuel level sending unit module, align the two mounting flanges with the hold-down tabs (see illustration). Then push down on the fuel filter/fuel pressure regulator and rotate it clockwise until the locking spring tab snaps into position in its locating slot. Reconnect the ground wire to the fuel filter/fuel pressure regulator.

9   Reconnect the fuel line to the fuel filter/fuel pressure regulator (refer to Section 4 if necessary).

10 Lower the vehicle and reconnect the cable to the negative battery terminal.

6 Fuel tank - removal and installation

※ WARNING:

Gasoline is extremely flammable, so take extra precautions when you work on any part of the fuel system. See the *Warning* in Section 2.

➡Note: If the vehicle still runs, it's best to use up as much fuel as possible before beginning this procedure.

REMOVAL

▶ Refer to illustrations 6.5, 6.6, 6.8a, 6.8b, 6.9 and 6.10

1 Slowly remove the fuel filler cap to relieve fuel tank pressure and then relieve the system fuel pressure (see Section 2).

2 Disconnect the negative battery cable.

3 Raise the rear of the vehicle and place it securely on jackstands.

4 Disconnect the vapor line from the EVAP canister and remove the EVAP canister (see "Evaporative Emissions Control (EVAP) system - general information and component replacement" in Chapter 6).

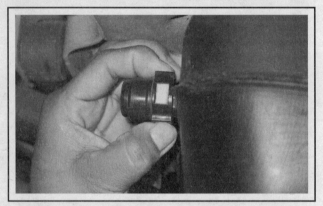

6.5 Before removing the fuel tank, unscrew the fuel tank drain cap and drain the fuel (after the tank straps have been detached, drain a little more fuel out of the tank if necessary)

5 Put an approved gasoline container under the drain cap, unscrew the fuel tank drain cap (see illustration) and then drain the tank. When the tank is empty, install the drain cap.

※ WARNING:

The fuel tank capacity is 14 gallons. Unless the fuel tank is almost empty, be prepared to collect a significant amount of fuel. Do not leave the fuel tank draining operation unattended. Be prepared to replace the drain cap in case the fuel tank capacity exceeds the container capacity. Remove the drain plug (see illustration) and allow the fuel to exit the fuel tank.

6 Disconnect the electrical connector (see illustration) from the fuel pump/fuel level sending unit module.

7 Disconnect the fuel line from the fuel filter/fuel pressure regulator (see illustrations 5.4a and 5.4b).

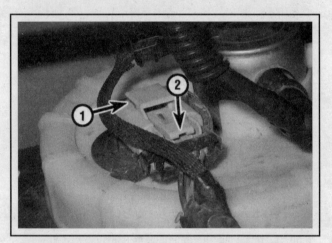

6.6 To disconnect the electrical connector from the fuel pump/fuel level sending unit module, push the red locking tab (1) to the right until it clicks, push down on the lock button (2) and pull firmly on the connector (not the wires!)

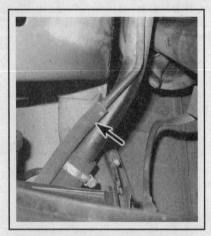

6.8a Disconnect the fuel filler neck vent hose . . .

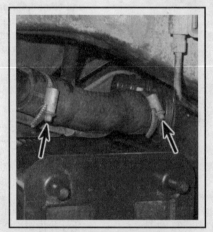

6.8b . . . loosen the fuel filler neck hose clamp screws and disconnect the fuel filler neck hose from the fuel inlet pipe and from the fuel tank

6.9 Support the tank with a transmission jack (shown) or use a floor jack; if you use a floor jack, put a piece of plywood between the jack head and the fuel tank to protect the tank

8   Disconnect the fuel filler neck vent hose from the fuel tank (see illustration), loosen the fuel filler-neck hose clamp and disconnect the fuel filler neck hose from the fuel tank inlet pipe (see illustration).

9   Place a transmission jack or a floor jack under the fuel tank to support it (see illustration). If you're going to use a floor jack, put a piece of plywood between the jack head and the tank to protect the tank.

10   Remove the nuts that secure the fuel tank straps (see illustration). Remove the EVAP canister mounting bracket from the right fuel tank mounting stud and push it aside.

11   Carefully lower the fuel tank from the vehicle.

12   Installation is the reverse of removal. Be sure to tighten the fuel tank strap nuts to the torque listed in this Chapter's Specifications. When you're done, start the engine and check for leaks at any fuel line connectors that you disconnected.

6.10 To detach the fuel tank assembly, remove these tank strap nuts; when installing the tank, don't forget to reattach the EVAP canister mounting bracket to the right mounting stud

7   Fuel tank cleaning and repair - general information

1   The fuel tank installed in these vehicles is plastic and is not repairable. If the tank becomes damaged it must be replaced.

2   If the fuel tank is removed from the vehicle, it should not be placed in an area where sparks or open flames could ignite the fumes coming out of the tank. Be especially careful inside garages where a gas-type appliance is located, because it could cause an explosion.

8   Fuel pump/fuel level sending unit module - removal and installation

▶ **Refer to illustration 8.2 and 8.3**

1   Relieve the system fuel pressure (see Section 2).

2   Before removing the fuel pump/fuel level sending unit module, make an alignment mark on top of the fuel tank, right underneath the fuel outlet pipe (see illustration). This mark ensures that the fuel pump/fuel level sending unit will be correctly realigned when installed again.

3   Before removing the fuel tank loosen the fuel pump/fuel level sending unit locknut with a pair of large water pump pliers (see illustration).

4   Remove the fuel tank (see Section 6).

5   Carefully withdraw the fuel pump/fuel level sending unit module from the fuel tank. Make sure that you don't damage the fuel pump inlet strainer or bend the sending unit float arm.

6   If you're replacing either the fuel pump or the fuel level sending unit, you'll have to separate the two components (see Section 9).

7   While the pump is removed, inspect the pump inlet strainer (see Section 9). Make sure that it's not clogged or damaged. If the filter is dirty, try washing it in clean solvent. If it's still clogged, replace it.

8   Installation is the reverse of removal. Be sure to tighten the fuel pump/fuel level sending unit locknut securely.

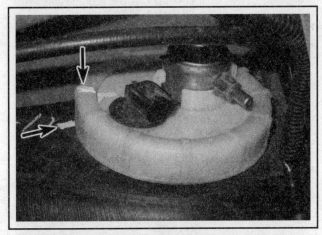

8.2 To ensure that the fuel pump/fuel level sending unit module will be correctly realigned during installation, mark the relationship of the module to the tank

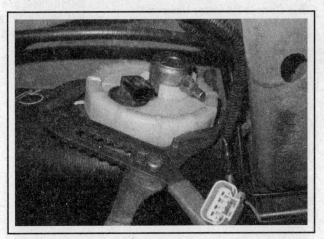

8.3 To remove the fuel pump/fuel level sending unit module, unscrew the locknut with a pair of large water pump pliers

9 Fuel pump/fuel level sending unit - component replacement

REMOVAL

▶ **Refer to illustrations 9.2, 9.3, 9.4, 9.5 and 9.6**

1 Remove the fuel pump/fuel level sending unit module from the

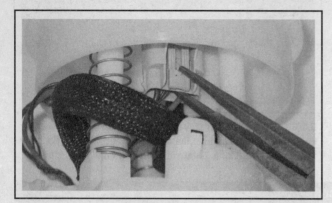

9.2 Depress the connector retaining tab and then remove the fuel pump/fuel level sending unit electrical connector from the underside of the fuel pump module electrical connector

fuel tank (see Section 8).

2 Depress the connector retaining tab (see illustration) and then remove the fuel pump/fuel level sending unit electrical connector from the underside of the fuel pump module electrical connector.

→**Note: The fuel pump module harness on top of the flange is NOT serviceable or removable.**

3 Remove the wedge lock from the electrical connector (see illustration).

4 Using a special terminal remover tool (C-4334) or a suitable similar tool, remove the terminals from the fuel level sending unit connector (see illustration).

5 Depress the tab and then slide the fuel level sending unit toward the bottom of the fuel pump module (see illustration).

6 Slide the fuel level sending unit out of the channel in the fuel pump module (see illustration).

INSTALLATION

▶ **Refer to illustrations 9.7, 9.8. 9.9, 9.10 and 9.11**

7 Install the electrical leads in the groove in the back of the fuel

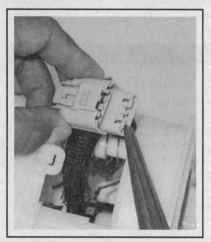

9.3 Remove the wedge lock from the electrical connector

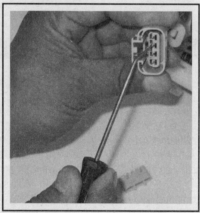

9.4 Using a special terminal remover tool (C-4334) or a suitable similar tool, remove the terminals from the fuel level sending unit connector

9.5 Depress this locking tab and then slide the fuel level sending unit off the fuel pump module

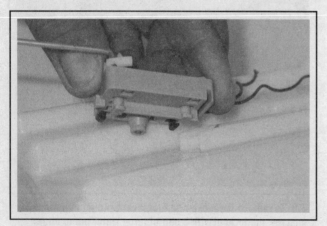

9.6 Remove the fuel level sending unit from the fuel pump module

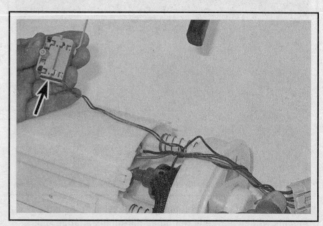

9.7 Install the leads in the groove in the back of the fuel level sending unit

level sending unit (see illustration).

8   Slide the fuel level sending unit up the channel until it snaps into place (see illustration). Make sure that the tab at the bottom of the sending unit locks into place.

9   Push the sending unit wires up through the connector (see illustration) until they lock into place. Make sure that the signal and ground wires are installed in the correct position.

10 Install the locking wedge in the connector (see illustration).

11 Push the connector back up into the underside of the fuel pump module electrical connector (see illustration).

12 Install the fuel pump/fuel level sending unit assembly (see Section 8).

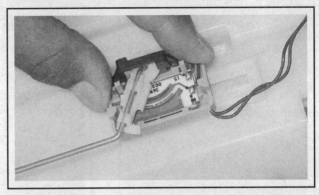

9.8 Slide the fuel level sending unit up the channel until it snaps into place; make sure that the tab at the bottom of the sending unit locks into place

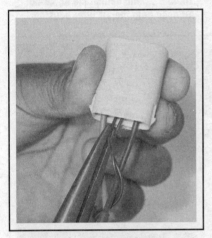

9.9 Push the sending unit wires up through the connector until they lock into place

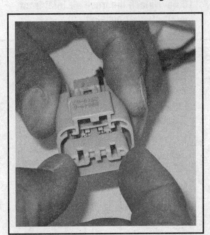

9.10 Install the locking wedge in the connector

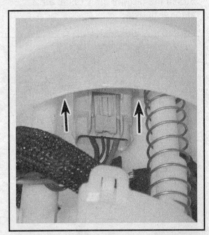

9.11 Push the electrical connector back up into the underside of the fuel pump module

10 Air filter housing - removal and installation

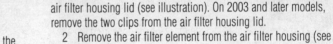

▶ Refer to illustrations 10.1, 10.2, 10.3, 10.4a, 10.4b, 10.5 and 10.6

1   On 2000 through 2002 models, remove the five screws from the air filter housing lid (see illustration). On 2003 and later models, remove the two clips from the air filter housing lid.

2   Remove the air filter element from the air filter housing (see illustration).

10.1 On 2000 through 2002 models, remove the five screws from the air cleaner lid (on 2003 and later models, remove the two clips from the air filter housing lid)

10.2 Remove the air filter element from the air filter housing

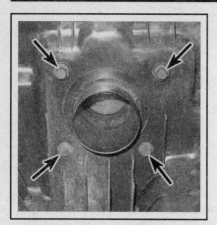

10.3 Remove the four bolts that attach the throttle body to the air cleaner housing

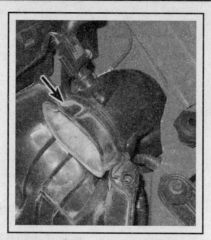

10.4a Detach the fresh air inlet from the air cleaner housing . . .

10.4b . . . and detach the make-up air hose from the air filter housing

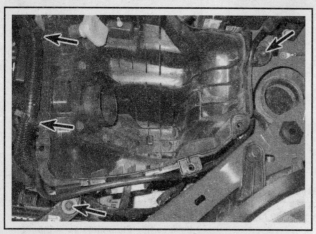

10.5 Disconnect the two wiring harness clips from the inner edge of the air cleaner housing and then remove the bolt and nut that secure the air cleaner housing to the vehicle body

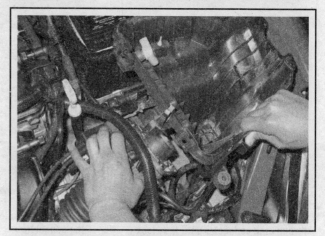

10.6 Remove the air filter housing assembly

3   Remove the four bolts that attach the throttle body to the air filter housing (see illustration).
4   Detach the fresh air inlet and the make-up air hose from the air filter housing (see illustrations).

5   Disconnect the wiring harness clips from the front of the air filter housing and remove the air filter housing mounting bolt and nut (see illustration).
6   Remove the air filter housing (see illustration).
7   Installation is the reverse of removal.

11 Accelerator cable - removal and installation

11.1a To remove the accelerator cable cover, squeeze this locking tab . . .

▶ Refer to illustrations 11.1a, 11.1b, 11.2a, 11.2b, 11.3, 11.5a, 11.5b and 11.6

➡ Note: 2004 and earlier engines use a conventional accelerator cable. On later models, the accelerator cable connects the accelerator pedal to the Accelerator Pedal Position Sensor (APPS) which is located in the right rear corner of the engine compartment. The throttle plate is electronically controlled by the Powertrain Control Module (PCM). To replace the sensor, refer to Chapter 6. If you want to replace the APPS cable, it's available separately and it is removed and installed just like any other accelerator cable (except that you disconnect it from the APPS instead of the throttle body).

1   Remove the accelerator cable cover (see illustrations).
2   Release the cable retaining tab from the cable bracket with a small screwdriver and then disengage the cable from the cable bracket (see illustrations).

11.1b . . . then slide the cover sideways to disengage the locking tab from its slot in the metal bracket and pull off the cover

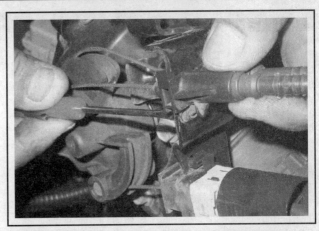

11.2a To disengage the accelerator cable from the cable bracket, pry the cable retaining tab loose from its hole in the cable bracket with a small screwdriver . . .

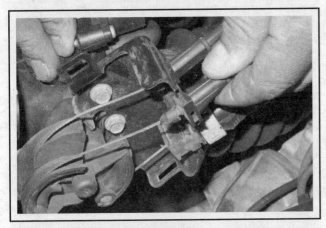

11.2b . . . and then slide the cable out of the bracket

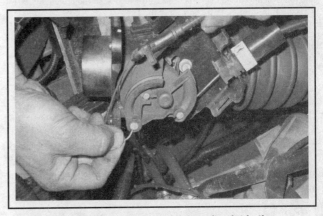

11.3 Disengage the cable end plug from its slot in the throttle cam

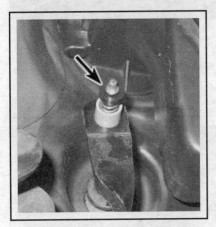

11.5a Working under the dash with a flashlight, locate the end of the accelerator cable at the top of the accelerator pedal

11.5b To disconnect the accelerator cable from the accelerator pedal, reach behind the pedal, pop the cable retainer from the slot in the top of the pedal, and then disengage the cable from the pedal

11.6 To remove the horseshoe-shaped clip that retains the cable grommet at the hole in the firewall, slide the cable straight up

3   Disengage the accelerator cable from its slot in the throttle lever cam (see illustration).

4   Disengage the accelerator cable from any clamps or clips in the engine compartment. Note the routing of the cable before proceeding.

5   Inside the vehicle, under the dash, remove the cable retainer from the top of the accelerator pedal (see illustrations) and then disengage the accelerator cable from the slot in the top of the accelerator pedal assembly.

6   Remove the clip that retains the cable grommet to its hole in the firewall (see illustration), pop the cable grommet out of the hole and push the cable through the hole.

7   Installation is the reverse of removal.

12 Sequential Multi-Port Electronic fuel injection (MPI) system - general information

The Sequential Multi-Port Electronic fuel injection (MPI) system consists of three sub-systems: air intake, electronic control and fuel delivery. The MPI system uses a Powertrain Control Module (PCM) and various information sensors to calculate the correct air/fuel ratio under all operating conditions. The MPI system and the emission control systems are closely linked. For more information on the emission control systems, refer to Chapter 6.

AIR INTAKE SYSTEM

The air intake system consists of the air filter housing, the air intake duct, the throttle body, the idle control system and the intake manifold. The intake manifold on most models is a one-piece design. RT models with the high output (HO) engine use a special two-piece intake manifold, which is equipped with a Manifold Tune Valve (MTV). See Chapter 2A for the removal and installation procedures for the intake manifolds and see Chapter 6 for more information about the MTV.

The throttle body, which is attached to the air filter housing, is a single-barrel, side-draft design. A Throttle Position (TP) sensor is attached to the throttle shaft to monitor changes in the throttle opening. An air intake duct connects the throttle body to the air intake manifold.

When the engine is idling, the air/fuel ratio is controlled by the idle air control system, which regulates the amount of airflow bypassing the throttle plate. The idle air control system consists of the PCM, the Idle Air Control (IAC) valve and various sensors such as the Engine Coolant Temperature (ECT) sensor, the Throttle Position (TP) sensor and the Manifold Absolute Pressure (MAP) sensor. The IAC valve is controlled by the PCM in accordance with the running conditions of the engine (air conditioning on, power steering demand, cold or warm temperature, etc.).

ELECTRONIC CONTROL SYSTEM

For information on the electronic control system, the Powertrain Control Module (PCM) and sensors, refer to Chapter 6.

FUEL DELIVERY SYSTEM

The fuel delivery system consists of the fuel filter/fuel pump/fuel pressure regulator assembly, the fuel supply line connecting the fuel pump to the fuel rail, the fuel pulsation damper, the fuel rail and the fuel injectors.

The electric fuel pump is located inside the fuel tank. Fuel is drawn through a filter into the pump, flows through the fuel line and then to the fuel rail, from which it's sprayed by the injectors into the intake ports. The fuel pressure regulator, which is an integral part of the fuel pump/fuel level sending unit module inside the tank, maintains a constant fuel pressure to the fuel rail and injectors. The fuel pulsation damper, which is located at the connection between the metal fuel supply line and the engine compartment fuel supply hose at the firewall, dampens the pressure pulses from the fuel pump.

Each injector consists of a solenoid coil, a pintle valve and the housing. When current is applied to the solenoid by the PCM, the pintle valve raises off its seat and the pressurized fuel inside the housing squirts out the nozzle. The amount of fuel injected is determined by the length of time that the pintle valve is open, which is determined by the length of time during which current is supplied to the solenoid. Because the injector on-time determines the air-fuel mixture ratio, injector timing must be very precise.

13 Fuel injection system - check

▶ Refer to illustrations 13.7 and 13.9

✳✳ WARNING:

Gasoline is extremely flammable, so take extra precautions when you work on any part of the fuel system. See the *Warning* in Section 2.

1   Inspect all system electrical connectors, especially the ground connections. Loose connectors and poor grounds are a common cause of many engine control system problems.

2   Verify that the battery is fully charged. The Powertrain Control Module (PCM) and sensors don't operate correctly without adequate supply voltage.

3   Inspect the air filter element (see Chapter 1). A dirty or partially blocked filter reduces performance and economy.

4   Check fuel pump operation (see Section 3). If the fuel pump fuse is blown, replace it and note whether it blows again. If it does, look for a short in the wiring harness to the fuel pump (see wiring diagrams in Chapter 12).

5   Inspect all vacuum hoses connected to the intake manifold for damage, deterioration and leakage.

6   Remove the air intake duct from the throttle body and look for

dirt, carbon, varnish, or other residue inside the throttle body bore, particularly around the throttle plate. If it's dirty, refer to Chapter 6 and troubleshoot the PCV system for the cause of the excessive residue

7   With the engine running, place an automotive stethoscope against each injector (see illustration) and listen for a clicking sound,

13.7 Use a stethoscope to listen for the clicking sound that indicates that each injector is working correctly; the clicking sound should rise and fall with changes in engine speed

which indicates that the injector is operating. If you don't have a stetho-scope, you can place the tip of a long screwdriver against the injector and listen through the handle.

8   Remove the intake manifold (see Chapter 2A).

9   With the engine off and the fuel injector electrical connectors dis-connected, measure the resistance of each injector with an ohmmeter (see illustration). Refer to the Specifications at the end of this chapter for the correct resistance.

10 Refer to Chapter 6 for other system checks.

13.9 Using an ohmmeter, measure the resistance across the terminals of each injector (intake manifold removed for clarity)

14 Throttle body - inspection, removal and installation

❋❋ WARNING:

Gasoline is extremely flammable, so take extra precautions when you work on any part of the fuel system. See the *Warning* in Section 2.

INSPECTION

▶ **Refer to illustrations 14.3a and 14.3b**

1   Remove the accelerator cable cover (see illustrations 11.1a and 11.1b).

2   Verify that the throttle linkage operates smoothly. If it doesn't, examine the routing of the accelerator cable.

3   Remove the air intake duct from the throttle body and then inspect the area around and behind the throttle plate for carbon and residue build-up. If it's dirty, clean it with aerosol carburetor cleaner and a toothbrush. Make sure that the can specifically states that it is safe with oxygen sensor systems and catalytic converters (see illustra-tions).

❋❋ CAUTION:

Do not clean the Throttle Position (TP) sensor or the Idle Air Control (IAC) valve with solvent.

REMOVAL AND INSTALLATION

▶ **Refer to illustration 14.6**

❋❋ WARNING:

Wait until the engine is completely cool before beginning this procedure.

4   Remove the air filter housing (see Section 10).

5   Disconnect the accelerator cable from the throttle body (see Sec-tion 11). If the vehicle is equipped with cruise control, disconnect the cruise control cable as well (see Chapter 12). And if the vehicle is equipped with an automatic transaxle, disconnect the kickdown cable (see Chapter 7B).

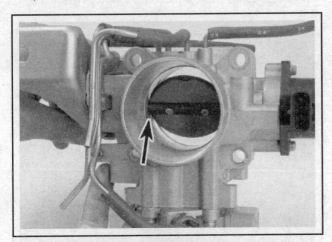

14.3a The area inside the throttle body behind the throttle plate (arrow) suffers from sludge build-up because of vapors vented into the throttle body by the PCV hose from the crankcase (throttle body removed for clarity)

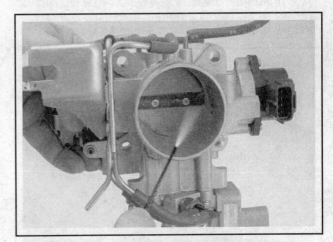

14.3b With the engine off, use aerosol carburetor cleaner (make sure it's approved for use with catalytic converters and oxygen sensors), a toothbrush and a rag to clean the throttle body; be sure to open the throttle plate and clean behind it

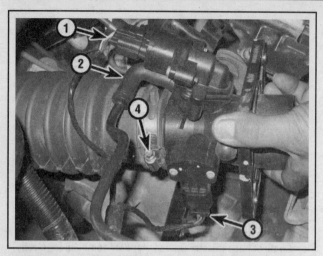

14.6 To remove the throttle body assembly, disconnect or remove the following items:

1   Idle Air Control (IAC) valve electrical connector
2   EVAP hose
3   Throttle Position (TP) sensor electrical connector
4   Air intake duct hose clamp screw

6   Disconnect the electrical connectors from the Idle Air Control (IAC) valve and from the Throttle Position (TP) sensor and disconnect the EVAP hose from the throttle body (see illustration).
7   Loosen the hose clamp (see illustration 14.6) and disconnect the air intake duct from the throttle body.
8   Remove the throttle body. If you're going to replace the throttle body, remove the IAC valve and the TP sensor from the old throttle body (see Chapter 6) and install them on the new unit.
9   Installation is the reverse of removal. Be sure to tighten the throttle body bolts to the torque listed in this Chapter's Specifications.

15   Fuel rail and injectors - removal and installation

15.2 To disconnect the fuel supply hose from the fuel rail, squeeze the two tabs on the quick-connect fitting together and pull up on the fitting (refer to Section 4 if necessary)

Refer to illustrations 15.2, 15.3a, 15.3b, 15.4, 15.5, 15.6a, 15.6b and 15.7

✳✳ WARNING:

Gasoline is extremely flammable, so take extra precautions when you work on any part of the fuel system. See the *Warning* in Section 2.

1   Relieve system fuel pressure (see Section 2) and disconnect the negative battery cable.
2   Disconnect the fuel supply line from the fuel rail (see illustration).
3   Disconnect the fuel injector electrical connectors (see illustrations).

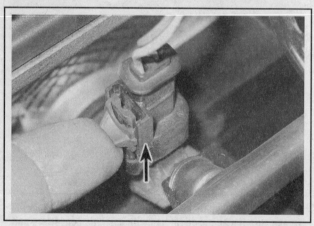

15.3a To disconnect each fuel injector electrical connector, push up on the lock . . .

15.3b . . . squeeze the tab (on the same side of the connector as the lock) and then pull up

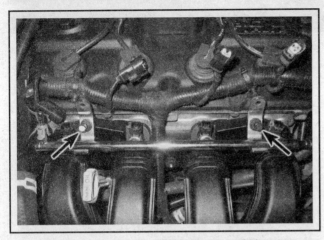

15.4 To detach the fuel rail, remove these two bolts

15.5 To remove the fuel rail and injectors, carefully pull up on the fuel rail to disengage the injectors from their bores in the intake manifold

4   Remove the fuel rail bolts (see illustration).

5   *Carefully* pull up on the fuel rail to disengage the injectors from their respective bores in the intake manifold and then remove the fuel rail and injectors as a single assembly (see illustration). The injectors might initially stick in their bores, but they'll pull free when sufficient pressure is applied. Just be careful not to damage the injectors - they're expensive!

6   Remove the injector retaining clips and remove the injectors from the fuel rail (see illustrations). Again, the injectors might initially stick to their bores, but if you pull and wiggle each injector, it will come out.

7   Remove the O-rings from each injector (see illustration). Discard these pieces and replace them with new O-rings.

8   Installation is the reverse of removal. Be sure to tighten the fuel rail retaining bolts to the torque listed in this Chapter's Specifications.

9   When you're done, start the engine and check for leaks at the fuel supply hose fitting at the fuel rail and at each injector.

15.6a To disassemble the fuel rail assembly, pull off each retaining clip . . .

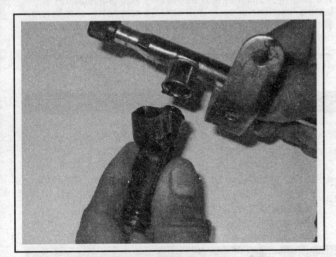

15.6b . . . and pull out the injector

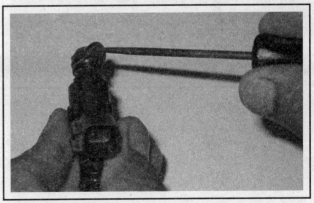

15.7 Remove the old O-rings from each injector and replace them with new ones (each of which must be coated with a little clean fuel prior to installing the injectors in the fuel rail)

16 Exhaust system servicing - general information

▶ **Refer to illustration 16.1**

✳✳ WARNING:

Inspection and repair of exhaust system components should be done only after the system components have cooled completely.

1 The exhaust system consists of the exhaust manifolds, the catalytic converters, the muffler, the tailpipe and all connecting pipes, brackets, hangers and clamps. The exhaust system is attached to the body with mounting brackets and rubber hangers (see illustration). If any of these parts are damaged or deteriorated, excessive noise and vibration will be transmitted to the body.

2 Conducting regular inspections of the exhaust system will keep it safe and quiet. Look for any damaged or bent parts, open seams, holes, loose connections, excessive corrosion or other defects which could allow exhaust fumes to enter the vehicle. Pay particular attention to rubber exhaust hangers. Because they're subjected to constant exhaust heat, they dry out, break down, develop cracks and then tear apart, allowing the exhaust system to hang unsupported. Sometimes an unsupported exhaust system "clunks" when the vehicle is driven over bumps and dips in the road; sometimes the exhaust touches the axle or other metal parts, which produces an annoying vibration at certain speeds.

3 If the exhaust system components are extremely corroded or rusted together, they will probably have to be cut from the exhaust system. The convenient way to accomplish this is to have a muffler repair shop remove the corroded sections with a cutting torch. If, however, you want to save money by doing it yourself and you don't have an oxy/acetylene welding outfit with a cutting torch, simply cut off the old components with a hacksaw. If you have compressed air, special pneumatic cutting chisels can also be used. If you do decide to tackle the job at home, be sure to wear eye protection to protect your eyes from metal chips and work gloves to protect your hands.

4 Here are some simple guidelines to apply when repairing the exhaust system:

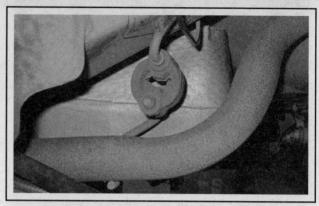

16.1 A typical rubber exhaust hanger; inspect all rubber hangers regularly, and anytime you have to raise the vehicle to service any under-vehicle component

a) *Work from the back to the front when removing exhaust system components.*

b) *Apply penetrating oil to the exhaust system component fasteners to make them easier to remove.*

c) *Use new gaskets, hangers and clamps when installing exhaust system components.*

d) *Apply anti-seize compound to the threads of all exhaust system fasteners during reassembly. Be sure to allow sufficient clearance between newly installed parts and all points on the underbody to avoid overheating the floor pan and possibly damaging the interior carpet and insulation. Pay particularly close attention to the catalytic converter and its heat shield.*

✳✳ WARNING:

The catalytic converter operates at very high temperatures and takes a long time to cool. Wait until it's completely cool before attempting to remove the converter. Failure to do so could result in serious burns.

Specifications

Fuel system

Fuel system pressure	
2000	44 to 54 psi
2001 on	53 to 63 psi
Injector resistance	12 ohms

Torque specifications

	Ft-lbs (unless otherwise indicated)	Nm
Fuel rail bolts	200 in-lbs	22.5
Fuel tank strap nuts	200 in-lbs	22.5

Section

Reference to other Chapters

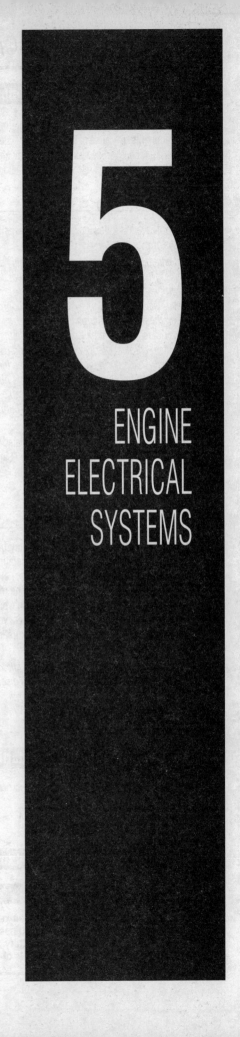

5

ENGINE ELECTRICAL SYSTEMS

1 General information, precautions and battery disconnection

The engine electrical systems include all ignition, charging and starting components. Because of their engine-related functions, these components are discussed separately from body electrical devices such as the lights, the instruments, etc. (which are included in Chapter 12).

PRECAUTIONS

Always observe the following precautions when working on the electrical system:

a) *Be extremely careful when servicing engine electrical components. They are easily damaged if checked, connected or handled improperly.*

b) *Never leave the ignition switched on for long periods of time when the engine is not running.*

c) *Never disconnect the battery cables while the engine is running.*

d) *Maintain correct polarity when connecting battery cables from another vehicle during jump starting - see the "Booster battery (jump) starting" section at the front of this manual.*

e) *Always disconnect the negative battery cable from the battery before working on the electrical system, but read the following battery disconnection procedure first.*

It's also a good idea to review the safety-related information regarding the engine electrical systems located in the "Safety first!" section at the front of this manual, before beginning any operation included in this Chapter.

BATTERY DISCONNECTION

Several systems on the vehicle require battery power to be available at all times, either to ensure their continued operation (such as the radio, alarm system, power door locks, windows, etc.) or to maintain control unit memories (such as that in the engine management system's Powertrain Control Module [PCM]) which would be lost if the battery were to be disconnected. Therefore, whenever the battery is to be disconnected, first note the following to ensure that there are no unforeseen consequences of this action:

a) *The engine management system's ECM will lose the information stored in its memory when the battery is disconnected. This includes idling and operating values, any fault codes detected and system monitors required for emissions testing. Whenever the battery is disconnected, the computer may require a certain period of time to "re-learn" the operating values.*

b) *On any vehicle with power door locks, it is a wise precaution to remove the key from the ignition and to keep it with you, so that it does not get locked inside if the power door locks should engage accidentally when the battery is reconnected!*

Devices known as "memory-savers" can be used to avoid some of the above problems. Precise details vary according to the device used. Typically, it is plugged into the cigarette lighter and is connected by its own wires to a spare battery; the vehicle's own battery is then disconnected from the electrical system, leaving the "memory-saver" to pass sufficient current to maintain audio unit security codes and ECM memory values, and also to run permanently live circuits such as the clock and radio memory, all the while isolating the battery in the event of a short-circuit occurring while work is carried out.

※ WARNING 1:

Some of these devices allow a considerable amount of current to pass, which can mean that many of the vehicle's systems are still operational when the main battery is disconnected. If a "memory-saver" is used, ensure that the circuit concerned is actually "dead" before carrying out any work on it!

※ WARNING 2:

If work is to be performed around any of the airbag system components, the battery must be disconnected. If a memory-saver device is used, power will be supplied to the airbag and personal injury may result if the airbag is accidentally deployed.

The battery on all vehicles is located in the front left corner of the engine compartment. To disconnect the battery for service procedures requiring power to be cut from the vehicle, loosen the negative cable clamp nut and detach the negative cable from the negative battery post (see Section 3). Isolate the cable end to prevent it from accidentally coming into contact with the battery post.

2 Battery - emergency jump starting

Refer to the Booster battery (jump) starting procedure at the front of this manual.

3 Battery - check, removal and installation

※ WARNING:

Hydrogen gas is produced by the battery, so keep open flames and lighted cigarettes away from it at all times. Always wear eye protection when working around a battery. Rinse off spilled electrolyte immediately with large amounts of water.

CHECK

▶ **Refer to illustrations 3.1a, 3.1b and 3.1c**

1   A battery cannot be accurately tested until it is at or near a fully charged state. Disconnect the negative battery cable from the battery and perform the following tests:

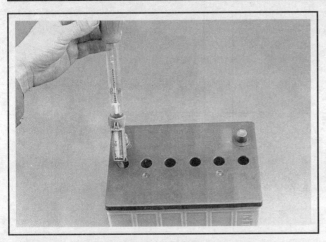

3.1a Use a battery hydrometer to draw electrolyte from the battery cell - this hydrometer is equipped with a thermometer to make temperature corrections

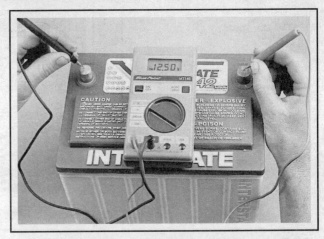

3.1b To test the open circuit voltage of the battery, connect the black probe of the voltmeter to the negative terminal and the red probe to the positive terminal of the battery - a fully charged battery should indicate approximately 12.5 volts depending on the outside air temperature

a) *Battery state of charge test* - Visually inspect the indicator eye (if equipped) on the top of the battery. If the indicator eye is dark in color, charge the battery as described in Chapter 1. If the battery is equipped with removable caps, check the battery electrolyte. The electrolyte level should be above the upper edge of the plates. If the level is low, add distilled water. DO NOT OVERFILL. The excess electrolyte may spill over during periods of heavy charging. Test the specific gravity of the electrolyte using a hydrometer (see illustration). Remove the caps and extract a sample of the electrolyte and observe the float inside the barrel of the hydrometer. Follow the instructions from the tool manufacturer and determine the specific gravity of the electrolyte for each cell. A fully charged battery will indicate approximately 1.270 (green zone) at 68-degrees F (20-degrees C). If the specific gravity of the electrolyte is low (red zone), charge the battery as described in Chapter 1.

b) *Open circuit voltage test* - Using a digital voltmeter, perform an open circuit voltage test (see illustration). Connect the negative probe of the voltmeter to the negative battery post and the positive probe to the positive battery post. The battery voltage should be greater than 12.5 volts. If the battery is less than the specified voltage, charge the battery before proceeding to the next test. Do not proceed with the battery load test until the battery is fully charged.

c) *Battery load test* - An accurate check of the battery condition can only be performed with a load tester (available at most auto parts stores). This test evaluates the ability of the battery to operate the starter and other accessories during periods of heavy amperage draw (load). Install a special battery load testing tool onto the battery terminals (see illustration). Load test the battery according to the tool manufacturer's instructions. This tool utilizes a carbon pile to increase the load demand (amperage draw) on the battery. Maintain the load on the battery for 15 seconds and observe that the battery voltage does not drop below 9.6 volts. If the battery condition is weak or defective, the tool will indicate this condition immediately.

➡Note: Cold temperatures will cause the minimum voltage requirements to drop slightly. Follow the chart given in the tool manufacturer's instructions to compensate for cold climates. Minimum load voltage for freezing temperatures (32 degrees F/0-degrees C) should be approximately 9.1 volts.

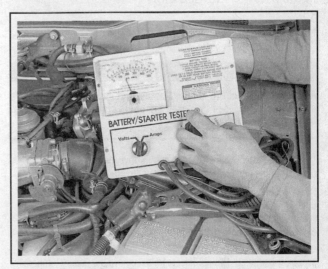

3.1c Some battery load testers are equipped with an ammeter which enables the battery load to be precisely dialed in, as shown - less expensive testers have a load switch and a voltmeter only

d) *Battery drain test* - This test will indicate whether there's a constant drain on the vehicle's electrical system that can cause the battery to discharge. Make sure all accessories are turned Off. If the vehicle has an underhood light, verify it's working properly, then disconnect it. Connect one lead of a digital ammeter to the disconnected negative battery cable clamp and the other lead to the negative battery post. A drain of approximately 100 milliamps or less is considered normal (due to the Powertrain Control Module, digital clocks, digital radios and other components that normally cause a key-off battery drain). An excessive drain (approximately 500 milliamps or more) will cause the battery to discharge. The problem circuit or component can be located by removing the fuses, one at a time, until the excessive drain stops and normal drain is indicated on the meter.

3.2 To prevent arcing, always disconnect the negative battery cable first

3.5 Remove the battery thermo-wrap

3.6 To detach the battery from the battery tray, remove this hold down clamp bolt

REMOVAL AND INSTALLATION

▸ Refer to illustrations 3.2, 3.5, 3.6 and 3.8

❊❊ CAUTION:

Always disconnect the negative cable first and hook it up last, or the tool being used to loosen the cable clamps may short the battery.

2 Loosen the cable clamp nut and remove the negative battery cable from the negative battery post (see illustration). Isolate the cable end to prevent it from accidentally coming into contact with the battery post.

3 Loosen the cable clamp nut and remove the positive battery cable from the positive battery post.

4 Disconnect the electrical connector from the battery temperature sensor, if equipped. (It's not necessary to remove the battery temperature sensor from the battery thermo-wrap.)

5 Remove the battery thermo-wrap (see illustration).

6 Remove the battery hold-down clamp bolt (see illustration).

7 Lift out the battery. Be careful - it's heavy.

➡**Note: Battery straps and handlers are available at most auto parts stores for a reasonable price. They make it easier to remove and carry the battery.**

8 While the battery is out, remove the battery tray (see illustration). (On 2000 models, the tray is attached by the four bolts shown in the accompanying illustration; on 2001 and later models, it's attached by these same four bolts and by a fifth bolt, which is located under the

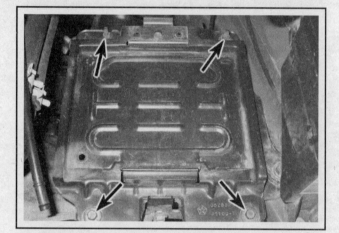

3.8 To remove the battery tray, remove these two bolts and two nuts (2000 model shown; on 2001 and later models, there's another bolt located under the center of the tray)

center of the tray.) Inspect the support bracket underneath for corrosion. If corrosion is present, remove it and then clean the deposits with a mixture of baking soda and water to prevent further corrosion. Flush the area with plenty of clean water and dry thoroughly.

9 If you are replacing the battery, make sure you replace it with a battery with the identical dimensions, amperage rating, cold cranking rating, etc.

10 Installation is the reverse of removal. Be sure to tighten the hold-down clamp bolt securely, but don't overtighten it, or you will crack the plastic tray.

4 Battery cables - replacement

ALL CABLES

1 Periodically inspect the entire length of each battery cable for damage, cracked or burned insulation and corrosion. Poor battery cable connections can cause starting problems and decreased engine performance.

2 Check the cable-to-terminal connections at the ends of the cables for cracks, loose wire strands and corrosion. The presence of white, fluffy deposits under the insulation at the cable terminal connection is a sign that the cable is corroded and should be replaced. Check the terminals for distortion, missing mounting bolts and corrosion.

3 When removing the cables always disconnect the negative cable

4.4 To detach the smaller negative battery cable from the body, remove this bolt

4.5 To detach the larger negative battery cable from the engine, remove this upper transaxle-to-engine bolt

from the negative battery post first and hook it up last, or the tool used to loosen the cable clamps could accidentally short the battery. Even if you're only replacing the positive cable, be sure to disconnect the negative cable first (see Chapter 1 for further information regarding battery cable maintenance).

NEGATIVE BATTERY CABLES

▶ **Refer to illustrations 4.4 and 4.5**

➡**Note: There are two negative battery cables.**

4   The smaller negative battery cable is grounded to the left side of the engine compartment, near the battery (see illustration).

5   The larger negative battery cable is grounded to an upper transaxle-to-engine bolt (see illustration). Before removing either cable, note its routing to ensure correct installation.

POSITIVE BATTERY CABLE

▶ **Refer to illustrations 4.9a and 4.9b**

6   Disconnect the positive cable from the battery (see Section 3).

7   Trace the positive and negative battery cables (they're routed together down to the engine ground wire bolt, at which point the positive cable branches off to the starter motor) and disengage them from all cable clips and guides.

8   Remove the bolt that grounds the battery negative cable to the engine (see illustration 4.5).

9   After removing the engine ground bolt, trace the positive battery cable to the red plastic housing on top of the starter solenoid (see illustration). Remove the nut on the starter solenoid terminal (see illustration) and then disconnect the positive battery cable from the starter solenoid.

10   If you have difficulty removing the starter solenoid terminal nut from above, raise the vehicle and place it securely on jackstands. Then remove the starter solenoid terminal nut from below.

ALL CABLES

11   If you are replacing a battery cable, take it with you when buying a new cable. It is vitally important that you replace battery cables with identical parts. Battery cables have easily identifiable characteristics: Positive cables are usually red and larger in cross-section; ground cables are usually black and smaller in cross-section.

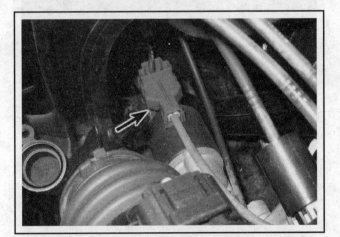

4.9a Trace the lower end of the positive battery cable to this red plastic housing on top of the solenoid, then find the nut that attaches the cable to the solenoid terminal, which is located on the right end of the solenoid (not visible in this photo, but shown in the next photo)

4.9b To disconnect the lower end of the positive battery cable from the starter solenoid, remove the nut on the starter solenoid terminal and then work the red plastic housing off the terminal and lift up (the red housing protects the end of the starter cable and it also houses a fusible link)

12 Clean the threads of the starter solenoid terminal or ground connection with a wire brush to remove rust and corrosion. Apply a light coat of battery terminal corrosion inhibitor or petroleum jelly to the threads to prevent future corrosion.

13 Reconnecting the lower end of any of the three battery cables is the reverse of removal.

14 After attaching the lower end of the new cable(s), install the battery (see Section 3).

15 Before connecting the new cable(s) to the battery, make sure that the cable reaches the battery post without having to be stretched.

16 Installation is otherwise the reverse of removal.

5   Ignition system - general information and precautions

1 All models are equipped with a distributorless ignition system, which is referred to as a Direct Ignition System (DIS). As the name implies, there is no distributor. The DIS system includes the Powertrain Control Module (PCM), the Camshaft Position (CMP) sensor, the Crankshaft Position (CKP) sensor, the knock sensor, the ignition coil pack, the spark plug wires and the spark plugs. For more information on the spark plug wires and the spark plugs, refer to Chapter 1. For more information on the CMP sensor, the CKP sensor, the knock sensor and the PCM, refer to Chapter 6.

2 The coil pack, which consists of two ignition coils molded into a single unit, is mounted on top of the valve cover. The Auto Shutdown Relay (ASD) controls battery voltage to the coil. A pair of "drivers" inside the PCM control the ground circuit for each coil. The ignition system is a "waste spark" design. Each time the PCM "fires" an ignition coil (each time a driver opens the ground circuit of a coil), the coil fires two spark plugs. One of these plugs is at the cylinder under compression, in which the piston is about to begin its power stroke. The other plug is at the cylinder in which the piston is on its exhaust

stroke. The rear coil fires cylinder numbers 1 and 4; the front coil fires cylinder numbers 2 and 3. So, for example, when the ignition coil for the No. 1 cylinder fires the No. 1 spark plug, it also fires the spark plug for the No. 4 cylinder, which is also at Top Dead Center (TDC), but which is on its exhaust stroke, not its power stroke. Hence the spark is "wasted."

3 When working on the ignition system, take the following precautions:

a) *Do not keep the ignition switch on for more than 10 seconds if the engine will not start.*

b) *Always connect a tachometer in accordance with the manufacturer's instructions. Some tachometers may be incompatible with this ignition system. Consult an auto parts counterperson before buying a tachometer for use with this vehicle.*

c) *Never allow the ignition coil terminals to touch ground. Grounding the coil could result in damage to the igniter and/or the ignition coil.*

d) *Do not disconnect the battery when the engine is running.*

6   Ignition system - check

▶ **Refer to illustration 6.2**

⁂ **WARNING:**

Because of the high voltage generated by the ignition system, extreme care should be taken whenever an operation is performed involving ignition components. This not only includes the igniter, coil and spark plug wires, but related components such as plug connectors, tachometer and other test equipment as well.

1 If the engine turns over but won't start, disconnect the spark plug wire from each of the spark plugs and attach it to a calibrated ignition tester, which is available at most auto parts stores.

2 Connect the clip on the tester to a bolt or metal bracket on the engine (see illustration).

3 Relieve the fuel system pressure (see Chapter 4). Keep the fuel system disabled while testing the ignition system.

4 Crank the engine and watch the end of the tester to see if bright blue, well-defined sparks occur.

5 If sparks occur, sufficient voltage is reaching the two companion spark plugs (Nos. 1 and 4 or Nos. 2 and 3) to fire them. Repeat this test at each of the spark plug wires to verify that both ignition coils are functioning.

6 If there is no spark, or only an intermittent spark, verify that there is battery voltage to the ignition coils (see wiring diagrams at the end of Chapter 12).

7 If there is battery voltage to the coils, inspect the plug wires (see Chapter 1).

6.2 To install a calibrated ignition tester, simply connect it to the spark plug wire boot, clip the tester to a convenient ground and with the fuel system pressure relieved and the fuel system disabled, crank the engine. If there's enough power to fire the plug, sparks will be visible between the electrode tip and the tester body

8 If the spark plug wires are okay, remove and then check the spark plugs for fouling (see Chapter 1). If necessary, install new plugs.

9 If the spark plugs are good, check the coil resistance (see Section 7).

10 If the ignition system checks out, the CMP sensor, the CKP sensor or the PCM might be defective (see Chapter 6).

7   Ignition coil pack - check and replacement

CHECK

▶ **Refer to illustrations 7.2a, 7.2b and 7.3**

➡**Note: The following checks should be made with the engine cold. If the engine is hot, the resistance will be greater.**

1   Remove the ignition coil pack (see below).

2   Check the primary resistance of the ignition coil. With the ignition key turned to OFF, disconnect the electrical harness connector from each coil. Connect an ohmmeter across the coil primary terminals (see illustrations) and then compare your measurement to the resistance listed in this Chapter's Specifications. If the primary resistance is incorrect, replace the coil.

3   Check the secondary resistance of each ignition coil. With the ignition key turned to OFF, label and detach the spark plug wires from each coil. Connect an ohmmeter across the two secondary terminals of each coil (see illustration). Compare your measurement to the resistance listed in this Chapter's Specifications. If the secondary resistance is incorrect, replace the coil.

REPLACEMENT

Refer to illustration 7.6

4   Disconnect the negative cable from the battery.

5   Label and disconnect the spark plug wires from the ignition coil (see Chapter 1). Pay close attention to the routing of the spark plug wires to ensure that they're correctly routed when reinstalled.

6   Disconnect the ignition coil electrical connector (see illustration).

7   Remove the ignition coil mounting bolts (see illustration 7.6) and remove the coil pack from the valve cover.

8   Installation is the reverse of removal.

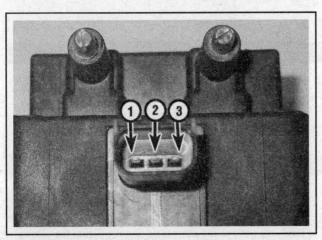

7.2a Ignition coil terminal guide

1   *Primary terminal No. 2 (to ignition coil No. 2 driver)*
2   *Battery voltage terminal (from Auto Shutdown Relay)*
3   *Primary terminal No. 1 (to ignition coil No. 1 driver)*

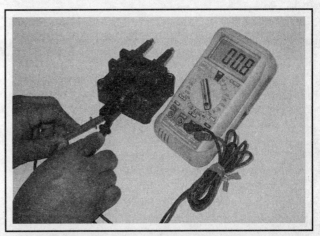

7.2b To check the primary resistance of the ignition coil, measure the resistance between the battery voltage terminal and each primary terminal with an ohmmeter

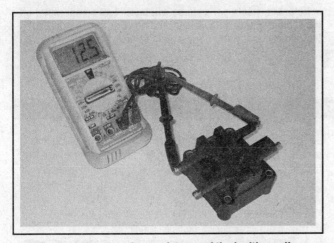

7.3 To check the secondary resistance of the ignition coil, remove the spark plug wires and measure the resistance between each pair of high-tension terminals with an ohmmeter

7.6 To detach the ignition coil pack from the valve cover, disconnect the electrical connector from the coil pack, then remove the coil pack mounting bolts

8   Charging system - general information and precautions

The charging system includes the alternator, an internal voltage regulator, a charge indicator, the battery and the wiring between all the components. The charging system supplies electrical power for the ignition system, the lights, the radio, etc. The alternator is driven by a drivebelt at the front of the engine.

The purpose of the voltage regulator is to limit the alternator's voltage to a preset value. This prevents power surges, circuit overloads, etc., during peak voltage output.

The charging system doesn't ordinarily require periodic maintenance. However, the drivebelt, battery and wires and connections should be inspected at the intervals outlined in Chapter 1.

The dashboard warning light should come on when the ignition key is turned to START, and then should go off immediately. If it remains on, there is a malfunction in the charging system. These vehicles are also equipped with a voltage gauge. If the voltage gauge indicates abnormally high or low voltage, check the charging system (see Section 9).

Be very careful when making electrical circuit connections to a vehicle equipped with an alternator and note the following:

a) *When reconnecting wires to the alternator from the battery, be sure to note the polarity.*
b) *Before using arc welding equipment to repair any part of the vehicle, disconnect the wires from the alternator and the battery terminals.*
c) *Never start the engine with a battery charger connected.*
d) *Always disconnect both battery leads before using a battery charger.*
e) *The alternator is driven by an engine drivebelt, which could cause serious injury if your hand, hair or clothes become entangled in it with the engine running.*
f) *Because the alternator is connected directly to the battery, it could arc or cause a fire if overloaded or shorted out.*

9   Charging system - check

1   If a malfunction occurs in the charging circuit, do not immediately assume that the alternator is causing the problem. First, check the following items:

a) *Make sure the battery cable clamps, where they connect to the battery, are clean and tight.*
b) *Test the condition of the battery (see Section 3). If it does not pass all the tests, replace it with a new battery.*
c) *Check the external alternator wiring and connections.*
d) *Check the drivebelt condition and tension (see Chapter 1).*
e) *Check the alternator mounting bolts for tightness.*
f) *Run the engine and check the alternator for abnormal noise.*
g) *Check the charge light on the dash. It should illuminate when the ignition key is turned ON (engine not running). If it does not, check the circuit from the alternator to the charge light on the dash.*
h) *Check all the fuses that are in series with the charging system circuit. The location of these fuses may vary from year and model but the designations are generally the same. Refer to the wiring schematics at the end of Chapter 12 for additional information.*

2   With the ignition key off, check the battery voltage with no accessories operating. It should be approximately 13.5 to 14.7 volts.

3   Start the engine and check the battery voltage again. It should now be greater than the voltage recorded in Step 2, but not more than 14.7 volts. Turn on all the vehicle accessories (air conditioning, rear window defogger, blower motor, etc.) and increase the engine speed to 2,000 rpm - the voltage should not drop below the voltage recorded in Step 2.

4   If the indicated voltage is greater than the specified charging voltage, replace the voltage regulator.

➡**Note: The voltage regulator is an integral part of the alternator and cannot be replaced separately.**

5   If the indicated voltage reading is less than the specified charging voltage, the alternator is probably defective. Have the charging system checked at a dealer service department or other properly equipped repair facility.

➡**Note: Many auto parts stores will bench test an alternator off the vehicle. Check with your local auto parts store regarding their policy, many will perform this service free of charge.**

10   Alternator - removal and installation

▶ **Refer to illustrations 10.2, 10.4, 10.5, 10.7, 10.8 and 10.9**

1   Disconnect the negative battery cable.
2   Loosen the upper alternator adjustment bolt and mounting bolt nut (see illustration).
3   Loosen the right front wheel lug nuts and then raise the vehicle. Remove the right front wheel.
4   Remove the accessory drivebelt splash shield (see illustration).
5   Loosen the alternator lower mounting bolt (see illustration).
6   Remove the alternator drivebelt (see Chapter 1, if necessary).
7   Disconnect the electrical connectors from the alternator (see illustration).

8   Remove the upper adjustment bolt (see illustration 10.2), then remove the upper and lower mounting bolts (see illustration) and set the alternator aside to give you some room to unbolt the alternator mounting bracket. If the alternator gets in the way during the next step, wire it up temporarily.
9   Remove the alternator mounting bracket (see illustration).
10 Remove the alternator from beneath the vehicle.
11 Installation is the reverse of removal. Be sure to tighten the alternator bolts securely.

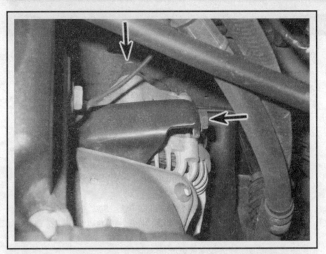

10.2 Loosen the upper alternator mounting bolt nut (upper arrow) and loosen the upper adjustment bolt (right arrow)

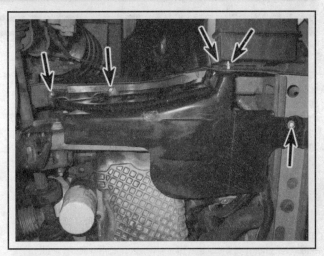

10.4 To remove the accessory drivebelt splash shield, remove these three bolts and two push fasteners

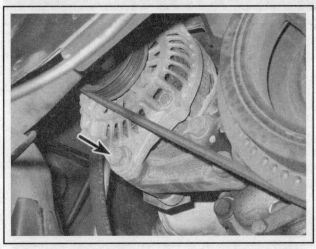

10.5 Loosen the alternator lower pivot bolt and then remove the alternator drivebelt (see Chapter 1)

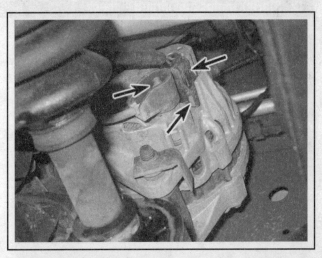

10.7 Push the red locking tab (lower arrow) to release the electrical connector for the alternator field circuit, then disconnect the connector; to disconnect the battery lead, simply remove the nut from the B+ terminal

10.8 To detach the alternator, remove the upper and lower mounting bolts (and the upper adjustment bolt, which is shown in illustration 10.2)

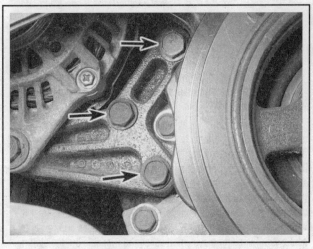

10.9 To detach the alternator mounting bracket from the block, remove these bolts

11 Starting system - general information and precautions

The starting system consists of the battery, the ignition switch, the starter relay, the starter motor, the starter solenoid and the electrical circuit connecting the components. The solenoid is mounted directly on the starter motor. The starting system on models with a manual transaxle also includes a clutch start switch, which is located at the top of the clutch pedal (see Chapter 8). The starting system on models with an automatic transaxle includes a Park/Neutral position switch, which is located on the transaxle (see Chapter 6).

The solenoid/starter motor assembly is installed on the front of the engine block, next to the transaxle bellhousing.

When the ignition key is turned to the START position, the starter solenoid is actuated through the starter control circuit. The starter solenoid then connects the battery to the starter. The battery supplies the electrical energy to the starter motor, which does the actual work of cranking the engine.

The starter motor on a vehicle equipped with a manual transaxle can be operated only when the clutch pedal is depressed; the starter on a vehicle equipped with an automatic transaxle can be operated only when the transaxle selector lever is in Park or Neutral.

Always observe the following precautions when working on the starting system:

a) *Excessive cranking of the starter motor can overheat it and cause serious damage. Never operate the starter motor for more than 15 seconds at a time without pausing to allow it to cool for at least two minutes.*

b) *The starter is connected directly to the battery and could arc or cause a fire if mishandled, overloaded or short circuited.*

c) *Always detach the cable from the negative terminal of the battery before working on the starting system.*

12 Starter motor and circuit - check

◆ **Refer to illustration 12.3**

1   If a malfunction occurs in the starting circuit, do not immediately assume that the starter is causing the problem. First, check the following items:

a) *Make sure the battery cable clamps, where they connect to the battery, are clean and tight.*

b) *Check the condition of the battery cables (see Section 4). Replace any defective battery cables with new parts.*

c) *Test the condition of the battery (see Section 3). If it does not pass all the tests, replace it with a new battery.*

d) *Check the starter solenoid wiring and connections. Refer to the wiring diagrams at the end of Chapter 12.*

e) *Check the starter mounting bolts for tightness.*

f) *Check the ignition switch circuit for correct operation (see Chapter 12).*

g) *Check the operation of the Park/Neutral Position switch (automatic transaxle) or clutch-start switch (manual transaxle). Make sure the shift lever is in PARK or NEUTRAL (automatic transaxle) or the clutch pedal is depressed (manual transaxle). Refer to Chapter 7 for the Park/Neutral Position switch check and adjustment procedure. Refer to the Chapter 12 wiring diagrams, if necessary, when performing circuit checks. These systems must operate correctly to provide battery voltage to the ignition solenoid.*

h) *Check the operation of the starter relay. The starter relay is located in the fuse/relay box inside the engine compartment. Refer to Chapter 12 for the testing procedure.*

2   If the starter does not actuate when the ignition switch is turned to the start position, check for battery voltage to the solenoid. This will determine if the solenoid is receiving the correct voltage signal from the ignition switch. Connect a test light or voltmeter between the starter solenoid positive terminal and ground. While an assistant turns the ignition switch to the start position the test light should light or the voltmeter should indicate voltage. If voltage is not available, refer to the wiring diagrams in Chapter 12 and check all the fuses and relays in series with the starting system. If voltage is available but the starter motor does not operate, remove the starter (see Section 13) and bench test it (see Step 4).

3   If the starter turns over slowly, check the starter cranking voltage

12.3 To use an inductive ammeter, simply hold the ammeter over the positive or negative cable (whichever is easier in terms of clearance)

and the current draw from the battery. This test must be performed with the starter assembly on the engine. Crank the engine over (for 10 seconds or less) and observe the battery voltage. It should not drop below 8.0 volts on manual transaxle models or 8.5 volts on automatic transaxle models. Also, observe the current draw using an amp meter (see illustration). It should not exceed 400 amps or drop below 250 amps.

❊❊ CAUTION:

The battery cables may be excessively heated because of the large amount of amperage being drawn from the battery. Discontinue the testing until the starting system has cooled down. If the starter motor cranking amp values are not within the correct range, replace it with a new unit. There are several conditions that may affect the starter cranking potential. The battery must be in good condition and the battery cold-cranking rating must not be under-rated for the particular application. Be sure to check the battery specifications carefully. The battery terminals and cables must be clean and not corroded. Also, in cases of extreme cold temperatures, make sure the battery and/or engine block is warmed before performing the tests.

4   If the starter is receiving voltage but does not activate, remove and check the starter/solenoid assembly on the bench. Most likely the solenoid is defective. In some rare cases, the engine may be seized so be sure to try and rotate the crankshaft pulley (see Chapter 2A) before proceeding. With the starter/solenoid assembly mounted in a vise on the bench, install one jumper cable from the negative battery terminal to the body of the starter. Install the other jumper cable from the positive battery terminal to the B+ terminal on the starter. Install a starter switch

and apply battery voltage to the solenoid S terminal (for 10 seconds or less) and see if the solenoid plunger, shift lever and overrunning clutch extends and rotates the pinion drive. If the pinion drive extends but does not rotate, the solenoid is operating but the starter motor is defective. If there is no movement but the solenoid clicks, the solenoid and/or the starter motor is defective. If the solenoid plunger extends and rotates the pinion drive, the starter/solenoid assembly is working properly.

13 Starter motor - removal and installation

▸ **Refer to illustration 13.4 and 13.5**

1   Detach the cable from the negative terminal of the battery.
2   On High Output (HO) models remove the intake manifold (see Chapter 2A). On all other models, remove the air intake duct between the throttle body and the intake manifold (see Chapter 4). Removing the air intake duct should provide you with adequate space to remove the starter. If you still need more room to work, remove the throttle body and the air filter housing, too (see Chapter 4).

3   Raise the front of the vehicle and place it securely on jackstands.
4   Remove the starter motor mounting bolts (see illustration).
5   Disconnect the electrical connector for the battery lead (see illustrations 4.9a and 4.9b) and disconnect the electrical connector for the solenoid assembly (see illustration).
6   Remove the starter motor assembly.
7   Installation is the reverse of removal. Be sure to tighten the starter mounting bolts to the torque listed in this Chapter's Specifications.

13.4 To detach the starter motor assembly, remove these two bolts (note that one bolt is on the starter side and one is on the bellhousing side); don't forget to reattach the ground wire, if it was attached to a starter bolt

13.5 To disconnect the electrical connector from the solenoid, release this red locking tab and then pull off the connector (to disconnect the connector for the battery lead, refer to illustrations 4.9a and 4.9b)

Specifications

Ignition coil

Primary resistance	
Weastec (steel towers)	0.45 to 0.65 ohms
Diamond (brass towers)	0.53 to 0.65 ohms
Secondary resistance	
Weastec (steel towers)	11.5 to 13.5 k-ohms
Diamond (brass towers)	10.9 to 14.7 k-ohms

Charging system

Charging voltage	13.5 to 14.7 volts

Torque specifications

	Ft-lbs (unless otherwise noted)	Nm
Ignition coil pack mounting bolts	105 in-lbs	

Notes

6

EMISSIONS
AND ENGINE
CONTROL
SYSTEMS

Section

1 General information

▶ **Refer to illustration 1.6**

To prevent pollution of the atmosphere from incompletely burned and evaporating gases, and to maintain good driveability and fuel economy, a number of emission control systems are incorporated. They include the:

On-Board Diagnostic II (OBD-II) system
Multi-Port Electronic Fuel Injection (MPI) system
Evaporative Emissions Control (EVAP) system
Positive Crankcase Ventilation (PCV) system
Catalytic converter

The Sections in this Chapter include general descriptions, checking procedures within the scope of the home mechanic and component replacement procedures (when possible) for each of the systems listed above.

Before assuming that an emissions control system is malfunctioning, check the fuel and ignition systems carefully. The diagnosis of some emission control devices requires specialized tools, equipment and training. If checking and servicing become too difficult or if a procedure is beyond your ability, consult a dealer service department or other repair shop. Remember, the most frequent cause of emissions problems is simply a loose or broken wire or vacuum hose, so always check the hose and wiring connections first.

This doesn't mean, however, that emissions control systems are particularly difficult to maintain and repair. You can quickly and easily perform many checks and do most of the regular maintenance at home with common tune-up and hand tools.

➡**Note: Because of a Federally mandated warranty which covers the emissions control system components, check with your dealer about warranty coverage before working on any emissions-related systems. Once the warranty has expired, you may wish to perform some of the component checks and/or replacement procedures in this Chapter to save money.**

Pay close attention to any special precautions outlined in this

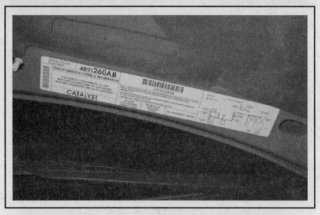

1.6 The Vehicle Emission Control Information (VECI) label contains such essential information as the types of emission control systems installed on the engine and the idle speed and ignition timing specifications; the vacuum hose routing diagram indicates important emission-control devices and components, and it shows how they're connected together by vacuum hoses

Chapter. It should be noted that the illustrations of the various systems might not exactly match the system installed on your vehicle because of changes made by the manufacturer during production or from year-to-year.

A Vehicle Emissions Control Information (VECI) label is attached to the underside of the hood (see illustration). This label contains important emissions specifications and adjustment information. Part of this label, the Vacuum Hose Routing Diagram, provides a vacuum hose schematic with emissions components identified. When servicing the engine or emissions systems, the VECI label and the vacuum hose routing diagram in your particular vehicle should always be checked for up-to-date information.

2 On-Board Diagnostic (OBD) system and trouble codes

2.1 Scanners like these from Actron and AutoXray are powerful diagnostic aids - programmed with comprehensive diagnostic information, they can tell you just about anything you want to know about your engine management system

SCAN TOOL INFORMATION

▶ **Refer to illustrations 2.1 and 2.2**

1   Hand-held scanners are the most powerful and versatile tools for analyzing engine management systems used on later model vehicles (see illustration). Early model scanners handle codes and some diagnostics for many systems. Each brand scan tool must be examined carefully to match the year, make and model of the vehicle you are working on. Often, interchangeable cartridges are available to access the particular manufacturer (Ford, GM, Chrysler, Toyota, etc.). Some manufacturers will specify by continent (Asia, Europe, USA, etc.).

➡**Note: An aftermarket generic scanner should work with any model covered by this manual. However, some early OBD-II models, although technically classified as OBD-II compliant by the manufacturer and by the federal government, might not be fully compliant with all SAE standards for OBD-II. Some generic scanners are unable to extract all the codes from these early OBD-II models. Before purchasing a generic scan tool, contact the manufacturer of the scanner you're planning to buy and ver-**

ify that it will work properly with the OBD-II system you want to scan. If necessary, of course, you can always have the codes extracted by a dealer service department or an independent repair shop with a professional scan tool.

2   With the arrival of the Federally mandated emission control system (OBD-II), a specially designed scanner has been developed. Several tool manufacturers have released OBD-II scan tools for the home mechanic (see illustration). Ask the parts salesman at a local auto parts store for additional information concerning dates and costs.

OBD SYSTEM GENERAL DESCRIPTION

3   All models are equipped with the second generation OBD-II system. This system consists of an on-board computer known as the Powertrain Control Module (PCM), and information sensors, which monitor various functions of the engine and send data to the PCM. This system incorporates a series of diagnostic monitors that detect and identify fuel injection and emission control systems faults and store the information in the computer memory. This updated system also tests sensors and output actuators, diagnoses drive cycles, freezes data and clears codes.

4   This powerful diagnostic computer must be accessed using the new OBD-II scan tool and 16 pin Data Link Connector (DLC) located under the driver's dash area. The PCM is the "brain" of the electronically controlled fuel and emissions system. It receives data from a number of sensors and other electronic components (switches, relays, etc.). Based on the information it receives, the PCM generates output signals to control various relays, solenoids (i.e., fuel injectors) and other actuators. The PCM is specifically calibrated to optimize the emissions, fuel economy and driveability of the vehicle.

✳ CAUTION:

It isn't a good idea to attempt diagnosis or replacement of the PCM or emission control components at home while the vehicle is under warranty. A Federally mandated warranty covers the emissions system components, and any owner-induced damage to the PCM, the sensors and/or the control devices may void this warranty. Take the vehicle to a dealer service department if the PCM or a system component malfunctions within the warranty period.

INFORMATION SENSORS

5   The PCM receives inputs from various information sensors and switches and uses the information it receives to control the operation of the engine in an appropriate and optimal manner. The following list provides a brief description of the function and location of each of the important information sensors.

6   **Battery temperature sensor -** The battery temperature sensor, which is located on the battery thermo-wrap, monitors the temperature of the battery so that the PCM can control charging voltage. When the ambient temperature is colder, the PCM maintains charging voltage at a higher level; as the ambient temperature goes up, the PCM decreases the charging voltage level.

7   **Camshaft Position (CMP) sensor -** The CMP sensor is one of two sensors (the CKP sensor is the other) that the PCM uses for cylinder identification and fuel injection synchronization. The CMP sensor is located on the left end of the cylinder head.

8   **Crankshaft Position (CKP) sensor -** The CKP sensor, which provides information on crankshaft position and engine speed to the

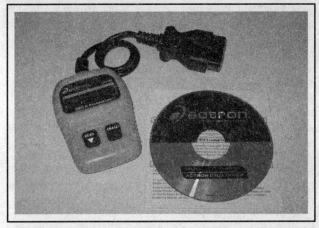

2.2 Trouble code readers like the Actron OBD-II PocketScan simplify the task of extracting the trouble codes

PCM, is also used (along with the CMP sensor) by the PCM to determine which spark plug to fire and which cylinder to inject with fuel. On 2000 through 2002 models, the CKP sensor is located on the rear side of the engine block, right above the oil filter adapter. On 2003 models, it's located on the front side of the block, right below the starter motor (on these models, you'll have to remove the structural collar to access the CMP sensor - see Chapter 2A).

9   **Engine Coolant Temperature (ECT) sensor -** The ECT sensor monitors engine coolant temperature and sends the PCM a voltage signal that affects PCM control of the fuel mixture, ignition timing and EGR operation. The ECT sensor is located on the left end of the cylinder head, next to the CMP sensor.

10   **Intake Air Temperature (IAT) sensor -** The IAT sensor is used by the PCM to calculate air density, one of the variables it must determine in order to calculate injector pulse width and adjust ignition timing. The IAT sensor is located on the air intake duct.

11   **Knock sensor -** The knock sensor monitors engine "knock" (pre-ignition or detonation) and then signals the PCM when knock occurs, so that the PCM can retard ignition timing accordingly. The knock sensor is located on the front side of the engine block, near the starter motor.

12   **Manifold Absolute Pressure (MAP) sensor -** The MAP sensor measures the intake manifold vacuum that draws the air/fuel mixture into the combustion chambers. The PCM uses this information to help it calculate injector pulse width and spark advance. The MAP sensor is located on top of the air cleaner housing.

13   **Oxygen sensors -** The oxygen sensors generate a voltage signal that varies in accordance with the difference between the oxygen content of the exhaust and the oxygen in the surrounding air. There are two oxygen sensors: an upstream, or pre-converter, oxygen sensor, and a downstream, or post-converter, oxygen sensor. The PCM uses the data from the upstream sensor to calculate the injector pulse width. The downstream oxygen sensor monitors the content of the exhaust gases as they exit the downstream catalytic converter. This information is used by the PCM to predict catalyst deterioration and/or failure. The upstream sensor is located on top of the exhaust manifold, just above the exhaust manifold flange. The location of the downstream oxygen sensor depends on the emissions package: On Federal (or 49-State) vehicles, the downstream sensor is located behind the catalytic converter. On Low Emission Vehicles (LEV), it's installed in the catalytic converter. On Ultra Low Emission Vehicles (ULEV), which have two catalytic converters, the oxygen sensor is located on the exhaust pipe between the two converters.

14 **Park/Neutral position switch** - The Park/Neutral position switch prevents the driver from starting the engine unless the automatic transaxle is in Park or Neutral. It also tells the PCM what gear the automatic transaxle is in.

15 **Power steering pressure switch** - The power steering pressure switch signals the PCM to increase the flow of air through the Idle Air Control (IAC) motor to prevent engine stalling when power steering pump pressure is excessive when parking and during other low-speed maneuvers. The power steering pressure switch is located on the power steering gear assembly.

16 **Throttle Position (TP) Sensor** - The TP sensor, which is located on the throttle body, produces a variable voltage signal in proportion to the opening angle of the throttle valve. This signal, which is monitored by the PCM, enables the PCM to calculate throttle position.

17 **Vehicle Speed Sensor (VSS)** - The VSS provides information to the PCM to indicate vehicle speed. The PCM uses this information to adjust the idle air control motor to maintain MAP values within an acceptable range during deceleration and to maintain an acceptable engine speed when the engine is idling. The VSS is also used to operate the speedometer, as an input for the cruise control system and, on Canadian vehicles, the Daytime Running Lights. The VSS is located above the inner CV joint for the right driveaxle on models with a manual transaxle; on models with an automatic, its located on the rear of the transaxle.

18 **Accelerator Pedal Position Sensor (APPS)** - The APPS, which is used on 2005 and later models, is located in the right rear corner of the engine compartment. The APPS provides the PCM with a variable voltage signal that's proportional to the position (angle) of the accelerator pedal. The PCM uses this data to control the position of the throttle plate inside the electronically controlled throttle body. There is an "accelerator cable" between the accelerator pedal and the APPS, but there's no cable to the throttle body.

OUTPUT ACTUATORS

19 Based on the information it receives from the information sensors described above, the PCM adjusts fuel injector pulse width, idle speed, ignition spark advance, ignition coil dwell and EVAP canister purge operation. It does so by controlling the output actuators. The following list provides a brief description of the function and location of each of the important output actuators.

20 **Auto Shutdown (ASD) relay** - When energized by the PCM, the ASD relay provides battery voltage to the fuel injectors, the ignition coil, the alternator field and the oxygen sensor heaters. The ASD relay is located inside the Power Distribution Center (PDC) in the engine compartment.

21 **Fuel injectors** - The PCM opens the fuel injectors sequentially (in firing order sequence). The PCM also controls the "pulse width," the interval of time during which each injector is open. The pulse width of an injector (measured in milliseconds) determines the amount of fuel delivered. For more information on the fuel delivery system and the fuel injectors, including injector replacement, refer to Chapter 4.

22 **Fuel pump relay** - When energized by the PCM, the fuel pump relay connects battery voltage to the fuel pump. The fuel pump relay is located inside the Power Distribution Center (PDC) in the engine compartment.

23 **Ignition coils** - The ignition coils are triggered by the PCM. The ignition coils are mounted on the valve cover. Refer to Chapter 5 for more information on the ignition coils.

2.28 The 16-pin Data Link Connector (DLC) is located under the left side of the dash

24 **Idle air control (IAC) motor** - The IAC motor, which is mounted on the throttle body, allows a certain amount of air to bypass the throttle plate when the throttle valve is closed or at idle position. The IAC motor is controlled by the PCM.

25 **Manifold Tuning Valve (MTV)** - The MTV, which is located on the right end of the intake manifold on High Output engines, is a PCM-controlled solenoid that opens and closes valves inside the lower half of the intake manifold (other engines are not equipped with an MTV). The MTV system optimizes the inside diameter of the intake manifold runners in response to engine operating conditions. At lower speeds, the MTV system decreases the size of the intake runners to maintain intake air velocity, which promotes better combustion. At higher speeds, the MTV system begins to increase the size of the intake runner in response to the engine's higher demand for intake air.

26 **Proportional purge solenoid** - When the engine is cold or still warming up, no captive fuel vapors are allowed to escape from the EVAP canister. After the engine is warmed up, the PCM energizes the proportional purge solenoid, which regulates the flow of these vapors from the canister to the intake manifold. The rate of the flow of vapors is regulated by the proportional purge solenoid in accordance with the current level, which is controlled by the PCM. The proportional purge solenoid is located at the lower left rear corner of the engine compartment, to the left of the power steering pressure switch.

OBTAINING OBD-II SYSTEM TROUBLE CODES

◗ **Refer to illustration 2.28**

27 The PCM will illuminate the CHECK ENGINE light (also called the Malfunction Indicator Light) on the dash if it recognizes a component fault for two consecutive drive cycles. It will continue to set the light until the PCM does not detect any malfunction for three or more consecutive drive cycles.

28 The diagnostic codes for the OBD-II system can be extracted from the PCM by plugging a generic OBD-II scan tool (see illustrations 2.1 and 2.2) into the PCM's data link connector (see illustration), which is located under the left side of the dash.

29 Plug the scan tool into the 16-pin data link connector (DLC), and then follow the instructions included with the scan tool to extract all the diagnostic codes.

DIAGNOSTIC TROUBLE CODES

Trouble code	Code identification
P0016	Camshaft position out of phase with crankshaft
P0030	Shorted condition in upstream oxygen sensor heater control feedback sensing circuit
P0031	Shorted condition in upstream oxygen sensor heater control feedback sensing circuit
P0036	Shorted condition in downstream oxygen sensor heater control feedback sensing circuit
P0037	Shorted condition in downstream oxygen sensor heater control feedback sensing circuit
P0038	Shorted condition in downstream oxygen sensor heater control feedback sensing circuit
P0068	Manifold absolute pressure sensor signal doesn't match throttle position sensor signal
P0071	Ambient temperature sensor value out of range
P0072	Ambient temperature sensor circuit, low voltage
P0073	Ambient temperature sensor circuit, high voltage
P0106	Manifold absolute pressure sensor input voltage out of range
P0107	Manifold absolute pressure sensor input below acceptable minimum voltage
P0108	Manifold absolute pressure sensor input above acceptable maximum voltage
P0110	Intake air temperature sensor stuck
P0111	Intake air temperature sensor circuit problem
P0112	Intake air temperature sensor input below minimum acceptable range
P0113	Intake air temperature sensor input above maximum acceptable range
P0116	Engine coolant temperature sensor range or performance problem
P0117	Engine coolant temperature sensor input below minimum acceptable voltage
P0118	Engine coolant temperature sensor input above maximum acceptable voltage
P0121	Throttle position sensor signal does not correlate to manifold absolute pressure signal
P0122	Throttle position sensor input below acceptable voltage range
P0123	Throttle position sensor input above acceptable voltage range
P0125	Time it takes to enter closed-loop fuel control is excessive
P0128	Thermostat rationality error detected
P0129	Manifold absolute pressure sensor input voltage out of range
P0130	Upstream oxygen sensor heater relay circuit malfunction
P0131	Upstream oxygen sensor input voltage below normal operating range
P0132	Upstream oxygen sensor input voltage above normal operating range
P0133	Upstream oxygen sensor response slower than minimum required switching frequency
P0134	Neither rich nor lean condition is detected from upstream oxygen sensor input
P0135	Upstream oxygen sensor heater or circuit fault

Trouble code	Code identification
P0136	Downstream oxygen sensor heater relay circuit malfunction
P0137	Downstream oxygen sensor input voltage below normal operating range.
P0138	Downstream oxygen sensor input voltage above normal operating range.
P0139	Downstream oxygen sensor response not as expected.
P0140	Neither rich nor lean condition is detected from downstream oxygen sensor input
P0141	Downstream oxygen sensor heater or circuit fault
P0165	Open or shorted condition in starter relay control circuit
P0171	Fuel injection system too lean
P0172	Fuel injection system too rich
P0201	Open or shorted condition detected in control circuit for injector No. 1
P0202	Open or shorted condition detected in control circuit for injector No. 2
P0203	Open or shorted condition detected in control circuit for injector No. 3
P0204	Open or shorted condition detected in control circuit for injector No. 4
P0300	Multiple cylinder misfires detected
P0301	Cylinder No. 1 misfire detected
P0302	Cylinder No. 2 misfire detected
P0303	Cylinder No. 3 misfire detected
P0304	Cylinder No. 4 misfire detected
P0315	Unable to learn crankshaft position sensor signal in preparation for misfire diagnostics
P0320	No reference signal from crankshaft position sensor detected during cranking
P0325	Knock sensor signal above or below minimum acceptable threshold voltage
P0335	Rationality error detected for loss of crankshaft position sensor circuit
P0339	Rationality error detected for intermittent loss of crankshaft position sensor circuit
P0340	No camshaft position sensor signal at powertrain control module
P0344	Rationality error detected for intermittent loss of camshaft position sensor circuit
P0350	Ignition coil draws too much current
P0351	Ignition coil No. 1 peak primary circuit current not achieved with maximum dwell time
P0352	Ignition coil No. 2 peak primary circuit current not achieved with maximum dwell time
P0420	Upstream catalytic converter efficiency below threshold
P0432	Downstream catalytic converter efficiency below threshold
P0440	General failure of evaporative emissions control system
P0441	Insufficient or excessive purge vapor flow detected in EVAP system
P0442	Medium leak detected in EVAP system

Trouble code	Code identification
P0443	Open or shorted condition detected in EVAP purge control solenoid circuit
P0452	Natural vacuum leak detection pressure switch stuck closed
P0453	Natural vacuum leak detection pressure switch stuck open
P0455	Large leak detected in EVAP system
P0456	Small leak detected in EVAP system
P0460	No movement of fuel level sending unit detected
P0461	No level of fuel level sending unit detected
P0462	Fuel level sending unit input below acceptable voltage
P0463	Fuel level sending unit input above acceptable voltage
P0480	Open or shorted low-speed fan relay control circuit
P0481	Open or shorted high-speed fan relay control circuit
P0498	Natural vacuum leak detection canister vent solenoid circuit low
P0499	Natural vacuum leak detection canister vent solenoid circuit high
P0500	No vehicle speed sensor signal detected during road load conditions
P0501	Rationality error detected in vehicle speed sensor circuit
P0505	Idle air control motor circuit fault
P0508	Idle air control motor circuit input below acceptable current
P0509	Idle air control motor circuit input above acceptable current
P0516	Battery temperature sensor input below minimum acceptable voltage
P0517	Battery temperature sensor input above maximum acceptable voltage
P0519	Idle speed performance - rationality error detected for target rpm during idle
P0522	Oil pressure sending unit input below acceptable voltage
P0523	Oil pressure sending unit input above acceptable voltage
P0532	Air conditioning pressure switch input below minimum acceptable voltage
P0533	Air conditioning pressure switch input above maximum acceptable voltage
P0551	Incorrect power steering switch input detected
P0562	Battery voltage low
P0563	Battery voltage high
P0572	Brake switch, low voltage
P0573	Brake switch, high voltage
P0579	Cruise control switch stuck in a valid voltage range
P0580	Cruise control switch input below minimum acceptable voltage
P0581	Cruise control switch input above maximum acceptable voltage

Trouble code	Code identification
P0582	Open or shorted condition detected in cruise control vacuum solenoid control circuit
P0586	Open or shorted condition detected in cruise control vent solenoid control circuit
P0594	Open or shorted condition detected in cruise control servo power control circuit
P0600	No communication between detected between co-processors in control module
P0601	Internal control module fault condition detected
P0604	Transmission control module RAM self-test fault detected
P0605	Transmission control module ROM self-test fault detected
P0615	Open or shorted condition detected in starter relay control circuit
P0622	Open or short condition detected in alternator field control circuit
P0627	Open or shorted condition detected in fuel pump relay control circuit
P0630	Vehicle identification number not programmed in control module memory
P0632	Odometer not programmed in control module memory
P0633	Sentry key immobilize module key not programmed in control module memory
P0645	Open or short condition detected in air conditioning clutch relay control circuit
P0660	Open or shorted condition detected in manifold tune valve solenoid control circuit
P0685	Open or shorted condition detected in automatic shutdown relay control circuit
P0688	Automatic shutdown relay control circuit low
P0700	Transaxle control system malfunction
P0703	Incorrect voltage input condition detected in brake switch circuit
P0711	Transaxle fluid temperature sensor or circuit malfunction or failure
P0712	Transaxle fluid temperature sensor input below acceptable voltage
P0713	Transaxle fluid temperature sensor input above acceptable voltage
P0720	Relationship between output shaft speed sensor and vehicle speed out of range
P0740	Failure of torque converter clutch lock-up system
P0743	Open or shorted condition detected in torque converter clutch solenoid control circuit
P0748	Open or shorted condition detected in governor pressure solenoid or trans relay circuit
P0751	Overdrive override switch input in a prolonged depressed state
P0753	Open or shorted condition detected in overdrive solenoid control or trans relay circuit
P0756	Shift solenoid B (2-3 shift) malfunction
P0783	Overdrive solenoid unable to engage gear change from third gear to overdrive gear
P0801	Open or short condition detected in transaxle reverse gear lockout solenoid control circuit
P0803	Rationality error detected in clutch start switch performance
P0833	Clutch release switch circuit and/or switch
P0850	Rationality error detected in Park/Neutral position switch performance

3   Camshaft Position (CMP) sensor - replacement

▶ **Refer to illustrations 3.2, 3.4, 3.5, 3.6 and 3.7**

1   Disconnect the negative battery cable.

2   The CMP sensor (see illustration) is located on the left end of the cylinder head. If necessary, move the brake booster hose and any electrical wires out of the way to access the CMP sensor.

3   Disconnect the CMP sensor electrical connector (see illustration 3.2).

4   Remove the CMP sensor retaining bolts (see illustration) and then remove the CMP sensor.

5   Remove the Torx bolt that attaches the target magnet to the camshaft (see illustration) and then remove the target magnet.

6   When installing the target magnet make sure that the locating dowels on the back of the magnet are aligned with the machined locating holes in the end of the camshaft (see illustration). Tighten the target magnet retaining screw to the torque listed in this Chapter's Specifications.

7   If you're planning to install the old CMP sensor, inspect the O-ring on the sensor mounting plate (see illustration). If it's cracked, torn or deteriorated, replace it. If you're installing a new CMP sensor, make sure a new O-ring is already installed.

8   Installation is otherwise the reverse of removal. Be sure to tighten the CMP sensor retaining bolts to the torque listed in this Chapter's Specifications.

3.2   The CMP sensor is located on the left end of the cylinder head; to release the locking tab on the CMP sensor electrical connector, push it out (toward the firewall)

3.4   To detach the CMP sensor from the cylinder head, remove these four bolts

3.5   To detach the CMP sensor target magnet from the end of the camshaft, remove this Torx bolt

3.6   When installing the CMP sensor target magnet, make sure that the locating dowels on the back of the magnet are aligned with the machined locating holes in the end of the camshaft

3.7   Before installing an old CMP sensor, inspect the O-ring on the mounting base; if it's torn, cracked or otherwise deteriorated, replace it

4 Crankshaft Position (CKP) sensor - replacement

▶ **Refer to illustration 4.4**

1 Disconnect the negative battery cable.

2 Raise the front of the vehicle and place it securely on jackstands.

3 On 2003 and later models remove the structural collar (see "Oil pan - removal and installation" in Chapter 2A).

4 Disconnect the CKP sensor electrical connector (see illustration).

5 Remove the CKP sensor retaining bolt (see illustration 4.4).

6 Installation is the reverse of removal. Be sure to tighten the CKP sensor retaining bolt to the torque listed in this Chapter's Specifications.

4.4 To remove the Crankshaft Position (CKP) sensor, unplug the electrical connector and then remove the retaining bolt

5 Engine Coolant Temperature (ECT) sensor - replacement

▶ **Refer to illustrations 5.3, 5.5 and 5.6**

✳✳ **WARNING:**

Wait until the engine has cooled completely before beginning this procedure.

✳✳ **CAUTION:**

Handle the Engine Coolant Temperature (ECT) sensor with care. Damage to the ECT sensor will affect the operation of the entire fuel injection system.

1 Disconnect the negative battery cable.

2 Drain the engine coolant (see Chapter 1).

3 The ECT sensor (see illustration) is located on the left end of the cylinder head, next to, and in front of, the CMP sensor. There is no easy way to access the ECT sensor. To get a wrench on the ECT sensor, you'll have to remove the CMP sensor. If you want to remove the ECT sensor with a deep socket, you'll have to remove the battery and the battery tray (see Chapter 5). We recommend removing the CMP sensor (see Section 3) because it's easier.

4 Disconnect the electrical connector from the ECT sensor (see illustration 5.3).

5 Unscrew the sensor from the cylinder head (see illustration).

6 Before installing the new ECT sensor, wrap the threads of the sensor with Teflon tape to prevent coolant leakage (see illustration).

7 Installation is otherwise the reverse of removal. Be sure to tighten the ECT sensor to the torque listed in this Chapter's Specifications. Refill the cooling system as described in Chapter 1.

5.3 Disconnect the electrical connector from the Engine Coolant Temperature (ECT) sensor . . .

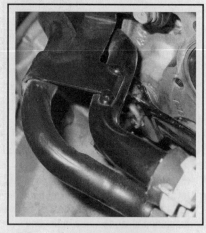

5.5 . . . then unscrew the sensor

5.6 Wrap the threads of the ECT sensor with Teflon tape to prevent coolant leakage

6 Intake Air Temperature (IAT) sensor - replacement

▶ **Refer to illustrations 6.2 and 6.4**

1   Disconnect the negative battery cable.

2   Disconnect the electrical connector from the IAT sensor (see illustration).

3   Remove the air filter housing cover and then remove the fresh air inlet duct (see "Air filter housing - removal and installation" in Chapter 4).

4   Using a pair of diagonal cutters, cut off the small steel band that secures the IAT sensor to the fresh air inlet duct (see illustration).

5   Remove the IAT sensor by pulling it out of the fresh air intake duct.

6   Install a new steel band on the pipe for the IAT sensor, install the new IAT sensor in the fresh air inlet duct and then crimp the band with a pair of CV joint band crimping pliers designed for this purpose. If you don't have a pair of CV joint band crimping pliers, use a conventional hose clamp instead and tighten it securely.

7   Installation is otherwise the reverse of removal.

6.2 Disconnect the electrical connector from the Intake Air Temperature (IAT) sensor

6.4 To remove the IAT sensor from the fresh air inlet duct, cut off this steel band with a pair of diagonal cutters and then pull the IAT sensor straight out of its mounting pipe

7 Knock sensor - replacement

▶ **Refer to illustration 7.3**

❋❋ WARNING:

Wait for the engine to cool completely before performing this procedure.

1   Disconnect the negative battery cable.

2   Raise the vehicle and place it securely on jackstands.

3   The knock sensor (see illustration) is located on the engine block, near the starter motor. On 2000 through 2002 models, it's screwed into the block. On 2003 models, it's bolted to the block.

4   Disconnect the electrical connector from the knock sensor (see illustration 7.3).

5   On 2000 through 2002 models unscrew the knock sensor with a deep socket (see illustration 7.3). On 2003 and later models, remove the knock sensor retaining bolt.

6   Installation is the reverse of removal. Be sure to tighten the knock sensor (2000 through 2002 models) or the knock sensor bolt (2003 and later models) to the torque listed in this Chapter's Specifications.

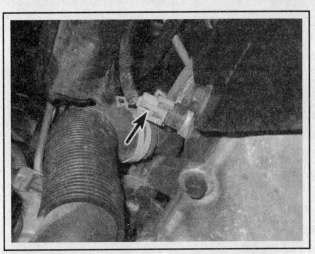

7.3 To remove the knock sensor, disconnect the electrical connector and unscrew the sensor (2000 through 2002 models, shown) or remove the retaining bolt (2003 models, not shown)

---

8 Manifold Absolute Pressure (MAP) sensor - replacement

▶ **Refer to illustrations 8.2, 8.4a and 8.4b.**

1 Disconnect the negative battery cable.

2 The MAP sensor (see illustration) is located on the front right corner of the intake manifold.

3 Disconnect the MAP sensor electrical connector (see illustration 8.2).

4 Remove the MAP sensor retaining screws and then remove the sensor (see illustrations). Remove and discard the two old sensor O-rings.

5 Whether you're planning to install the old MAP sensor or a new unit, install new O-rings on the sensor.

6 Installation is otherwise the reverse of removal. Be careful not to damage the O-rings when installing the MAP sensor.

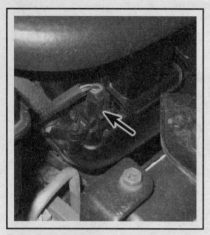 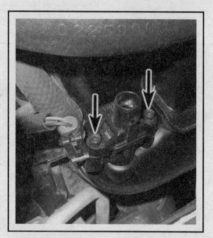 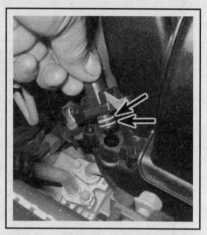

8.2 Disconnect the electrical connector from the Manifold Absolute Pressure (MAP) sensor

8.4a To detach the MAP sensor from the intake manifold, remove these two retaining screws . . .

8.4b . . . and then remove the sensor from the manifold; be sure to discard the old O-rings and replace them

---

9 Oxygen sensors - general information and replacement

GENERAL INFORMATION

1 Use special care when servicing an oxygen sensor:

a) *Oxygen sensors have a permanently attached pigtail and electrical connector which cannot be removed from the sensor. Damage or removal of the pigtail or electrical connector will ruin the sensor.*

b) *Grease, dirt and other contaminants should be kept away from the electrical connector and the louvered end of the sensor.*

c) *Do not use cleaning solvents of any kind on an oxygen sensor or air/fuel ratio sensor.*

d) *Do not drop or roughly handle an oxygen sensor or air/fuel ratio sensor.*

e) *Be sure to install the silicone boot in the correct position to prevent the boot from melting and to allow the sensor to operate properly.*

REPLACEMENT

➡**Note: Because it is installed in the exhaust manifold or catalytic converter, both of which contract when cool, an oxygen sensor might be very difficult to loosen when the engine is cold. Rather than risk damage to the sensor, start and run the engine for a minute or two, then shut it off. Be careful not to burn yourself during the following procedure.**

2 Disconnect the cable from the negative terminal of the battery.

3 Raise the vehicle and place it securely on jackstands.

Upstream oxygen sensor

▶ **Refer to illustrations 9.4, 9.5 and 9.6**

4 The upstream oxygen sensor (see illustration) is located on top of the exhaust manifold, just above the flange.

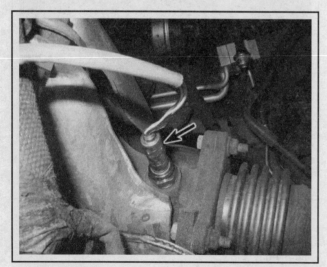

9.4 The upstream oxygen sensor is located on top of the exhaust manifold

5 Trace the electrical lead from the sensor to the electrical connector (see illustration) and then disconnect it.

6 Using a special oxygen sensor socket, unscrew the sensor (see illustration).

7 After removing the old upstream oxygen sensor, clean the threads of the sensor bore in the exhaust manifold with an 18 mm tap with the appropriate thread pitch.

8 If you're going to install the old sensor, apply anti-seize compound to the threads of the sensor to facilitate future removal. If you're going to install a new oxygen sensor, it's not necessary to apply anti-seize compound to the threads. The threads on new sensors already have anti-seize compound on them.

9 Install the upstream oxygen sensor and then tighten it to the torque listen in this Chapter's Specifications.

10 Installation is otherwise the reverse of removal.

Downstream oxygen sensor

▶ Refer to illustrations 9.12 and 9.13

11 The location of the downstream oxygen sensor depends on the emissions package: On Federal (or 49-State) vehicles, the downstream sensor is located on the exhaust pipe behind the catalytic converter. On Low Emission Vehicles (LEV), it's installed on the catalytic converter. On Ultra Low Emission Vehicles (ULEV), which have *two* catalytic converters, the oxygen sensor is located on the exhaust pipe *between* the two converters.

12 Trace the electrical lead from the sensor to the electrical connector and then disconnect it (see illustration).

13 Unscrew and remove the downstream oxygen sensor (see illustration).

14 After removing the old oxygen sensor, clean the threads of the sensor bore in the catalytic converter with an 18 mm tap with the appropriate thread pitch.

9.5 To find the electrical connector for the upstream oxygen sensor, trace the lead from the sensor to the connector, which is attached to a bracket bolted to the bellhousing right behind the left end of the cylinder head; to unplug the connector, pull it off the bracket, lift up the locking tab (not visible in this photo - it's on the side that's attached to the bracket) and pull the connector halves apart

15 If you're going to install the old sensor, apply anti-seize compound to the threads of the sensor to facilitate future removal.

16 If you're going to install a new oxygen sensor, it's not necessary to apply anti-seize compound to the threads. The threads on new sensors already have anti-seize compound on them.

17 Install the oxygen sensor and then tighten it to the torque listed in this Chapter's Specifications.

18 Installation is otherwise the reverse of removal.

9.6 Use a special oxygen sensor socket to unscrew the upstream oxygen sensor

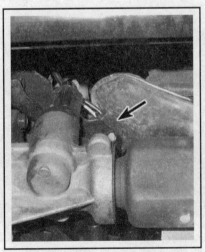

9.12 The electrical connector for the downstream oxygen sensor is attached to a bracket located above the right inner CV joint; to unplug the connector, pull it off the bracket, lift up the locking tab (not visible in this photo) and then pull the connector halves apart

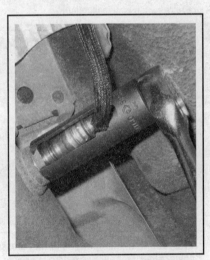

9.13 To remove the downstream oxygen sensor, unscrew it with a special socket

10 Park/Neutral position switch - replacement

▶ **Refer to illustration 10.2**

2000 AND 2001 MODELS

1   Raise the front of the vehicle and place it securely on jackstands.

2   The Park/Neutral position switch (see illustration) is located at the front lower left corner of automatic transaxles.

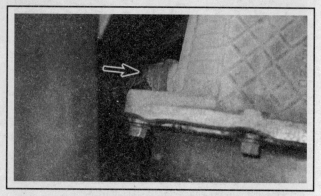

10.2 The Park/Neutral position switch is located on the front lower left corner of 2000 and 2001 automatic transaxles

3   Disconnect the electrical connector from the Park/Neutral position switch.

4   Unscrew the Park/Neutral position switch. Discard the old O-ring.

5   Install a new switch O-ring. Even if you're planning to reuse the old Park/Neutral position switch, replace the O-ring.

6   Install the Park/Neutral position switch and tighten it to the torque listed in this Chapter's Specifications.

7   Plug in the electrical connector.

8   Lower the vehicle.

9   Check the automatic transaxle fluid level (see Chapter 1) and add fluid as necessary.

2002 AND LATER MODELS

10   2002 and later models are equipped with an updated transaxle range sensor (Park/Neutral switch) that is mounted to the valve body. The transaxle range sensor has four switch contacts that monitor shift lever position. The assembly is also equipped with a transaxle fluid temperature sensor that relays temperature data to the TCM and PCM. The transaxle range sensor electrical connector is accessible from the side of the transaxle but sensor replacement will require valve body removal. Have the sensor replaced by a dealer service department or other qualified automotive repair facility.

11 Power steering fluid pressure switch - replacement

▶ **Refer to illustrations 11.3 and 11.4**

1   Disconnect the negative battery cable.

2   Raise the vehicle and place it securely on jackstands.

3   Disconnect the electrical connector from the power steering fluid pressure switch (see illustration).

4   Using a 7/8-inch deep socket and an extension (see illustration), unscrew and remove the power steering fluid pressure switch.

5   Seal the threads of the new switch with Teflon tape.

6   Installation is the reverse of removal. Be sure to tighten the switch to the torque listed in this Chapter's Specifications.

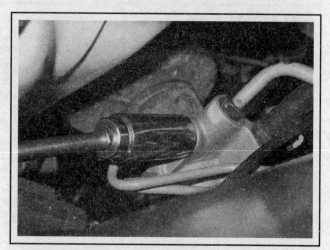

11.3 You'll find the power steering pressure switch on the power steering gear assembly; to unplug the electrical connector, pry up on the locking tab (not visible in this photo - it's on the other side of the connector) to release it and then pull off the connector

11.4 Use a 7/8-inch deep socket to unscrew the power steering pressure switch (a deep socket will protect the plastic connector)

12 Throttle Position (TP) sensor - replacement

▶ **Refer to illustrations 12.2, 12.5 and 12.7**

1   Disconnect the negative battery cable.

2   The TP sensor (see illustration) is located on the front side of the throttle body, which is bolted to the air filter housing.

3   Disconnect the electrical connector from the TP sensor (see

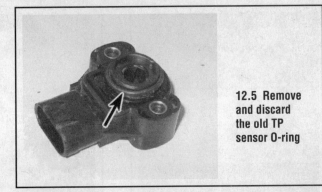

12.5 Remove and discard the old TP sensor O-ring

12.2 The Throttle Position (TP) sensor is located on the front side of the throttle body, which is bolted to the air filter housing; to detach the TP sensor from the throttle body, unplug the electrical connector and remove the sensor retaining screws (it's easier to replace the TP sensor if you unbolt the throttle body from the air filter housing and disconnect it from the air intake duct)

illustration 12.2).

4   Remove the two TP sensor mounting screws (see illustration 12.2) and then remove the TP sensor.

5   Remove and discard the old TP sensor O-ring (see illustration).

6   Before installing the TP sensor, install a new O-ring.

7   When installing the TP sensor, make sure that the blade on the end of the throttle shaft fits into the hole in the backside of the TP sensor (see illustration). There are two tabs inside the hole. The blade must fit between the two tabs, i.e., each side of the blade should be touching one of the tabs.

8   Install the TP sensor with the mounting holes of the sensor a few degrees to the left of the mounting holes in the throttle body and then rotate it clockwise until the holes in the TP sensor and the holes in the throttle body are aligned. The TP sensor should offer a little resistance when rotated into position. But if it's difficult to rotate, then you've got the tip of the throttle shaft blade on the wrong sides of the tabs. Rein-

12.7 When installing the TP sensor, make sure that the blade on the end of the throttle shaft fits *between* the two tabs inside the hole in the backside of the TP sensor; in other words, each side of the blade must be touching one of the tabs

stall the TP sensor with the throttle shaft blade on the opposite sides of the sensor tabs.

9   When the TP sensor holes are lined up, install the TP sensor screws and tighten them securely.

10   Installation is otherwise the reverse of removal.

13   Vehicle Speed Sensor (VSS) - replacement

ALL MODELS

1   Disconnect the negative battery cable.

2   The VSS, which is located on the top of the transaxle extension housing, near one of the inner CV joints, is difficult, but not impossible, to reach from above or below. If you want to remove it from above, remove the air filter housing (see Chapter 4) and then remove the battery and the battery tray bracket (see Chapter 5). Or, if you'd prefer not to remove those components, raise the vehicle, place it securely on jackstands and then remove the VSS from below.

MODELS WITH A MANUAL TRANSAXLE

♦ **Refer to illustration 13.3**

➡**Note: Refer to illustrations 13.11 through 13.15 for disassembly details. The VSS used on manual transaxles is virtually identical to the unit used on 2000 and 2001 models with an automatic transaxle.**

3   Unplug the VSS electrical connector (see illustration).

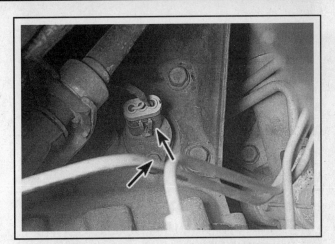

13.3 On vehicles with a manual transaxle, the Vehicle Speed Sensor (VSS) is located on top of the transaxle, near the left or right inner CV joint; to detach the VSS from the transaxle, disconnect the electrical connector and remove the hold-down bolt

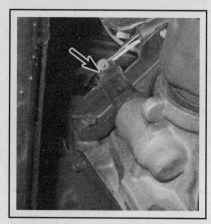

13.11 Disconnect the VSS electrical connector (2000 and 2001 models with automatic transaxle)

13.12 To detach the VSS from the transaxle, remove this hold-down bolt (2000 and 2001 models with automatic transaxle)

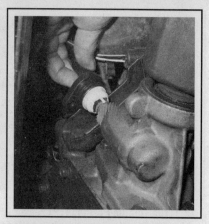

13.13 Remove the VSS assembly from the transaxle (2000 and 2001 models with automatic transaxle)

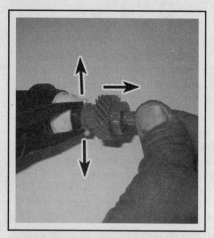

13.14 To remove the driven pinion gear from the VSS, spread the two white plastic retainers apart and pull off the pinion gear (2000 and 2001 models with automatic transaxle)

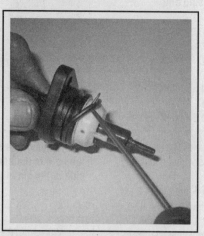

13.15 Even if you're going to reuse the old VSS, remove and discard the old O-ring from the VSS and replace it with a new O-ring (2000 and 2001 models with automatic transaxle)

13.19 To unplug the electrical connector from the input shaft VSS, pry open the locking tang on the underside of the connector (not visible in this photo) and then pull off the connector; to remove the VSS, simply unscrew it (2002 and later models with an automatic transaxle)

4 Remove the VSS retaining bolt and then remove the VSS from the transaxle.

5 Remove the VSS pinion driven gear from the VSS.

6 Remove and discard the old O-ring.

7 Install a new O-ring on the VSS and coat it with a little clean oil.

8 Install the drive gear and then install the VSS.

9 Install the VSS retaining bolt and then tighten it to the torque listed in this Chapter's Specifications.

10 Installation is otherwise the reverse of removal.

MODELS WITH AN AUTOMATIC TRANSAXLE

2000 and 2001 models

▸ **Refer to illustrations 13.11, 13.12, 13.13, 13.14 and 13.15**

11 Disconnect the VSS electrical connector (see illustration).

12 Remove the VSS hold-down bolt (see illustration).

13 Remove the VSS (see illustration).

14 Remove the pinion driven gear from the VSS (see illustration). (The new VSS will not necessarily include a new pinion driven gear, so you'll need to swap this part onto the new VSS.)

15 If you're planning to reuse the old VSS, remove the old O-ring from the VSS (see illustration) and discard it. If you're replacing the VSS, make sure that the new VSS has a new O-ring installed. Coat the new O-ring with clean oil to protect it during installation.

16 Install the VSS pinion driven gear. Make sure that the white plastic retainers snap into place and that the pinion gear is fully seated.

17 Installation is otherwise the reverse of removal. Be sure to tighten the VSS hold-down bolt to the torque listed in this Chapter's Specifications.

2002 and later models

Input shaft VSS

▸ **Refer to illustrations 13.19 and 13.21**

18 On 2002 and later models, remove the air filter housing and the throttle body to access the input shaft speed sensor (see Chapter 4).

19 Disconnect the input shaft VSS electrical connector (see illustration).

20 Unscrew and then remove the input shaft VSS from the transaxle.

13.21 Remove and discard the old O-ring from the VSS (2002 and later models with an automatic transaxle)

21 Remove and discard the old O-ring (see illustration).
22 Install a new O-ring on the VSS.
23 Install the VSS and then tighten it to the torque listed in this Chapter's Specifications.

Output shaft VSS

▶ **Refer to illustration 13.24**

24 Unplug the output shaft VSS electrical connector (see illustration).

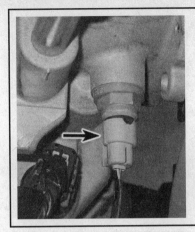

13.24 To unplug the electrical connector from the output shaft VSS, pry open the locking tang on the underside of the connector (not visible in this photo) and pull off the connector; to remove the VSS, simply unscrew it (2002 and later models with an automatic transaxle)

25 Unscrew and then remove the output shaft VSS from the transaxle.
26 Remove and discard the old O-ring (see illustration 13.20).
27 Install a new O-ring on the VSS.
28 Install the VSS and then tighten it to the torque listed in this Chapter's Specifications.

14 Accelerator Pedal Position Sensor (APPS) - replacement

➡ **Note: This procedure applies to 2005 and later models only.**

1 Disconnect the cable from the negative terminal of the battery (see Chapter 5, Section 1). Disconnect the APPS cable from the throttle lever at the throttle body (see Chapter 4).
2 Working in the right rear corner of the engine compartment, remove the two APPS mounting bolts.

3 Disconnect the APPS electrical connector from the module housing.
4 Pry the locating tab and lift the top of the APPS assembly to access the cable.
5 Disconnect the cable end from the APPS lever arm.
6 Installation is the reverse of removal.

15 Powertrain Control Module (PCM) - replacement

▶ **Refer to illustrations 15.3, 15.5a and 15.5b**

✳✳ WARNING:

The models covered by this manual are equipped with Supplemental Restraint systems (SRS), more commonly known as airbags. Always disable the airbag system before working in the vicinity of any airbag system components to avoid the possibility of accidental deployment of the airbag, which could cause personal injury (see Chapter 12).

✳✳ CAUTION:

To avoid electrostatic discharge damage to the PCM, handle the PCM only by its case. Do not touch the electrical terminals during removal and installation. If available, ground yourself to the vehicle with an anti-static ground strap, available at computer supply stores.

➡ **Note: On 2002 and later models, if the PCM is replaced with a new unit, it will be necessary to calibrate the speedometer using a professional-grade scan tool capable of performing this operation. Have the vehicle calibrated by a dealer service department or other qualified automotive repair facility.**

1 The Powertrain Control Module (PCM) is located in the front lower left corner of the engine compartment.

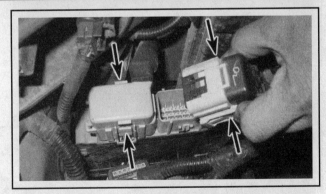

15.3 To disconnect the electrical connectors from the Powertrain Control Module (PCM), depress the large locking tabs on the sides of the connectors and pull firmly (on the connector, *not* on the wires!)

2 Remove the air filter housing, the throttle body and air intake duct (see Chapter 4).

➡ **Note: Remove these components as a single assembly. It's not necessary to separate the intake duct from the throttle body, or the throttle body from the air filter housing.**

3 Disconnect the electrical connectors from the PCM (see illustration).

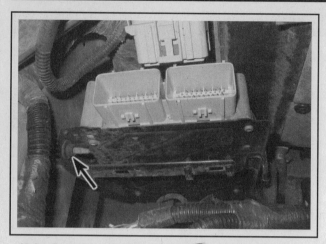

15.5a To remove the PCM, remove the upper nut and . . .

15.5b . . . remove the two lower mounting bolts

4   Unclip the PCM wiring harness from the PCM mounting plate and set it aside so that you have clear access to the PCM mounting fasteners.

5   Remove the PCM upper mounting nut and the two PCM lower mounting bolts (see illustrations) and remove the PCM.

❋❋ CAUTION:

Avoid any static electricity damage to the computer by grounding yourself to the body before touching the PCM and using a special anti-static pad to store the PCM on once it is removed.

6   Installation is the reverse of removal.

16 Idle Air Control (IAC) motor - replacement

16.2 To disconnect the electrical connector from the IAC motor, depress the locking tab and pull firmly on the connector (not the wires)

▶ Refer to illustrations 16.2, 16.3, 16.4 and 16.6

1   Disconnect the negative battery cable.
2   Disconnect the IAC motor electrical connector (see illustration).
3   Remove the IAC motor mounting screws (see illustration) and then detach the IAC motor from the throttle body.
4   Remove and discard the old O-ring from the IAC motor (see illustration).
5   Install a new O-ring on the IAC motor.
6   If you're going to install a new IAC motor, measure the length of the IAC motor pintle valve (the plunger) before installing the IAC motor (see illustration). The pintle valve should not exceed one inch (25 mm) in length. If it does, carefully push the pintle in while wiggling it until it protrudes less than one inch.
7   Install the IAC motor and tighten the IAC motor mounting screws securely.
8   Installation is otherwise the reverse of removal.

16.3 To detach the IAC motor from the throttle body, remove these two mounting screws

16.4 Remove and discard the old O-ring from the IAC motor

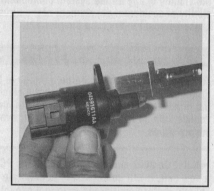

16.6 If you installing a new IAC motor, measure the length of the IAC motor pintle valve before installing the motor; the pintle valve should not exceed one inch in length

17 Manifold Tuning Valve (MTV) - general information and component replacement

GENERAL INFORMATION

1   The Manifold Tuning Valve (MTV), which is used only on High Output engines, optimizes the flow of intake air through the intake manifold runners by opening and closing butterfly valves inside the manifold in response to engine speed and load. When the engine is idling, its speed and load are low, and so is its demand for intake air. In order to keep the velocity of intake air high enough to promote optimal combustion at idle, the PCM orders the MTV solenoid to close the butterfly valves. As the engine speeds up, the valves are gradually opened to allow as much air into the engine as possible without sacrificing intake air velocity.

2   The MTV system consists of a special two-piece aluminum manifold, which contains the butterfly valves, and a PCM-controlled solenoid, which in turn controls the butterfly valves inside the manifold. The replacement procedure for the solenoid is in this section. For more information on the MTV intake manifold, see "Intake manifold - removal and installation" in Chapter 2A.

COMPONENT REPLACEMENT

3   Remove the three power steering bracket bolts.
4   Remove the power steering bracket.
5   Remove the MTV solenoid mounting bolts.
6   Remove the MTV solenoid.
7   Disconnect the MTV solenoid electrical connector.
8   Installation is the reverse of removal.

18 Catalytic converter - general information, check and replacement

➡**Note: Because of the Federally mandated extended warranty which covers emissions-related components such as the catalytic converter, check with a dealer service department before replacing the converter at your own expense.**

GENERAL INFORMATION

1   The catalytic converter is an emission control device installed in the exhaust system that reduces pollutants from the exhaust gas stream. There are two types of converters: The oxidation catalyst reduces the levels of hydrocarbon (HC) and carbon monoxide (CO) by adding oxygen to the exhaust stream. The reduction catalyst lowers the levels of oxides of nitrogen (NOx) by removing oxygen from the exhaust gases. These two types of catalysts are combined into a three-way catalyst that reduces all three pollutants.

CHECK

2   The equipment for testing a catalytic converter is expensive. If you suspect that the converter on your vehicle is malfunctioning, take it to a dealer or authorized emissions inspection facility for diagnosis and repair.

3   Whenever the vehicle is raised for servicing underbody components, inspect the converter for leaks, corrosion, dents and other damage. Inspect the welds/flange bolts that attach the front and rear ends of the converter to the exhaust system. If damage is discovered, the converter should be replaced.

4   Although catalytic converters don't break too often, they can become plugged. The easiest way to check for a restricted converter is to use a vacuum gauge to diagnose the effect of a blocked exhaust on intake vacuum.

a) Connect a vacuum gauge to an intake manifold vacuum source (see Chapter 2B).
b) Warm the engine to operating temperature, place the transaxle in Park (automatic) or Neutral (manual) and apply the parking brake.
c) Note and record the vacuum reading at idle.

d) Quickly open the throttle to near full throttle and release it shut. Note and record the vacuum reading.
e) Perform the test three more times, recording the reading after each test.
f) If the reading after the fourth test is more than one in-Hg lower than the reading recorded at idle, the exhaust system may be restricted (the catalytic converter could be plugged or an exhaust pipe or muffler could be restricted).

REPLACEMENT

▶ **Refer to illustrations 18.8a and 18.8b**

5   Raise the vehicle and place it securely on jackstands.
6   Disconnect the electrical connector for the downstream oxygen sensor and remove the downstream oxygen sensor (see Section 9).
7   Be sure to spray the nuts on the exhaust flange studs with penetrant before removing them from the catalytic converter.
8   Loosen the clamp behind the catalytic converter and then remove the nuts that attach the catalyst to the exhaust manifold (see illustrations). Separate the catalytic converter from the exhaust system.

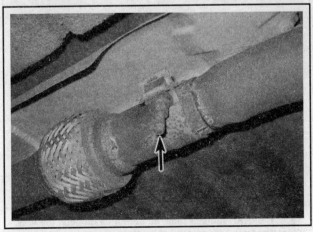

18.8a To remove the catalytic converter, loosen this clamp bolt behind the converter . . .

18.8b ... and then remove these four nuts from the exhaust manifold studs (one nut not visible)

9   Before installing the converter, coat the threads of the exhaust manifold studs and the clamp bolt with anti-seize compound. Be sure to tighten the fasteners to the torque listed in this Chapter's Specifications.

10 Installation is otherwise the reverse of removal.

19 Evaporative emissions control (EVAP) system - general information and component replacement

GENERAL DESCRIPTION

▶ **Refer to illustration 19.2**

1   The fuel evaporative emissions control (EVAP) system absorbs fuel vapors (unburned hydrocarbons) and, during engine operation, releases them into the intake manifold from which they're drawn into the intake ports where they mix with the incoming air-fuel mixture. The charcoal canister is mounted at the right front corner of the fuel tank.

2   When the engine is not operating, fuel vapors migrate through a system of hoses from the fuel tank and intake manifold to the EVAP canister (see illustration), where they are stored until the next time the vehicle is operated. Vapors are routed through the control valve and then through the flow management valve on their way to the EVAP canister. The flow management valve also serves as a liquid separator to prevent fuel from contaminating the EVAP canister. The flow management valve is replaceable, but the control valve is not. If the control valve malfunctions, the fuel tank must be replaced. A rollover valve, which is located in the top of the fuel tank, prevents the flow of fuel through the vapor hoses if the vehicle rolls over. The rollover valve is also not serviceable. If it malfunctions, replace the fuel tank.

3   The Powertrain Control Module (PCM) purges the EVAP canister through the proportional purge solenoid. The solenoid regulates the rate of flow of vapors from the canister to the throttle body. The solenoid is not energized during cold starts and during warm-up. After the engine has been started and warmed up to its normal operating temperature, the PCM energizes the solenoid, which controls the rate of flow in proportion to the current level, which is controlled by the PCM.

4   On 2000 through 2002 California models, the EVAP system is equipped with a Leak Detection Pump (LDP), which is mounted in front of the fuel tank, near the EVAP canister. The LDP, which is controlled by the PCM, tests for, and can detect, leaks in the system. When a leak is detected, the PCM stores a diagnostic trouble code (see Section 2). The LDP is equipped with a filter, which is located on top of the fuel tank.

5   On 2003 and later California models, the EVAP system is equipped with a Natural Vacuum Leak Detection (NVLD) device instead of an LDP. The NVLD is capable of detecting any leak, from a large leak to one caused by a hole in the system as small as 0.020 inch (0.5 mm). To remove the NVLD device, refer to "Leak Detection Pump (LDP)" below.

REPLACEMENT

Proportional purge solenoid

▶ **Refer to illustration 19.6**

6   Raise the vehicle and place it securely on jackstands. The proportional purge solenoid (see illustration) is attached to a bracket which is bolted to the body, right below the power brake booster and to the left of the steering shaft.

7   Disconnect the solenoid electrical connector (see illustration 19.6) from the solenoid.

8   Clearly label the EVAP hoses to ensure correct reassembly and

19.2 The EVAP canister (and, on California models, the leak detection pump) is located behind the fuel tank

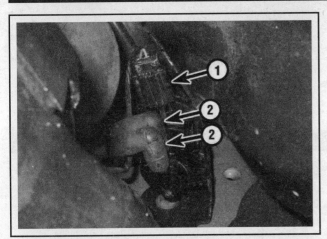

19.6 The proportional purge solenoid is located on a bracket which is mounted on the body below the power brake booster and to the left of the steering gear; to remove the proportional purge solenoid, disconnect the electrical connector (1), detach the two EVAP hoses (2) and push the solenoid toward the steering shaft to disengage it from the mounting bracket

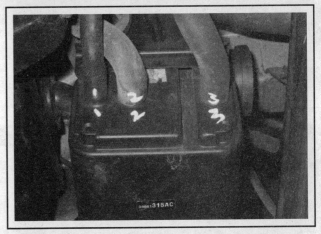

19.19 The EVAP canister is located under the vehicle, near the fuel tank; clearly label, then disconnect, the three vapor hoses

then detach the two EVAP hoses (see illustration 19.6) from the solenoid.

9   To detach the solenoid from its mounting bracket slide it off, in the direction of the steering shaft. If you have difficulty detaching the solenoid from its mounting bracket, remove the bracket mounting bolt, remove the bracket and solenoid as a single assembly and separate the solenoid from the bracket.

10 Installation is the reverse of removal.

Leak Detection Pump (LDP)

11 Raise the vehicle and place it securely on jackstands.

12 Disconnect the electrical connector from the LDP.

13 Loosen the rear stabilizer bar bracket (see Chapter 10) and remove the LDP pump and bracket as a single assembly.

14 Clearly label and disconnect the EVAP hoses from the LDP.

15 Remove the LDP filter.

16 Detach the LDP from its mounting bracket.

17 Installation is the reverse of removal.

EVAP canister

▶ **Refer to illustrations 19.19, 19.20a and 19.20b**

18 Raise the vehicle and place it securely on jackstands.

19 Clearly label the EVAP hoses (see illustration) and then disconnect them from the canister.

20 Remove the EVAP canister mounting nuts (see illustration) and detach the canister from the bracket (see illustration).

21 Installation is the reverse of removal.

EVAP canister breather (non-California models)

▶ **Refer to illustration 19.22**

22 The breather (see illustration) is located on top of the EVAP canister.

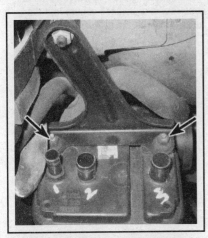

19.20a To detach the EVAP canister from its mounting bracket, remove these two nuts . . .

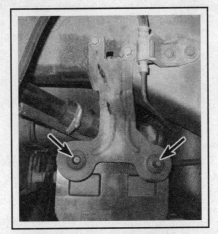

19.20b . . . and pull the canister locator pins out of their grommets in the mounting bracket; when installing the canister, make sure these two locator pins are fully seated in these grommets

19.22 To detach the breather from the EVAP canister/mounting bracket assembly, detach the hose (1) and remove the bolt (2) (bolt head faces toward the canister); when installing the breather, make sure the locator pin (3) is aligned with its slot in the mounting bracket

23 Detach the EVAP canister from its mounting bracket (see illustrations 19.20a and 19.20b). It's not necessary to actually remove the canister, just pull it down far enough so that you can remove the breather.

24 Detach the hose that connects the breather to the EVAP canister (see illustration 19.22).

25 Remove the single bolt that attaches the breather to the mounting bracket (see illustration 19.22).

26 Remove the breather.

27 When installing the breather make sure that the locator pin is correctly aligned with its slot in the mounting bracket (see illustration 19.22).

28 Installation is otherwise the reverse of removal.

20 Positive Crankcase Ventilation (PCV) system

▶ **Refer to illustrations 20.2a and 20.2b**

1   The Positive Crankcase Ventilation (PCV) system reduces hydrocarbon emissions by scavenging crankcase vapors. It does this by circulating fresh air from the air cleaner through the crankcase, where it mixes with blow-by gases, before being drawn through a PCV valve into the intake manifold.

2   The PCV system consists of the PCV valve and two hoses, one of which connects the air cleaner to the crankcase (see illustration) and the other (containing the PCV valve) which connects the crankcase to the air intake duct (see illustration).

3   To maintain idle quality, the PCV valve restricts the flow when the intake manifold vacuum is high. If abnormal operating conditions (such as piston ring problems) arise, the system is designed to allow excessive amounts of blow-by gases to flow back through the crankcase vent tube into the air cleaner to be consumed by normal combustion.

4   Checking and replacement of the PCV valve is covered in Chapter 1.

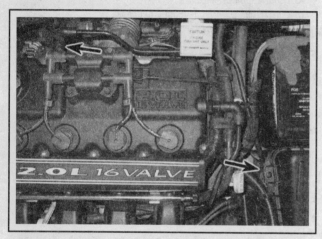

20.2a The Positive Crankcase Ventilation (PCV) system consists of the hose between the valve cover and the air filter housing, which draws clean air into the crankcase . . .

20.2b . . . the PCV valve itself, which is screwed into the valve cover, and the hose connecting the PCV valve to the air intake manifold; this hose routes crankcase vapors into the manifold so they can be burned with the air/fuel mixture during combustion

Specifications

Torque specifications	Ft-lbs (unless otherwise indicated)	Nm
Camshaft Position (CMP) sensor		
CMP sensor retaining bolts	80 in-lbs	9
CMP sensor target magnet retaining bolt	30 in-lbs	3.5
Catalytic converter-to-exhaust manifold flange bolts	250 in-lbs	28
Crankshaft Position (CKP) sensor retaining bolt		
2000 through 2002	90 in-lbs	10
2003	80 in-lbs	9
Engine Coolant Temperature (ECT) sensor	165 in-lbs	18.5
Exhaust pipe slip joint band clamps	35	47
Knock sensor		
Knock sensor (2000 through 2002)	84 in-lbs	9.5
Knock sensor retaining bolt (2003)	195 in-lbs	22
Oxygen sensors (upstream and downstream)	20	27
Park/Neutral position switch	24	32.5
Power steering pressure switch	70 in-lbs	8
Vehicle Speed Sensor (VSS) retaining bolt		
Manual transaxle	60 in-lbs	6.8
Automatic transaxle		
2000 and 2001	60 in-lbs	6.8
2002 and later (input shaft VSS and output shaft VSS)	20	27

Notes

Section

Reference to other Chapters

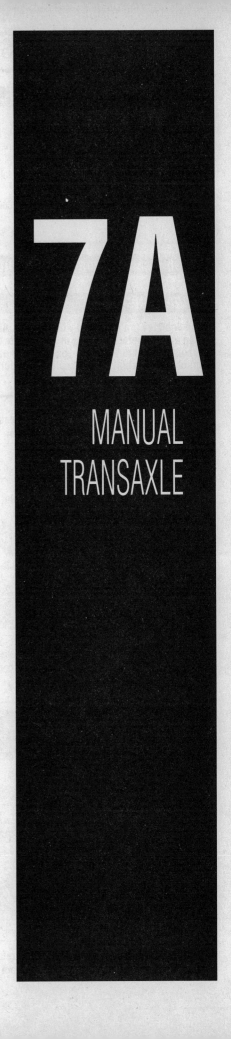

7A

MANUAL TRANSAXLE

1 General information

The vehicles covered in this manual are equipped with either a 5-speed manual, or a 3- or 4-speed automatic transaxle. Information on the manual transaxles are included in this Part of Chapter 7. Service procedures for the automatic transaxles are contained in Chapter 7, Part B.

The manual transaxle is a compact, two-piece, lightweight aluminum alloy housing containing both the transmission and differential assemblies.

Because of the complexity, unavailability of replacement parts and special tools necessary, internal repair procedures for the manual transaxle are beyond the scope of this manual. The bulk of information in this Chapter is devoted to removal and installation procedures.

2 Shift cables - removal, installation and adjustment

※※ WARNING:

These vehicles are equipped with airbags. Always disable the airbag system before working in the vicinity of any airbag system components to avoid the possibility of accidental deployment of the airbag(s), which could cause personal injury (see Chapter 12).

➡Note: The selector cable and crossover cable are available as an assembly and must be replaced together.

REMOVAL

▶ **Refer to illustrations 2.3a, 2.3b, 2.5, 2.8a, 2.8b, 2.9 and 2.11**

1   Remove the gearshift knob and lever boot (see Section 3).
2   Remove the center console (see Chapter 11).
3   Remove the clip securing the selector cable to the shift lever and slide the cable off of the lever (see illustrations).
4   Remove the clip securing the crossover cable to the shift lever and slide the cable off of the lever.
5   Remove the grommet plate-to-floorpan fasteners (see illustration).
6   Remove the air filter housing assembly (see Chapter 4).
7   Remove the accelerator cable and cruise control cable (if equipped) (see Chapter 4).

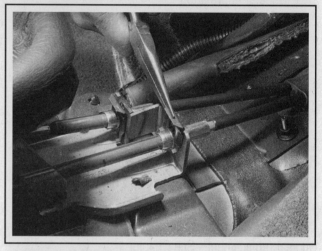

2.3a Use needle-nose pliers to remove both cable retaining clips at the shifter

8   Use two flat blade screwdrivers and carefully pry the crossover and selector cables from the shifter levers (see illustrations).
9   Remove both cable retaining clips and remove the cables from the bracket (see illustration).
10 Raise the vehicle and place it securely on jackstands.
11 Working under the vehicle, remove the catalytic converter heat shield (see illustration).

2.3b Use a flat-bladed screwdriver to carefully pry the selector cable and the crossover cable from the shifter

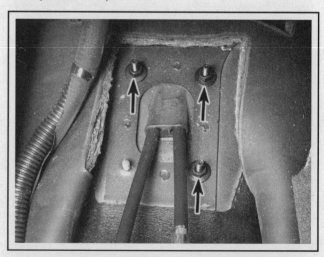

2.5 Remove the shift cable floor pan grommet mounting nuts

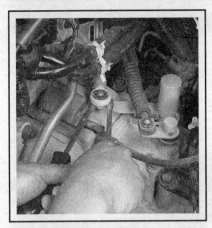

2.8a Use two-flat-bladed screwdrivers to carefully pry the selector cable from the select lever . . .

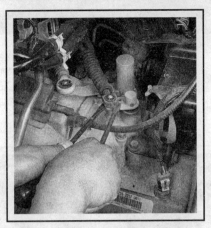

2.8b . . . and the crossover cable from the crossover lever

2.9 Using pliers, remove the cable bracket retaining clip and lift the cable from the bracket

12 Remove the last bolt and cable grommet from the floor pan and pull the cable assembly out from under vehicle.

INSTALLATION

13 Installation is the reverse of removal, but note the following points:

   a) Install new cable retaining clips.
   b) Adjust the crossover cable (see Steps 14 through 23).

ADJUSTMENT

➡**Note: Only the crossover cable is adjustable. The selector cable does not require adjustment.**

2000 models

▸ **Refer to illustrations 2.14 and 2.15**

14 Loosen the crossover cable adjustment screw at the shifter (see illustration).

15 Align the hole in the crossover lever with the hole in the transaxle raised boss. Insert a 1/4 inch drill bit through the lever and into the raised boss (see illustration). Make sure the drill bit goes into the transaxle hole at least 1/2 inch.

16 Place the shifter lever in the Neutral position. Allow the shift lever to self-center in this position.

17 Tighten the crossover cable adjustment screw, taking care not to let the shift lever move off-center.

18 Remove the 1/4 drill bit from the lever and the transaxle.

19 Verify proper shifting pattern. Readjust the cable if necessary.

2001 and later models

20 Loosen the crossover cable adjustment screw at the shifter (see illustration 2.14).

21 Allow the shift lever and crossover lever to relax in the neutral position.

22 Tighten the adjustment screw taking care not to let the shift lever move off-center.

23 Verify proper shifting pattern.

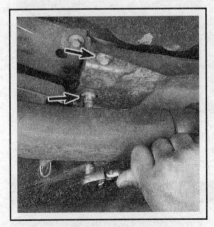

2.11 Working under the vehicle, remove the nuts from the catalytic converter heat shield

2.14 Loosen the crossover cable adjustment screw

2.15 Align the hole in the crossover lever with the hole in the transaxle raised boss. Insert a 1/4 inch drill bit through the lever and into the raised boss at least 1/2 inch

3 Shift lever - removal and installation

▶ **Refer to illustrations 3.2 and 3.5**

1 Push down on the gearshift knob, rotate it clockwise and remove it from the lever.

2 Remove the shifter lever boot from the center console (see illustration).

3 Remove the center console (see Chapter 11).

4 Disconnect the crossover and selector cable from the shifter assembly (see Section 2).

5 Unbolt the shift lever bracket bolts (see illustration) and remove the shift lever.

6 Installation is the reverse of removal.

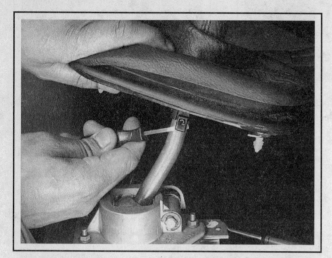

3.2 Use a screwdriver to release the clamp and remove the shifter boot

3.5 Remove the four shift lever mounting nuts (only two are visible) then remove the shift lever assembly

4 Back-up light switch - replacement

▶ **Refer to illustration 4.1**

1 The back-up light switch is located on top of the transaxle (see illustration).

2 Disconnect the electrical connector from the back-up light switch.

3 Unscrew the switch from the case.

4 Wrap the threads of the new switch with Teflon tape, or equivalent.

5 Screw in the new switch and tighten it securely.

6 Connect the electrical connector.

7 Check the operation of the back-up lights.

4.1 Back-up light switch location - T350 transaxle shown, other models similar

5 Transaxle mounts - check and replacement

1 Insert a large screwdriver or prybar between the mount and the transaxle and pry up.

2 The transaxle should not move excessively away from the mount. If it does, replace the mount.

3 To replace a mount, support the transaxle with a jack, remove the nuts and bolts and remove the mount. It may be necessary to raise the transaxle slightly to provide enough clearance to remove the mount.

4 Installation is the reverse of removal.

6   Manual transaxle - removal and installation

REMOVAL

▶ **Refer to illustrations 6.8, 6.14, 6.17, 6.19 and 6.23**

1   Open the hood and place protective covers on the front fenders and cowl. Special fender covers are available, but an old bedspread or blankets will also work.

2   Disconnect both cables from the battery terminals (see Chapter 5).

✻✻✻ CAUTION:

Always disconnect the negative cable first and hook it up last or the battery may be shorted by the tool being used to loosen the cable clamps.

3   Remove the air filter housing assembly (see Chapter 4).

4   Remove the battery and the battery tray (see Chapter 5). Disconnect the negative battery cable ground stud from the battery tray.

5   Disconnect the shift cables from the transaxle (see Section 2), then disconnect the harness connectors from the vehicle speed sensor and back-up light switch.

6   On 2000 models, disconnect the clutch cable from the transaxle (see Chapter 8).

7   On 2001 and later models, disconnect the clutch release cylinder from the manual transaxle (see Chapter 8). Do not disconnect the hydraulic line from the cylinder, but position the assembly off to the side.

8   Remove the transaxle-to-engine upper bolts (see illustration).

9   Loosen the driveaxle hub nuts (see Chapter 8) and front wheel lug nuts.

10   Raise the vehicle and place it securely on jackstands. Remove both front wheels.

11   Support the engine from above with a hoist, or place a floor jack under the oil pan. Place a wood block on the jack head to spread the load on the oil pan.

12   Drain the transaxle fluid (see Chapter 1).

13   Remove the driveaxles (see Chapter 8).

14   Remove the structural collar (see illustration).

6.8 Remove the bolts that secure the upper transaxle to the engine

15   Remove the bolts securing the intake manifold support bracket (see Chapter 2A).

16   Remove right, lateral bending brace and the starter (see Chapter 5).

17   Remove the lower inspection cover (see illustration).

18   If the modular clutch assembly is to be reinstalled, match-mark the clutch assembly to the driveplate at the slotted tolerance hole, and remove the four modular clutch assembly-to-driveplate bolts. To gain access to each bolt, rotate the engine from the drivebelt end of the engine using the crankshaft damper/pulley bolt. Remove all four bolts and discard them. Use a screwdriver placed in the ring gear of the driveplate to keep the crankshaft from turning during removal of the bolts.

19   Support the transaxle with a transmission jack (see illustration), if available, or use a floor jack. Secure the transaxle to the jack using straps or chains so it doesn't fall off during removal.

20   Remove the upper transaxle mount-to-bracket bolts.

21   Remove the transaxle upper mount bracket.

22   Remove the lower transaxle-to-engine bolts. Make sure all the engine bolts are removed.

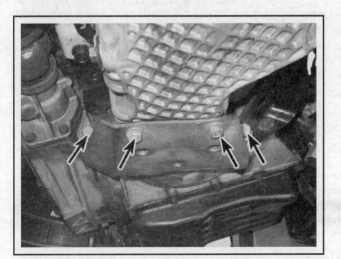

6.14 Remove the bolts from the structural collar

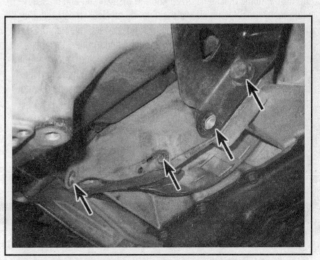

6.17 Remove the bolts that secure the inspection cover to the engine and slide the inspection cover away from the flywheel housing

6.19 Install a transaxle jack under the transaxle

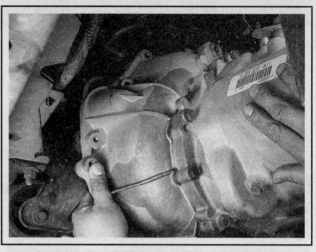

6.23 Carefully separate the transaxle from the engine

23 Make a final check that all wires, hoses and brackets have been disconnected from the transaxle, then with the engine properly supported, carefully lower the transaxle and remove it from under the vehicle (see illustration). Make sure you keep the transaxle level as you maneuver it or the modular clutch assembly may fall out.

➡**Note: If necessary have someone help with the removal procedure.**

INSTALLATION

24 With the transaxle secured to the hoist/floor jack as on removal, raise it into position, and then carefully slide it onto the engine.

25 Once the transaxle is successfully mated to the engine, insert as many of the flange bolts as possible, and tighten them progressively, to draw the transaxle fully onto the locating dowels.

26 Install the transaxle upper mount bracket.

27 Install the remaining transaxle-to-engine bolts, and tighten all of them to the specified torque.

28 Install the driveplate-to-clutch module bolts.

29 Once the transaxle mount has been replaced, the engine hoist or supporting jack can be removed.

30 The remainder of installation is a reversal of removal, noting the following points:

a) Install the structural collar and intake manifold support bracket as described in Chapter 2A.

b) Install the starter motor as described in Chapter 5.

c) Install the driveaxles as described in Chapter 8.

d) Fill the transaxle with lubricant (see Chapter 1).

e) Road test the vehicle and check for proper transaxle operation and check for fluid leaks.

7   Manual transaxle overhaul - general information

1 Overhauling a manual transaxle unit is a difficult and involved job for the home mechanic. In addition to dismantling and reassembling many small parts, clearances must be precisely measured and, if necessary, changed by selecting shims and spacers. Internal transaxle components are also often difficult to obtain and in many instances, extremely expensive. Because of this, if the transaxle develops a fault or becomes noisy, the best course of action is to have the unit overhauled by a transmission specialist or to obtain an exchange reconditioned unit.

2 Nevertheless, it is not impossible for the more experienced mechanic to overhaul the transaxle if the special tools are available and the job is carried out in a deliberate step-by-step manner, to ensure that nothing is overlooked.

3 The tools necessary for an overhaul include internal and external snap-ring pliers, bearing pullers, a slide hammer, a set of pin punches, a dial test indicator and possibly a hydraulic press. In addition, a large, sturdy workbench and a vise will be required.

4 During dismantling of the transaxle, make careful notes of how each component is installed to make reassembly easier and accurate.

5 Before disassembling the transaxle, it will help if you have some idea of where the problem lies. Certain problems can be closely related to specific areas in the transaxle which can make component examination and replacement easier. Refer to the *Troubleshooting* Section in this manual for more information.

Torque specifications

	Ft-lbs (unless otherwise indicated)	Nm
Transaxle-to-engine bolts	70	95
Drain plug	See Chapter 1	

Section

Reference to other Chapters

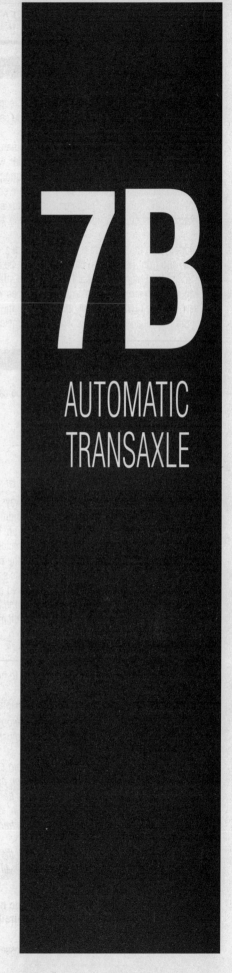

7B

AUTOMATIC
TRANSAXLE

1   General information

All information on the automatic transaxle is included in this Part of Chapter 7. Information for the manual transaxle can be found in Part A of this Chapter.

These models are equipped with several different automatic transaxles. 2000 and 2001 models are equipped with the 31TH automatic transaxle. 2002 and 2003 models are equipped with the 41TE automatic transaxle. 2004 and 2005 models are equipped with the 40TE automatic transaxle. The automatic transaxle and the differential are housed in a compact, lightweight, two-piece aluminum alloy housing.

Models with the 31TH transaxle are equipped with rear (low/-reverse) and front (kickdown) bands that must be adjusted on a periodic schedule. Refer to Chapter 1 for the band adjustment procedure.

Models with the 41TE transaxle are equipped with a Transmission Control Module (TCM) which is the "brain" of the transaxle. The TCM monitors engine and transaxle operating parameters through numerous sensors and then generates output signals to various relays and solenoids to regulate hydraulic pressures, optimize driveability, provide efficient torque management and maintain maximum fuel economy. The TCM is a separate component on 2002 models and is mounted behind the right, front fender splash shield. 2003 and later models incorporate the TCM into the PCM. The TCM is part of the On-Board Diagnostic system OBD-II. For more information, see Chapter 6.

Because of the complexity of the automatic transaxles and the specialized equipment necessary to perform most service operations, this Chapter contains only those procedures related to general diagnosis, adjustment and removal and installation procedures.

If the transaxle requires major repair work, it should be left to a dealer service department or an automotive or transmission repair shop. Once properly diagnosed you can, however, remove and install the transaxle yourself and save the expense, even if the repair work is done by a transmission shop.

2   Diagnosis - general

1   Automatic transaxle malfunctions may be caused by five general conditions:

a) *Poor engine performance*
b) *Improper adjustments*
c) *Hydraulic malfunctions*
d) *Mechanical malfunctions*
e) *Malfunctions in the computer or its signal network*

2   Diagnosis of these problems should always begin with a check of the easily repaired items: fluid level and condition (see Chapter 1), shift cable adjustment and shift lever installation. Next, perform a road test to determine if the problem has been corrected or if more diagnosis is necessary. If the problem persists after the preliminary tests and corrections are completed, additional diagnosis should be performed by a dealer service department or other qualified transmission repair shop. Refer to the *Troubleshooting* Section at the front of this manual for information on symptoms of transaxle problems.

PRELIMINARY CHECKS

3   Drive the vehicle to warm the transaxle to normal operating temperature.

4   Check the fluid level as described in Chapter 1:

a) *If the fluid level is unusually low, add enough fluid to bring the level within the designated area of the dipstick, then check for external leaks (see following).*
b) *If the fluid level is abnormally high, drain off the excess, then check the drained fluid for contamination by coolant. The presence of engine coolant in the automatic transmission fluid indicates that a failure has occurred in the internal radiator oil cooler walls that separate the coolant from the transmission fluid (see Chapter 3).*
c) *If the fluid is foaming, drain it and refill the transaxle, then check for coolant in the fluid, or a high fluid level.*

5   Check the engine idle speed.

➡**Note: If the engine is malfunctioning, do not proceed with the preliminary checks until it has been repaired and runs normally.**

6   Check and adjust the shift cable, if necessary (see Section 4).

7   If hard shifting is experienced, inspect the shift cable under the center console and at the manual lever on the transaxle (see Section 4).

FLUID LEAK DIAGNOSIS

8   Most fluid leaks are easy to locate visually. Repair usually consists of replacing a seal or gasket. If a leak is difficult to find, the following procedure may help.

9   Identify the fluid. Make sure it's transmission fluid and not engine oil or brake fluid (automatic transmission fluid is a deep red color).

10   Try to pinpoint the source of the leak. Drive the vehicle several miles, then park it over a large sheet of cardboard. After a minute or two, you should be able to locate the leak by determining the source of the fluid dripping onto the cardboard.

11   Make a careful visual inspection of the suspected component and the area immediately around it. Pay particular attention to gasket mating surfaces. A mirror is often helpful for finding leaks in areas that are hard to see.

12   If the leak still cannot be found, clean the suspected area thoroughly with a degreaser or solvent, then dry it thoroughly.

13   Drive the vehicle for several miles at normal operating temperature and varying speeds. After driving the vehicle, visually inspect the suspected component again.

14   Once the leak has been located, the cause must be determined before it can be properly repaired. If a gasket is replaced but the sealing flange is bent, the new gasket will not stop the leak. The bent flange must be straightened.

15   Before attempting to repair a leak, check to make sure that the following conditions are corrected or they may cause another leak.

➡**Note: Some of the following conditions cannot be fixed without highly specialized tools and expertise. Such problems must be referred to a qualified transmission shop or a dealer service department.**

Gasket leaks

16   Check the pan periodically. Make sure the bolts are tight, no bolts are missing, the gasket is in good condition and the pan is flat (dents in the pan may indicate damage to the valve body inside).

17 If the pan gasket is leaking, the fluid level or the fluid pressure may be too high, the vent may be plugged, the pan bolts may be too tight, the pan sealing flange may be warped, the sealing surface of the transaxle housing may be damaged, the gasket may be damaged or the transaxle casting may be cracked or porous. If sealant instead of gasket material has been used to form a seal between the pan and the transaxle housing, it may be the wrong type of sealant.

Seal leaks

18 If a transaxle seal is leaking, the fluid level may be too high, the vent may be plugged, the seal bore may be damaged, the seal itself may be damaged or improperly installed, the surface of the shaft protruding through the seal may be damaged or a loose bearing may be causing excessive shaft movement.

19 Make sure the dipstick tube seal is in good condition and the tube is properly seated. Periodically check the area around the sensors for leakage. If transmission fluid is evident, check the seals for damage.

Case leaks

20 If the case itself appears to be leaking, the casting is porous and will have to be repaired or replaced.

21 Make sure the oil cooler hose fittings are tight and in good condition.

Fluid comes out vent pipe or fill tube

22 If this condition occurs, the possible causes are: the transaxle is overfilled; there is coolant in the fluid; the dipstick is incorrect; the vent is plugged or the drain-back holes are plugged.

3 Driveaxle oil seals - replacement

▶ Refer to illustration 3.3 and 3.5

1 The driveaxle oil seals are located on the sides of the transaxle, where the inner ends of the driveaxles are splined into the differential side gears. If you suspect that a driveaxle oil seal is leaking, raise the vehicle and support it securely on jackstands. If the seal is leaking, you'll see lubricant on the side of the transaxle, below the seal.

2 Remove the driveaxle (see Chapter 8).

3 Using a screwdriver or prybar, carefully pry the oil seal out of the transaxle bore (see illustration).

➡Note: Driveaxle oil seals on the right side of the transaxle may require a slide hammer equipped with a hook-type tool for removal.

4 If the oil seal cannot be removed with a screwdriver or prybar, a special oil seal removal tool (available at auto parts stores) will be required.

5 Using a seal installer, install the new oil seal. Drive it into the bore squarely until it bottoms (see illustration).

6 Install the driveaxle (see Chapter 8).

3.3 Using a large screwdriver or prybar, carefully pry the oil seal out of the transaxle (if you can't remove the oil seal with a screwdriver or prybar, you may need to obtain a special seal removal tool - available at most auto parts stores - to do the job)

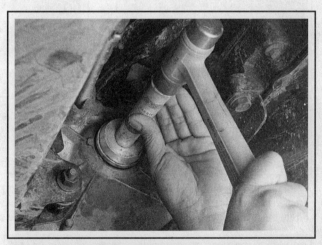

3.5 Using a seal installer, large section of pipe or a large deep socket as a drift, drive the new seal squarely into the bore and make sure that it's completely seated; lubricate the lip of the new seal with multi-purpose grease

4 Shift cable - removal, installation and adjustment

✷ WARNING:

These vehicles are equipped with airbags. Always disconnect the negative battery cable and wait two minutes before working in the vicinity of any airbag system components to avoid the possibility of accidental deployment of the airbag(s), which could cause personal injury (see Chapter 12).

✷ CAUTION:

Do not attempt this procedure until the vehicle has cooled completely. The exhaust system components must be cold to avoid physical harm.

REMOVAL

▶ **Refer to illustration 4.5, 4.7, 4.9, 4.10 and 4.11**

1   In the event of hard shifting, disconnect the cable at the transaxle and operate the shifter from the driver's seat. If the shift lever moves smoothly through all positions with the cable disconnected, then the cable should be adjusted as described at the end of this Section. To remove and install the shift cable, perform the following.

2   Raise the hood and place a blanket over the left (driver's) fender to protect it.

3   Remove the air filter housing (see Chapter 4).

4   Remove the battery and battery tray (see Chapter 5).

5   Working in the engine compartment, disconnect the shift cable from the shift lever (see illustration).

6   Working in the engine compartment, disconnect the shift cable from the bracket.

7   Working inside the vehicle remove the shift knob set screw and remove the knob from the shifter (see illustration).

8   Remove the center console (see Chapter 11).

9   Remove gearshift indicator lamp from the shift trim bezel (see illustration).

10 Remove the shift trim bezel screws and the bezel (see illustration).

11 Use a flat-blade screwdriver and carefully pry the shift cable end from the shift lever mounting stud (see illustration).

12 Carefully pull the cable conduit end out from the shift housing. Pull up and remove the cable.

13 Remove the grommet plate-to-floor-pan fasteners.

14 Raise the vehicle and support it securely on jackstands.

15 Working under the vehicle, remove the catalytic converter heat shield and the intermediate pipe heat shield (see Chapter 6).

16 Working in the passenger compartment, remove the last remaining grommet plate fastener and remove the cable from the vehicle.

INSTALLATION

17 Working in the engine compartment, install the shift cable into the bracket.

18 Connect the shift cable onto the shift lever post. Make sure it snaps into place.

19 Verify the shift lever is in the park position.

20 Raise the vehicle and support it securely on jackstands.

21 Route the shifter cable through to the floor pan and up to the shifter.

22 Install the grommet plate-to-floorpan fasteners and lower the vehicle.

23 Connect the shift cable to the shift lever pin.

24 The remainder of installation is the reverse of removal.

25 Adjust the shift cable as described in the following steps.

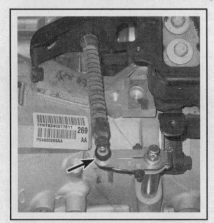

4.5 Use two flat blade screwdrivers to carefully pry the selector cable from the select lever

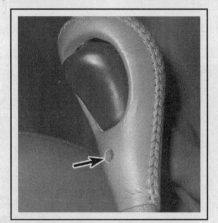

4.7 To remove the handle from the shift lever, loosen the set screw and pull the handle straight up

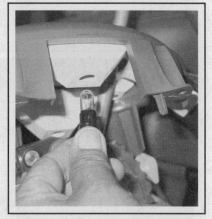

4.9 Remove the gearshift indicator lamp from the shift trim bezel

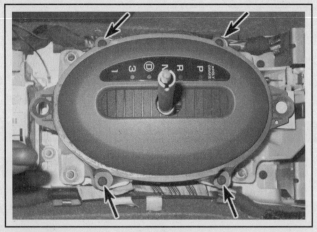

4.10 Remove all four of the shift lever bezel mounting screws

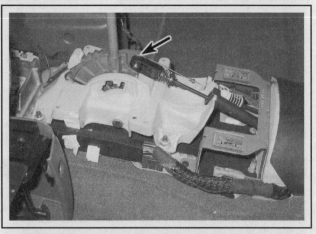

4.11 Carefully pry the shift cable from the shift lever mounting stud

ADJUSTMENT

▶ **Refer to illustration 4.29**

26 Working inside the vehicle remove the shift knob set screw and remove the knob from the shifter.

27 Remove the center console (see Chapter 11).

28 Place the shift lever in the Park position.

29 Loosen the gearshift cable adjustment screw at the shifter cable (see illustration).

30 Make sure the shift lever at the transaxle is in the Park position. The park sprag must be engaged when adjusting the cable. Rock the vehicle back and forth to ensure that the park sprag is fully engaged.

31 Tighten the shift cable adjustment screw.

32 Install the center console (see Chapter 11).

33 Check the shift lever for proper operation. It should operate smoothly without binding. The engine should start only in the Park or Neutral positions.

34 Shift the transaxle into all gear positions to make sure the cable is functioning properly. Readjust if necessary.

4.29 Gearshift cable adjustment screw

5   Shift interlock cable - removal, installation and adjustment

REMOVAL

▶ **Refer to illustrations 5.5, 5.8, 5.10 and 5.11**

1 The interlock cable slides into the housing behind the lock cylinder and attaches to the shift lever base. The floor-shift interlock system is adjusted by a nut at the shift lever assembly. If the system must be adjusted (but not replaced), adjust it as described in Steps 22 through 25.

2 Disconnect the cable from the negative battery terminal (see Chapter 5).

3 Remove the shift lever handle (see Section 4).

4 Remove the console bezel (see Chapter 11).

5 Detach the interlock cable from the shift lever base (see illustration).

6 Remove the driver's side knee bolster and reinforcement (see Chapter 11).

7 Remove the steering column covers (see Chapter 11).

8 Disconnect the Brake/Transmission Shift Interlock solenoid from the interlock cable connector (see illustration).

9 Turn the ignition key to the Run position.

10 Slide the cable end plug out of its groove in the interlock lever (see illustration).

11 Turn the ignition key to the Off or On/Run position. At the ignition key lock cylinder, squeeze the lock tab on the interlock cable and pull the cable out of the lock cylinder housing (see illustration).

12 Disconnect the interlock cable from the retaining clips and remove the cable.

INSTALLATION

13 Route the interlock cable through the instrument panel, below the steering column and down to the shift lever assembly.

14 Make sure the ignition switch is in the Off or On/Run position.

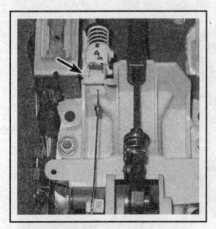

5.5 To detach the interlock cable, pry the cable up from the shifter base

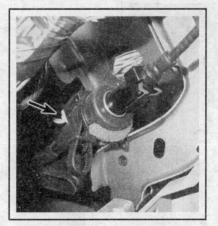

5.8 Disconnect the brake/transaxle shift interlock solenoid connector from the interlock cable

5.10 Slide the interlock cable end from the lever

15 Insert the forward end of the interlock cable into the lock cylinder housing and push it in until it snaps into place.

16 Turn the ignition key to the Off/Lock position.

17 Connect the Brake/Transmission Shift Interlock solenoid to the interlock cable connector.

18 Insert the cable end plug into its groove in the interlock lever. Make sure the plug is properly seated in the groove.

19 Reattach the cable to the shift lever base. The cable housing is fully seated when it snaps into place.

20 Adjust the interlock cable (see Step 22).

21 The remainder of installation is the reverse of removal.

ADJUSTMENT

22 Remove the ignition key from the lock cylinder with the switch in the Lock position.

23 When the adjustment clip on the interlock cable is disengaged, the cable automatically indexes itself to the correct position. Pry up on the adjuster lock and allow the cable to index itself. Snap the clip back into place.

➡Note: If you're installing a new cable, the cable will come with an adjustment pin; once the pin is removed, the cable will self-adjust. After that, snap the adjustment clip into place (and don't reinstall the pin).

24 With the ignition key in the Off (locked) position, the shift lever should be locked in the Park position. If it isn't, inspect the system for

5.11 Squeeze the tab on the forward end of the interlock cable, then pull the cable housing straight out of the lock cylinder

binding and repeat the adjustment procedure.

25 Without starting the engine, place the ignition switch in the Run position. Move the shift lever to the Reverse position. You should be unable to remove the ignition key from the lock cylinder. If you can remove the key at this point, inspect the system for binding and repeat the adjustment procedure.

6   Throttle valve cable (2000 and 2001 models) - adjustment and replacement

ADJUSTMENT

▶ **Refer to illustration 6.1**

1 Working in the engine compartment, release the adjustment locking clip on the throttle valve cable at the throttle body (see illustration).

2 Grip the throttle valve cable and push in toward the throttle body. This will allow the throttle valve cable to self-adjust.

3 Press the locking clip to secure the position of the throttle valve cable.

4 Road test the vehicle and test for proper kickdown actuation.

REPLACEMENT

▶ **Refer to illustration 6.7**

5 Raise the hood and place a blanket over the left (driver's) fender to protect it.

6 Remove the air filter housing (see Chapter 4).

7 Working in the engine compartment, disconnect the throttle valve cable from the throttle body assembly (see illustration).

8 Disconnect the throttle valve cable from the lever on the transaxle.

9 Remove the throttle valve cable.

10 Installation is the reverse of removal.

11 Adjust the throttle valve cable (see Steps 1 through 4).

6.1 Release the locking clip by lifting the tabs and sliding the clip horizontally

6.7 Pry the lock mechanism out (arrow) and slide the throttle valve cable horizontally from the bracket

7 Transmission Control Module (TCM) (2002 models) - removal and installation

▶ Refer to illustration 7.5

✳ CAUTION:

The TCM is an Electro-Static Discharge (ESD) sensitive electronic device, meaning a static electricity discharge from your body could possibly damage electrical components. Be sure to properly ground yourself and the TCM before handling it. Avoid touching the electrical terminals of the TCM.

➡ Note: After replacing a TCM, take the vehicle to your local dealer service department or other qualified transmission shop to have the TCM calibrated for your vehicle.

1   Disconnect the cable from the negative battery terminal (see Chapter 5).
2   Loosen the right front wheel lug nuts, then raise the vehicle and support it securely on jackstands.
3   Remove the right front wheel.
4   Remove the fenderwell splash shield.
5   Detach the electrical connector from the TCM (see illustration).
6   Remove the TCM bracket mounting screws and withdraw the

7.5 Unscrew the bolt securing the connector to the module

TCM from the vehicle.
7   Installation is the reverse of removal.

8 Automatic transaxle - removal and installation

▶ Refer to illustration 8.15

REMOVAL

1   Remove the battery (see Chapter 5) and the battery tray.

✳ CAUTION:

Always disconnect the negative cable first and hook it up last or the battery may be shorted by the tool being used to loosen the cable clamps. Loosen the front wheel lug nuts and the driveaxle/hub nut (see Chapter 8).

➡ Note: Depending on the type of wheels installed on the vehicle and the thickness of the socket you are using, you may have to loosen the driveaxle/hub nuts after the wheels have been removed.

2   Remove the air filter housing and air intake duct assembly (see Chapter 4).
3   Remove the transaxle oil cooler lines and plug them to prevent fluid from spilling.

➡ Note: It will be necessary to replace the transaxle lines once they are removed from the radiator. The inner wall of the transaxle lines will become damaged and prone to leaking. Consult with a dealer parts department for information.

4   Clearly label, then unplug, all electrical connectors.

➡ Note: On 2002 and later models, these electrical connectors will include the input speed sensor, the output speed sensor, the solenoid/pressure switch as well as the torque converter controls.

5   Disconnect the shift cable from the manual lever and bracket (see Section 4).
6   On 2000 and 2001 models, remove the throttle valve cable from

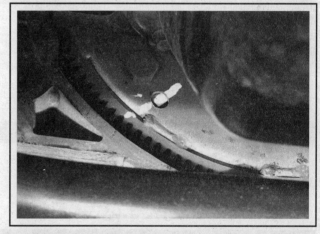

8.15 Mark the relationship of the torque converter to the driveplate

the bracket and separate the assembly from the transaxle (see Section 6).
7   Raise the vehicle and support it securely on jackstands. Remove the wheels.
8   Drain the transaxle fluid (see Chapter 1).
9   Remove both driveaxles (see Chapter 8).
10 Remove the bolts securing the power steering cooler to the crossmember (see Chapter 10).
11 Remove the intake manifold support bracket and structural collar (see Chapter 2A).
12 Remove the starter motor (see Chapter 5).
13 Support the engine from above with a hoist or place a jack and a block of wood under the oil pan to spread the load.
14 Remove the torque converter cover.
15 Mark the relationship of the torque converter to the driveplate so they can be installed in the same position (see illustration).

16 Remove the torque converter-to-driveplate bolts. Turn the crankshaft 120-degrees at a time for access to each bolt. After all three bolts are removed, push the torque converter into the bellhousing so it doesn't stay with the engine when the transaxle is removed.

17 Support the transaxle with a transmission jack, if available, or use a floor jack. Secure the transaxle to the jack using straps or chains so it doesn't fall off during removal.

18 Remove the transaxle upper mount-to-bracket bolts (see Chapter 2A) Remove the upper transaxle-to-engine bolts.

19 Make a final check that all wires and hoses have been disconnected from the transaxle, then move the transaxle jack toward the side of the vehicle until the transaxle is clear of the engine locating dowels. Make sure you keep the transaxle level as you do this.

INSTALLATION

20 Installation of the transaxle is a reversal of the removal procedure, but note the following points:

 a) As the torque converter is reinstalled, ensure that the drive tangs at the center of the torque converter hub engage with the recesses in the automatic transaxle fluid pump inner gear. This can be confirmed by turning the torque converter while pushing it towards the transaxle. If it isn't fully engaged, it will "clunk" into place.

 b) When installing the transaxle, make sure the matchmarks you made on the torque converter and driveplate line up.

 c) Install all of the driveplate-to-torque converter bolts before tightening any of them.

 d) Tighten the driveplate-to-torque converter bolts to the torque listed in this Chapter's Specifications.

 e) Tighten the transaxle mounting bolts to the torque listed in this Chapter's Specifications.

 f) Tighten the driveaxle/hub nuts to the torque value listed in the Chapter 8 Specifications.

 g) Tighten the wheel lug nuts to the torque listed in the Chapter 1 Specifications.

 h) Fill the transaxle with the correct type and amount of automatic transmission fluid as described in Chapter 1.

 i) On completion, adjust the shift cable (see Section 4).

9 Automatic transaxle overhaul - general information

In the event of a problem occurring, it will be necessary to establish whether the fault is electrical, mechanical or hydraulic in nature, before repair work can be contemplated. Diagnosis requires detailed knowledge of the transaxle's operation and construction, as well as access to specialized test equipment, and so is deemed to be beyond the scope of this manual. It is therefore essential that problems with the automatic transaxle are referred to a dealer service department or other qualified repair facility for assessment.

Note that a faulty transaxle should not be removed before the vehicle has been diagnosed by a knowledgeable technician equipped with the proper tools, as troubleshooting must be performed with the transaxle installed in the vehicle.

Specifications

General

Lubricant type and capacity	See Chapter 1

Torque specifications	Ft-lbs (unless otherwise indicated)	Nm
Torque converter-to-driveplate bolts	65	88
Transaxle-to-engine bolts	80	108
Structural collar	See Chapter 2A	

8

CLUTCH AND DRIVEAXLES

1 General information

The information in this Chapter deals with the components from the rear of the engine to the front wheels, except for the transaxle, which is dealt with in Chapter 7A and 7B. For the purposes of this Chapter, these components are grouped into two categories: clutch and driveaxles. Separate Sections within this Chapter offer general descriptions and checking procedures for both groups.

Since nearly all the procedures covered in this Chapter involve working under the vehicle, make sure it's securely supported on sturdy jackstands or a hoist where the vehicle can be easily raised and lowered.

2 Clutch - description and check

1   All vehicles with a manual transaxle use a single dry plate, diaphragm spring type clutch. The clutch disc has a splined hub which allows it to slide along the splines of the transaxle input shaft. The clutch and pressure plate are held in contact by spring pressure exerted by the diaphragm in the pressure plate. All models use a modular clutch assembly that includes a clutch assembly where the clutch pressure plate and friction disc are an integral unit.

2   The clutch release system is operated using a cable (2000 models) or hydraulic pressure (2001 and later models). The hydraulic system consists of the clutch pedal, a master cylinder, the hydraulic line, a release-cylinder which actuates the clutch release lever and the clutch release (or throw-out) bearing. 2001 and 2002 models are equipped with quick-release type hydraulic line fittings while 2003 and later models are equipped with pin-insert hydraulic line fittings.

3   When pressure is applied to the clutch pedal to release the clutch, hydraulic pressure is exerted against the outer end of the release lever, which pivots, moving the release bearing. The bearing pushes against the fingers of the diaphragm spring of the pressure plate assembly, which in turn releases the clutch plate.

4   Other than replacing components that have obvious damage, some preliminary checks should be performed to diagnose a clutch system failure:

a) *The first check should be of the clutch release system. On 2000 models, check the condition of the clutch cable. If the cable is stretched or damaged, replace it. On 2001 and later models, check the fluid level in the clutch master cylinder. If the fluid level is low, add fluid as necessary and inspect the hydraulic clutch system for leaks. If the master cylinder reservoir has run dry, bleed the system as described in Section 6 and re-test the clutch operation.*

b) *To check "clutch spin down time," run the engine at normal idle speed with the transaxle in Neutral (clutch pedal up - engaged). Disengage the clutch (pedal down), wait several seconds and shift the transaxle into Reverse. No grinding noise should be heard. A grinding noise would most likely indicate a problem in the pressure plate or the clutch disc.*

c) *To check for complete clutch release, run the engine (with the parking brake applied to prevent movement) and hold the clutch pedal approximately 1/2-inch from the floor. Shift the transaxle between 1st gear and Reverse several times. If the shift is not smooth, component failure is indicated.*

d) *Visually inspect the clutch pedal bushings at the top of the clutch pedal to make sure there is no sticking or excessive wear.*

3 Clutch cable (2000 models) - removal and installation

REMOVAL

▶ **Refer to illustrations 3.3, 3.4, 3.5a, 3.5b and 3.5c**

1   Raise the hood and place a blanket over the left (driver's) fender to protect it.

2   Remove the battery and the battery tray (see Chapter 5).

3   Remove the clutch cable inspection cover from the bellhousing (see illustration).

4   Pull back on the clutch cable housing and disengage it from the

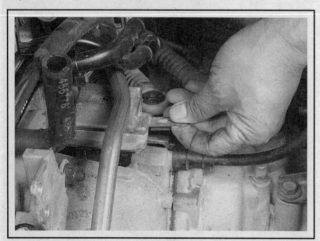

3.3 Pull off the rubber clutch inspection cover

3.4 Grab onto the cable housing and pull the clutch cable from the groove in the bellhousing, then disconnect it from the release lever

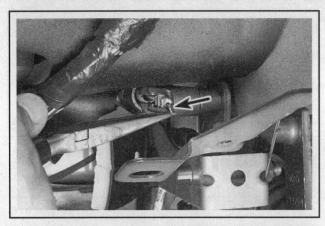

3.5a Working in under the driver's dash, slightly depress the clutch pedal to allow access to the clutch cable then remove the clip securing the spacer to the clutch pedal pivot ball

groove in the bellhousing, then disconnect it from the release lever (see illustration).

5   At the clutch pedal, slightly depress the clutch pedal to allow access to the clutch cable. Remove the clip securing the spacer to the clutch pedal pivot ball (see illustration). Wedge a narrow flat-blade screwdriver between the clutch pedal pivot ball and the spacer, then pull the spacer from the pivot ball (see illustration). Remove the spacer from the clutch cable end (see illustration).

✳✳ CAUTION:

Do not pull on the clutch cable while removing it from the dash panel as the cable self-adjuster may be damaged.

Within the engine compartment, hold onto the grommet and, using a slight twisting motion, carefully remove the clutch cable grommet from the firewall and clutch bracket. If necessary, carefully use a screwdriver to free the grommet from the firewall opening. Remove the clutch cable assembly.

INSTALLATION

▶ **Refer to illustration 3.6**

6   Within the engine compartment, using a slight twisting motion. insert the self-adjusting end of the clutch cable through the grommet in the firewall and the clutch bracket (see illustration).

7   In the passenger compartment, seat the cylindrical part of the grommet into the firewall opening and clutch bracket. Make sure the self-adjuster is firmly seated against the clutch bracket to ensure the adjuster will function properly.

8   Connect the clutch cable end onto the spacer and install the spacer onto the clutch pedal pivot ball. Install the clip and make sure it is properly seated.

9   Within the engine compartment, using slight pressure, pull the clutch cable end to draw the cable tight. Push the cable housing toward the firewall with less than 25-lbs of pressure. The cable housing should move about 1-inch - this indicates proper adjustment. If it does not adjust, determine if the mechanism is properly seated on the bracket.

10 Guide the cable through the slot in the bellhousing and connect it to the release lever seating the cupped washer securely in the release lever tangs.

11 Pull back on the clutch cable housing and insert it into the bell-housing.

12 Install the clutch cable inspection cover onto the bellhousing.

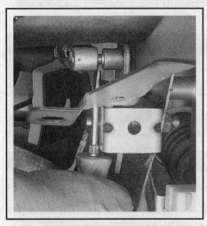

3.5b Wedge a flat-bladed screwdriver between the clutch pedal pivot ball and the spacer, then pull the spacer from the pivot ball

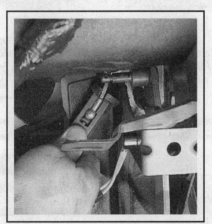

3.5c Remove the spacer from the clutch cable end

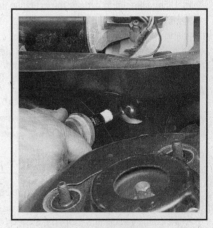

3.6 Working in the engine compartment, free the grommet from the firewall, then pull on the grommet and carefully remove the clutch cable and grommet from the firewall

4   Clutch master cylinder (2001 and later models) - removal and installation

➡**Note: On 2003 and later models equipped with the T350 transaxle, in the event the clutch master cylinder must be replaced, it will be necessary to replace the master cylinder, hydraulic fluid line and release cylinder as one complete assembly. If the release cylinder must be replaced, simply install a new release cylinder (see Section 5).**

REMOVAL

▶ **Refer to illustration 4.8**

1   Disconnect the cable from the negative battery terminal (see Chapter 5).

2   Remove the air filter housing (see Chapter 4).

3   Working in the passenger compartment, remove the knee bolster (see Chapter 11), remove the clip that secures the master cylinder pushrod to the clutch pedal and slide the pushrod off the clutch pedal pin.

➡️Note: Inspect the plastic retainer for damage. Replace the retainer with a new part if necessary.

4   Remove the clip securing the brake pedal to the brake booster push rod, and remove the brake booster mounting nuts (see Chapter 9).

➡️Note: On some models, it will be necessary to remove the Power Distribution Center bracket bolts and position the assembly off to the side (see Chapter 12).

5   Move the brake booster forward and out of the way to gain access to the clutch master cylinder.

6   Remove the clutch master cylinder mounting nuts from the firewall.

7   Raise the vehicle and support it securely on jackstands.

2001 and 2002 models, and 2004 and later models with the T850 transaxle

8   Working under the vehicle, separate the hydraulic line in between the release cylinder and the master cylinder using the special tool (see illustration).

9   Working in the engine compartment, remove the fasteners securing the clutch master cylinder and fluid reservoir to the firewall.

10   Using care not to damage the hydraulic line, work the clutch master cylinder, reservoir and line from the engine compartment.

2003 and later models with the T350 transaxle

11   Remove the clutch release cylinder mounting bolts (see Section 5).

12   Remove the hydraulic line mounting bracket.

13   Working in the engine compartment, remove the fasteners securing the clutch master cylinder and fluid reservoir to the firewall.

14   Using care not to damage the hydraulic line, work the clutch master cylinder, reservoir, the release cylinder and line from the engine compartment.

➡️Note: If necessary to make additional clearance, remove the ABS module mounting bolts and position the ABS module with the purge control solenoid off to the side. Do not disconnect the ABS module.

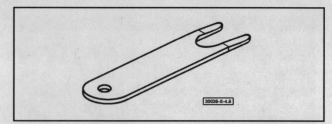

4.8 Tool for separating the quick-connect fitting on the clutch hydraulic line

INSTALLATION

15   Place the clutch master cylinder pushrod through the firewall and install the fasteners finger tight.

16   Working inside the vehicle, connect the master cylinder pushrod to the clutch pedal pin and install a new clip.

2001 and 2002 models, and 2004 and later models with the T850 transaxle

17   Connect the master cylinder hydraulic line to the release cylinder quick-connect fitting. Verify proper connection by an audible click and then pulling outward on the connection.

2003 and later models with the T350 transaxle

18   Working under the vehicle, install the release cylinder mounting bolts (see Section 5).

19   Install the hydraulic line mounting bracket.

All models

20   The remainder of installation is the reverse of removal. Tighten the mounting fasteners securely.

21   Bleed the clutch hydraulic system, if necessary (see Section 6).

22   Wash off any spilled brake fluid with water.

✳✳ CAUTION:

Don't allow brake fluid to come into contact with paint, as it will damage the finish.

---

5   Clutch release cylinder (2001 and later models) - removal and installation

REMOVAL

1   Disconnect the cable from the negative battery terminal (see Chapter 5).

2   Raise the vehicle and support it securely on jackstands.

2001 and 2002 models

3   Disconnect the hydraulic line quick disconnect fitting in between the release cylinder and the master cylinder.

4   Remove the lateral bending brace with the clutch release cylinder mounting bolts.

2003 and later models with the T350 transaxle

5   Remove the clip securing the hydraulic hose to the release cylinder. Using a punch, drive out the roll pin and disconnect the hose from the release cylinder.

✳✳ CAUTION:

Don't allow fluid to come in contact with paint as it will damage the finish.

2004 and later models with the T850 transaxle

6   Lift the nylon tab using a small screwdriver, press the release cylinder in toward the transaxle, then rotate the release cylinder 60-degrees counterclockwise. Remove the release cylinder from the transaxle.

7   Remove the bolts securing the release cylinder to the transaxle.

INSTALLATION

8   Installation is the reverse of the removal; refer to the Steps with the following additions:

   a) *Install the release cylinder and, if applicable, the lateral bending brace mounting bolts and tighten them securely.*

   b) *Connect the hydraulic line and check the hydraulic fluid level in the reservoir, adding fluid if necessary, until the level is correct.*

6 Clutch hydraulic system bleeding

→Note 1: This procedure applies only to 2001 and later models.

→Note 2: The clutch hydraulic system may require bleeding after servicing. Follow the preliminary bleeding procedure and if this does not produce a proper operating clutch pedal, bleed the release cylinder.

PRELIMINARY BLEEDING PROCEDURE

1   Make sure the clutch fluid reservoir is full. Sit in the driver's seat and actuate the clutch pedal between 60 to 100 times and check the operation of the clutch pedal. If the pedal remains spongy or the clutch does not engage properly, perform the release cylinder bleeding procedure.

BLEEDING THE RELEASE CYLINDER

2   Verify the clutch fluid reservoir is properly filled (see Chapter 1).

3   Raise the vehicle and secure it on jackstands.
4   Remove the release cylinder bolts, separate it from the transaxle (see Section 5) but do not disconnect the hydraulic line from the release cylinder.
5   Allow the release cylinder to hang down so that it's the lowest point in the system.
6   Depress the release cylinder pushrod until it bottoms then releases, forcing air out the top of the clutch fluid reservoir.
7   Continue bleeding until all the air has been eliminated from the hydraulic system.
8   Install the clutch release cylinder (see Section 5).
9   Sit in the driver's seat and actuate the clutch pedal 30 times and check the operation of the clutch pedal. If the pedal remains spongy or the clutch does not engage properly, perform Steps 2 through 6.
10 If several attempts bleeding the clutch hydraulic system are unsuccessful, replace the clutch master cylinder and the release cylinder.

7 Clutch components - removal and installation

✸ WARNING:

Dust produced by clutch wear and deposited on clutch components may contain asbestos, which is hazardous to your health. DO NOT blow it out with compressed air and DO NOT inhale it. DO NOT use gasoline or petroleum-based solvents to remove the dust. Brake system cleaner should be used to flush the dust into a drain pan. After the clutch components are wiped clean with a rag, dispose of the contaminated rags and cleaner in a labeled, covered container.

→Note: If the clutch pressure plate or flywheel requires replacement, the modular clutch assembly must be replaced as a set. Separate components are not available.

REMOVAL

→Note: Access to the clutch components is normally accomplished by removing the transaxle, leaving the engine in the vehicle. If, of course, the engine is being removed for major overhaul, then the opportunity should always be taken to check the clutch for wear and replace worn components as necessary. However, the relatively low cost of the clutch components compared to the time and labor involved in gaining access to them warrants their replacement any time the engine or transaxle is removed, unless they are new or in near-perfect condition. The following procedures assume that the engine will stay in place.

1   Remove the transaxle from the vehicle (see Chapter 7, Part A). Be sure to properly matchmark the location of the clutch assembly to the driveplate and remove the mounting bolts before removing the transaxle. Support the engine while the transaxle is out. Preferably, an engine hoist should be used to support it from above. However, if a jack is used underneath the engine, make sure a piece of wood is used between the jack and oil pan to spread the load.
2   Remove the modular clutch assembly from the transaxle input shaft.

✸ CAUTION:

Do not touch the clutch disc facing with your oily or dirty hands as the clutch surface will become contaminated.

INSTALLATION

3   Install the clutch release bearing and lever (see Section 7).
4   Install the modular clutch assembly onto the transaxle input shaft.
5   Install the transaxle (see Chapter 7A).

→Note: Be sure to install new bolts when attaching a modular clutch to the driveplate. Tighten the modular clutch-to-driveplate bolts to the torque listed in this Chapter's Specifications.

6   Using moly-base grease, lubricate the face of the release bearing where it contacts the fingers of the pressure plate diaphragm spring. Also lightly lubricate the splines of the transaxle input shaft.

✸ CAUTION:

Don't use too much grease.

7   Install the clutch release bearing, if removed (see Section 6).
8   Install the transaxle (see Chapter 7A) and all components removed previously, tightening all fasteners to the proper torque specifications.

8   Clutch release bearing and lever - removal, inspection and installation

❊ WARNING:

Dust produced by clutch wear and deposited on clutch components may contain asbestos, which is hazardous to your health. DO NOT blow it out with compressed air and DO NOT inhale it. DO NOT use gasoline or petroleum-based solvents to remove the dust. Brake system cleaner should be used to flush it into a drain pan. After the clutch components are wiped clean with a rag, dispose of the contaminated rags and cleaner in a labeled, covered container.

❊ CAUTION:

Don't allow brake fluid to come into contact with paint as it will damage the finish.

REMOVAL

▶ Refer to illustration 8.4

➡Note: Because of the difficulty involved in removing the transaxle for release bearing replacement, we recommend rou-

tinely replacing the release bearing when the clutch components are replaced.

1   Disconnect the cable from the negative terminal of the battery (see Chapter 5).
2   Remove the transaxle (see Chapter 7 Part A).
3   Using two hands, move the lever and bearing so the lever is at right angles to the input shaft. Secure the spring clip in place with needle-nose pliers.

❊ CAUTION:

Do not use a screwdriver or prybar to disengage the lever as the spring clip will be damaged.

4   Hold the release bearing in place and pull the release lever from the pivot stud and remove it (see illustration).
5   Slide the release bearing off the input shaft.

INSPECTION

▶ Refer to illustration 8.6

6   Hold the bearing by the outer race and rotate the inner race while

8.4 Hold the release bearing in place, disengage the clip and pull the release lever from the pivot stud. Then slide the release lever up to disengage it from the release bearing

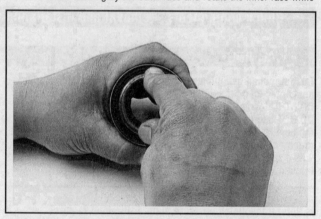

8.6 Hold the bearing by the outer race and rotate the inner race while applying pressure - if the bearing doesn't turn smoothly or if it's noisy, it must be replaced (it's a good idea to replace the bearing even if it's in good condition)

8.8a Using high-temperature grease, lubricate the release lever ends . . .

8.8b . . . the inner groove of the release bearing . . .

8.8c . . . and the sleeve around the input shaft

applying pressure (see illustration). If the bearing doesn't turn smoothly or if it's noisy, replace the bearing assembly with a new one. Wipe the bearing with a clean rag and inspect it for damage, wear and cracks. Don't immerse the bearing in solvent - it's sealed for life and to do so would ruin it. Also check the release lever and fork for cracks and bends.

7   Check the release lever for wear or damage especially in the area where the lever contacts the release bearing. Also be sure to clean any dirt off the pivot ball stud and stud pocket in the release fork.

INSTALLATION

▶ **Refer to illustrations 8.8a, 8.8b and 8.8c**

8   Lubricate the release lever ends, the inner diameter of the release bearing and the input shaft (see illustrations).
9   Install the release bearing onto the input shaft. Install the release lever fingers onto the release bearing and onto the pivot ball. Snap the lever into place on the pivot ball. Make sure it is properly seated and that the spring clip is inserted through the release lever slot.
10 Install the transaxle (see Chapter 7 Part A).

9   Clutch start switch - check and replacement

CHECK

1   Verify that the engine will not start when the clutch pedal is released.
2   Verify that the engine will start when the clutch pedal is depressed all the way.
3   If the engine won't start with the pedal depressed, or starts with the pedal released, unplug the electrical connector to the switch (located near the top of the clutch pedal) and check continuity between the connector terminals with the clutch pedal depressed.
4   If there's continuity between the terminals with the pedal depressed, the switch is okay; if there's no continuity between the terminals with the pedal depressed, replace the switch. If there's continuity between the terminals when the clutch pedal is released, replace the switch.

REPLACEMENT

5   Remove the knee bolster from the instrument panel (see Chapter 11).

6   Unplug the switch electrical connector, if you haven't already done so.
7   Depress the wing tabs on the switch and push the switch out of the mounting bracket.
8   Remove the clip securing the clutch master cylinder pushrod to the pedal pin.
9   Remove the switch and wires out of the slot in the bracket.
10 Installation is the reverse of removal.

ADJUSTMENT

11 Loosen the adjustment screw on the clutch master cylinder pushrod.
12 Carefully lift the pedal upward and connect the pushrod to the pedal pin.
13 Tighten the pushrod adjustment screw.
14 Verify the switch operates properly.

10   Driveaxles - general information and inspection

1   Power is transmitted from the transaxle to the wheels through a pair of driveaxles. The inner end of each driveaxle is splined to the differential side gears. The driveaxles can be pulled out to replace the oil seals (see Chapter 7A). The outer ends of the driveaxles are splined to the front hubs and locked in place by a large nut.
2   Each driveaxle assembly consists of an inner and outer constant velocity (CV) joint connected together by a driveaxle shaft. The inner ends of the driveaxles are equipped with a tripod joint on all models. The design is capable of both angular and axial motion. In other words, the inner CV joints are free to slide in-and-out as the driveaxle moves up-and-down with the wheel. These joints can be disassembled and cleaned in the event of a boot failure, but if any parts are damaged, the entire driveaxle assembly must be replaced as a unit.
3   The outer CV joints use a ball-and-socket design, capable of angular but not axial movement. These joints can be cleaned and repacked if an outer boot is torn, but if any parts are damaged, the entire driveaxle assembly must be replaced as a unit.

4   The boots should be inspected periodically for damage and leaking lubricant. Torn CV joint boots must be replaced immediately or the joints can be damaged. Boot replacement involves removal of the driveaxle (see Sections 10 and 11).

➡**Note: Some auto parts stores carry "split" type replacement boots, which can be installed without removing the driveaxle from the vehicle. This is a convenient alternative; however, the driveaxle should be removed and the CV joint disassembled and cleaned to ensure the joint is free from contaminants such as moisture and dirt which will accelerate CV joint wear.**

The most common symptom of worn or damaged CV joints, besides lubricant leaks, is a clicking noise in turns, a clunk when accelerating after coasting and vibration at highway speeds. To check for wear in the CV joints and driveaxle shafts, grasp each axle (one at a time) and rotate it in both directions while holding the CV joint housings, feeling for play indicating worn splines or sloppy CV joints. Also check the axleshafts for cracks, dents and distortion.

11   Driveaxles - removal and installation

REMOVAL

▶ **Refer to illustrations 11.1a, 11.1b, 11.7, 11.8 and 11.10**

1   Remove the hub nut cotter pin and discard it (see illustration). Discard the old cotter pin - you'll need a new one for reassembly. Remove the driveaxle hub nut lock and spring washer (see illustration).

2   Set the Parking brake firmly, then loosen the driveaxle/hub nut with a large socket and breaker bar.

3   Loosen the front wheel lug nuts, raise the vehicle and support it securely on jackstands. Remove the wheel.

4   It's not absolutely necessary that you drain the transaxle lubricant prior to removing a driveaxle, but if the mileage on the odometer indicates that the transaxle is nearing the lubricant-change interval prescribed in Chapter 1, now is a good time to do it.

5   Separate the tie-rod end from the steering knuckle (see Chapter 10).

6   Remove the bolt and nut securing the balljoint to the steering knuckle then pry the lower control arm down to separate the components (see Chapter 10).

7   To loosen the driveaxle from the hub splines, tap the end of the driveaxle with a soft-faced hammer (see illustration). If the driveaxle is stuck in the hub splines and won't move, it may be necessary to push it from the hub with a puller.

8   Pull out on the steering knuckle and detach the driveaxle from the hub (see illustration). Suspend the outer end of the driveaxle on a bungee cord or piece of wire.

9   Before you remove the driveaxle, look for lubricant leakage in the area around the differential seal. If there's evidence of a leak, you'll want to replace the seal after removing the driveaxle (see Chapter 7B).

10   To remove a right (passenger's side) driveaxle, position the prybar against the inner joint and carefully pry the joint off the transaxle side gear and retaining clip (see illustration). Do not use the driveaxle to pull on the inner joint. Doing so might damage the inner joint components. Pry straight out on the driveaxle to avoid damage to the transaxle oil seal. Remove the driveaxle assembly, being careful not to over extend the inner joint or damage the axleshaft boots.

11   To remove the left (driver's side) driveaxle, position a prybar against the inner joint and carefully pry the joint off the transaxle side gear and retaining clip. Do not use the driveaxle to pull on the inner

11.1a   Remove the hub nut cotter pin . . .

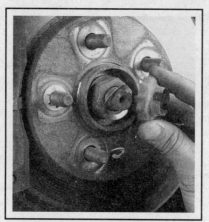

11.1b   . . . then remove the lock and spring washer

11.7   To loosen the driveaxle from the hub splines, tap the end of the driveaxle with a soft-faced hammer

11.8   Angle the steering knuckle as required and pull the driveaxle from the wheel hub

11.10   Using a large prybar, pry the inner CV joint out sharply to disengage it from the differential gears inside the transaxle

tripod joint. Doing so might damage the inner joint components. Pull straight out on the driveaxle to avoid damage to the transaxle oil seal. Remove the driveaxle assembly, being careful not to over extend the inner joint or damage the axleshaft boots.

12 Should it become necessary to move the vehicle while the driveaxle is out, place a large bolt with two large washers (one on each side of the hub) through the hub and tighten the nut securely.

13 If you noted evidence of a leaking driveaxle seal, refer to Chapter 7B for the seal replacement procedure.

INSTALLATION

14 Installation is the reverse of removal, but with the following additional points:

a) *Apply an even bead of multi-purpose grease around the splines of the inner joint.*

b) *When installing the driveaxle, hold the driveaxle straight out, push it in sharply to seat the driveaxle snap-ring into the groove in the transaxle side gear. To make sure the snap-ring is properly seated in the gear groove, attempt to pull the driveaxle out of the transaxle by hand. If the snap-ring is properly seated the inner joint will not move out.*

c) *Clean all foreign matter from the driveaxle outer CV joint threads. Install the spring washer and the nut. Tighten the hub nut to the torque listed in this Chapter's Specifications. Install the nut lock and a NEW cotter pin. Bend the ends over completely.*

d) *Install the wheel and lug nuts, then lower the vehicle. Tighten the lug nuts to the torque listed in the Chapter 1 Specifications.*

e) *Add transaxle lubricant as necessary (see Chapter 1).*

---

12 Driveaxle boot replacement and CV joint inspection

➡**Note 1:** If the CV joints or boots must be replaced, explore all options before beginning the job. Complete, rebuilt driveaxles may be available on an exchange basis, eliminating much time and work. Whichever route you choose to take, check on the cost and availability of parts before disassembling the vehicle.

➡**Note 2:** The inner and outer boots on the vehicles covered in this manual are constructed from different materials: The inner boots are made from either high-temperature application silicone material or Hytrel thermoplastic; the outer boot is made of Hytrel thermoplastic. Make sure you obtain a boot made of the correct material for the CV joint boot you're replacing.

INNER CV JOINT

1 Remove the driveaxle (see Section 11).

2 Mount the driveaxle in a vise with wood-lined jaws (to prevent damage to the axleshaft). Check the CV joints for excessive play in the radial direction, which indicates worn parts. Check for smooth operation throughout the full range of motion for each CV joint. If a boot is torn, the recommended procedure is to disassemble the joint, clean the components and inspect for damage due to loss of lubrication and possible contamination by foreign matter. If the CV joint is in good condition, lubricate it with CV joint grease and install a new boot.

Disassembly

◆ **Refer to illustrations 12.4, 12.5, 12.6, 12.7**

3 Cut the boot clamps with side-cutters, then remove and discard them.

4 Using a screwdriver, carefully pry up on the edge of the CV boot, pull it off the CV joint housing and slide it down the axleshaft, exposing the tripod spider assembly. To separate the axleshaft and spider assembly from the inner tripod joint housing, simply pull them straight out (see illustration).

➡**Note:** When removing the spider assembly, hold the rollers in place on the spider trunnion to protect the rollers and the needle bearings from falling free.

5 Remove the spider assembly snap-ring with a pair of snap-ring pliers (see illustration).

6 Mark the tripod to the axleshaft to ensure that they are reassembled properly (see illustration).

7 Use a hammer and a brass drift to drive the spider assembly from the axleshaft (see illustration).

8 Slide the boot off the shaft.

Inspection

9 Thoroughly clean all components with solvent until the old CV

12.4 Remove the boot from the inner CV joint and slide the tripod from the joint housing

12.5 Remove the snap-ring with a pair of snap-ring pliers

12.6 Mark the relationship of the tripod bearing assembly to the axleshaft

12.7 Drive the tripod joint off the axleshaft with a brass punch and hammer; be careful not to damage the bearing surfaces or the splines on the shaft

12.10a Wrap the axleshaft splines with electrical tape to prevent damaging the boot as it's slid onto the shaft

joint grease is completely removed. Inspect the bearing surfaces of the inner tripods and housings for cracks, pitting, scoring and other signs of wear. If any part of the inner CV joint is worn, you must replace the entire driveaxle assembly (inner tripod joint, axleshaft and outer CV joint). The only components that can be purchased separately are the boots themselves and the boot clamps.

Reassembly

▶ **Refer to illustrations 12.10a, 12.10b, 12.10c, 12.10.d, 12.11 and 12.13**

10 Wrap the splines on the inner end of the axleshaft with electrical or duct tape to protect the boots from the sharp edges of the splines and slide the clamps and boot onto the axleshaft (see illustration). Remove the tape and place the tripod spider on the axleshaft with the chamfer toward the shaft (see illustration). Tap the spider onto the shaft with a brass drift until it's seated and install the snap-ring. Apply grease to the tripod assembly and inside the housing (see illustration). Insert the tripod into the housing and pack the remainder of the grease around the tripod (see illustration).

11 Slide the boot into place, making sure the raised bead on the inside of the seal boot is positioned in the groove on the interconnect-

12.10b Install the tripod spider on the axleshaft (make sure your match mark is facing out)

ing shaft. If the driveaxle has multiple locating grooves on the shaft, position the boot so only one of the grooves (the thinnest) is exposed (see illustration). Position the sealing boot into the groove on the tri-

12.10c Place grease at the bottom of the CV joint housing

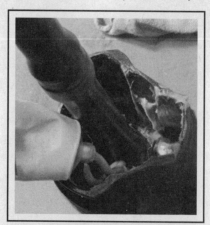

12.10d Install the boot and clamps onto the axleshaft, then insert the tripod into the housing, followed by the rest of the grease

12.11 Make sure that the thinnest groove on the axleshaft is the only one showing

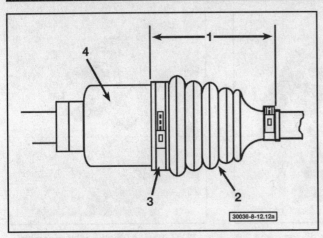

12.12a CV joint adjustment details - Hytrel boot

1	*See Specifications*	3	*Clamp*
2	*Boot*	4	*Inner CV joint*

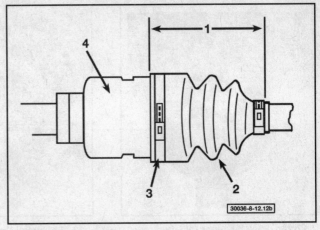

12.12b CV joint adjustment details - silicone boot

1	*See Specifications*	3	*Clamp*
2	*Boot*	4	*Inner CV joint*

pod housing retaining groove.

12 Adjust the length of the CV joint (see illustrations). There are two different types of boots equipped on these models. Hytrel boots are designed with more rigid material with four bellows while silicone boots are more flexible with two bellows. Refer to the Specifications listed in this Chapter for the correct CV boot adjustment length.

13 Equalize the pressure inside the boot by inserting a small, dull, flat screwdriver tip between the boot and the CV joint housing (see illustration). Then remove the screwdriver tip.

14 Make sure each end of the boot is seated properly, and the boot is not distorted.

15 Two types of clamps are used on the inner CV joint. If a crimp-type clamp is used, clamp the new boot clamps onto the boot with a special crimping tool (available at most automotive parts stores). Place the crimping tool over the bridge of each new boot clamp, then tighten the nut on the crimping tool until the jaws are closed. If a low profile latching type clamp is used, place the prongs of the clamping tool (available at most automotive parts stores) in the holes of the clamp and squeeze the tool together until the top band latches behind the tabs of the lower band.

16 The driveaxle is now ready for installation (see Section 11).

OUTER CV JOINT

Disassembly

▶ **Refer to illustrations 12.21a and 12.21b**

17 Remove the driveaxle (see Section 9).

18 Mount the driveaxle in a vise with wood-lined jaws (to prevent damage to the axleshaft). Check the CV joints for excessive play in the radial direction, which indicates worn parts. Check for smooth operation throughout the full range of motion for each CV joint. If a boot is torn, the recommended procedure is to disassemble the joint, clean the components and inspect for damage due to loss of lubrication and possible contamination by foreign matter. If the CV joint is in good condition, lubricate it with CV joint grease and install a new boot.

19 Cut the boot clamps with side-cutters, then remove and discard them.

20 Slide the boot away from the outer CV joint, and clean the CV joint and the interconnecting parts.

21 Strike the edge of the CV joint housing sharply with a soft-faced hammer to dislodge the outer CV joint housing from the axleshaft (see illustration). Remove the circlip from the shaft (see illustration).

22 Slide the outer CV joint and the sealing boot off the axleshaft.

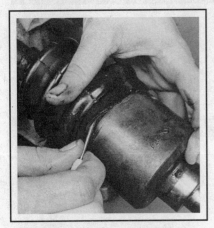

12.13 Equalize the pressure inside the boot by inserting a small, dull screwdriver between the boot and the CV joint housing

12.21a Strike the edge of the CV joint housing sharply with a soft-faced hammer to dislodge the CV joint from the shaft

12.21b Using a pair of snap-ring pliers, remove the circlip from the axleshaft

12.26a Pack the outer CV joint assembly with grease

12.26b Install the small clamp and the new boot on the driveaxle, then apply grease to the inside of the boot

12.27 Position the CV joint assembly on the driveaxle, aligning the splines

Inspection

23 Thoroughly clean all components with solvent until the old CV grease is completely removed. Inspect the bearing surfaces of the inner tripods and housings for cracks, pitting, scoring, and other signs of wear. If any part of the outer CV joint is worn, you must replace the entire driveaxle assembly (inner CV joint, axleshaft and outer CV joint).

Reassembly

▸ Refer to illustrations 12.26a, 12.26b and 12.27

24 Slide a new sealing boot clamp and sealing boot onto the axleshaft.

➡**Note: The sealing boot must be positioned on the shaft so the raised bead on the inside of the boot is in the groove on the shaft.**

25 Replace the circlip onto the axleshaft.

26 Place half the grease provided in the sealing boot kit into the outer CV joint assembly housing. Put the remaining grease into the sealing boot (see illustrations).

27 Align the splines on the axleshaft with the splines on the outer CV joint assembly and start the outer CV joint onto the axleshaft (see illustration).

28 Thread a nut loosely onto the stub axle and use a soft-faced hammer to strike the nut (the nut is installed to protect the threads).

29 Drive the CV joint onto the axleshaft until the CV joint is seated to the axleshaft.

30 Install the outer CV joint sealing boot to the axleshaft (see Steps 11 through 16).

➡**Note: The length of the outer joint isn't adjustable; just make sure that each end of the boot is seated properly and that there are no dimples in the folds of the boot.**

Specifications

CV boot adjustment length

Hytrel boot	4.22 inches (107 mm)
Silicone boot	4.14 inches (105 mm)

Torque specifications

	Ft-lbs (unless otherwise indicated)	Nm
Modular clutch-to-driveplate bolts	65	88
Driveaxle/hub nut	180	244
Wheel lug nuts	See Chapter 1	

Section

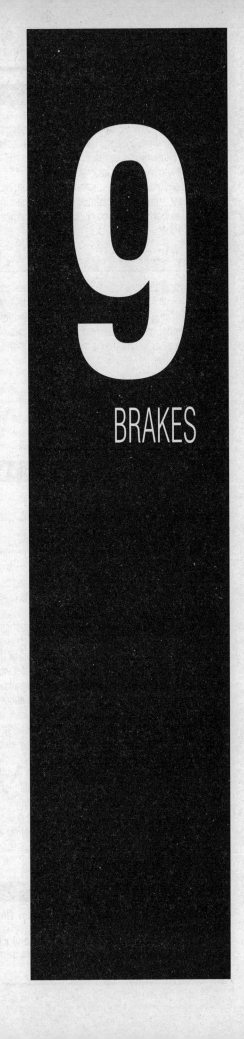

9

BRAKES

1 General information

The vehicles covered by this manual are equipped with a hydraulically operated brake system. The front brakes are discs and the rear brakes are either drums or discs. Both the front and rear disc brakes automatically compensate for disc and pad wear. As the pads wear down, the pistons gradually protrude farther from the calipers, but don't retract as far, automatically compensating for the thinner pads. Rear drum brakes have automatic adjusters, which compensate for wear of the brake shoes.

HYDRAULIC SYSTEM

The hydraulic system consists of two separate circuits that are diagonally split (one circuit operates the left front and right rear brakes, while the other circuit operates the right front and left rear brakes). The master cylinder has separate reservoirs for the two circuits, and, in the event of a leak or failure in one hydraulic circuit, the other circuit will remain operative.

POWER BRAKE BOOSTER

The power brake booster, which is mounted on the firewall, utilizes engine manifold vacuum and atmospheric pressure to provide assistance to the hydraulically operated brakes.

PARKING BRAKE

The parking brake operates the rear brakes only, through cable actuation. It's activated by a lever mounted in the center console. The parking brake on rear disc brake models uses brake shoes and small brake drums integral with the rear brake discs.

SERVICE

After completing any operation involving disassembly of any part of the brake system, always test-drive the vehicle to check for proper braking performance before resuming normal driving. When testing the brakes, perform the tests on a clean, dry, flat surface. Conditions other than these can lead to inaccurate test results.

Test the brakes at various speeds with both light and heavy pedal pressure. The vehicle should stop evenly without pulling to one side or the other. Avoid locking the brakes, because this slides the tires and diminishes braking efficiency and control of the vehicle.

Tires, vehicle load and wheel alignment are factors which also affect braking performance.

2 Antilock Brake System (ABS) - general information and speed sensor removal and installation

GENERAL INFORMATION

1   The Mark 20e Antilock Brake System (ABS) is designed to maintain vehicle steerability, directional stability and optimum deceleration under severe braking conditions on most road surfaces. It does so by monitoring the rotational speed of each wheel and controlling the brake line pressure to each wheel during braking. This prevents the wheels from locking up.

2   The ABS system has three principal components - the wheel speed sensors, the Hydraulic Control Unit (HCU) and the ABS computer, which is referred to as the Integrated Control Unit (ICU). Four wheel speed sensors - one at each wheel - send a variable voltage signal to the control unit, which monitors these signals, compares them to its program and determines whether a wheel is about to lock up. When a wheel is about to lock up, the control unit signals the hydraulic unit to reduce hydraulic pressure (or not increase it further) at that wheel's brake caliper. Pressure modulation is handled by electrically-operated solenoid valves.

3   If a problem develops within the system, an "ABS" warning light will glow on the dashboard. Sometimes, a visual inspection of the ABS system can help you locate the problem. Carefully inspect the ABS wiring harness. Pay particularly close attention to the harness and connections near each wheel. Look for signs of chafing and other damage caused by incorrectly routed wires. If a wheel sensor harness is damaged, the sensor must be replaced.

✳✳ WARNING:

Do NOT try to repair an ABS wiring harness. The ABS system is sensitive to even the smallest changes in resistance. Repairing the harness could alter resistance values and cause the system to malfunction. If the ABS wiring harness is damaged in any way, it must be replaced.

✳✳ CAUTION:

Make sure the ignition is turned off before unplugging or reattaching any electrical connections.

DIAGNOSIS AND REPAIR

4   If a dashboard warning light comes on and stays on while the vehicle is in operation, the ABS system requires attention. Although a special scan tool is necessary to properly diagnose the system, you can perform a few preliminary checks before taking the vehicle to a dealer service department.

   a) *Check the brake fluid level in the reservoir.*
   b) *Verify that the computer electrical connectors are securely connected.*
   c) *Check the electrical connectors at the hydraulic control unit.*
   d) *Check the fuses.*
   e) *Follow the wiring harness to each wheel and verify that all connections are secure and that the wiring is undamaged.*

5   If the above preliminary checks do not rectify the problem, the vehicle should be diagnosed by a dealer service department or other qualified repair shop. Due to the complex nature of this system, all actual repair work must be done by a qualified automotive technician.

WHEEL SPEED SENSOR - REMOVAL AND INSTALLATION

6   Loosen the wheel lug nuts, raise the vehicle and support it securely on jackstands. Remove the wheel.

7   Make sure the ignition key is turned to the Off position.

8   Trace the wiring back from the sensor, detaching all brackets

and clips while noting its correct routing, then disconnect the electrical connector.

9   Remove the mounting bolt and carefully pull the sensor out from the knuckle or brake backing plate.

10 Installation is the reverse of the removal procedure. Tighten the mounting bolt to the torque listed in this Chapter's Specifications.

11 Install the wheel and lug nuts, tightening them securely. Lower the vehicle and tighten the lug nuts to the torque listed in the Chapter 1 Specifications.

3   Disc brake pads - replacement

▶ Refer to illustrations 3.3, 3.4a through 3.4l

✳✳ WARNING:

Disc brake pads must be replaced on both front or rear wheels at the same time - never replace the pads on only one wheel. Also, the dust created by the brake system may contain asbestos, which is harmful to your health. Never blow it out with compressed air and don't inhale any of it. An approved filtering mask should be worn when working on the brakes. Do not, under any circumstances, use petroleum-based solvents to clean brake parts. Use brake system cleaner only!

➡Note: This procedure applies to both the front and the rear disc brakes.

1   Loosen the front wheel lug nuts, raise the front of the vehicle and support it securely on jackstands. Apply the parking brake. Remove the front wheels (or the rear wheels, if you're working on a vehicle with rear discs).

2   Remove about two-thirds of the fluid from the master cylinder reservoir and discard it. Position a drain pan under the brake assembly and clean the caliper and surrounding area with brake system cleaner.

3   Push the piston back into its bore to provide room for the new brake pads with a C-clamp (see illustration). As the piston is depressed to the bottom of the caliper bore, the fluid in the master cylinder will rise. Make sure it doesn't overflow. If necessary, siphon off some of the fluid.

4   To replace the brake pads, follow the accompanying photos, beginning with illustration 3.4a. Be sure to stay in order and read the caption under each illustration.

5   While the pads are removed, inspect the caliper for brake fluid

3.3 Use a C-clamp to depress the piston into its bore; this aids in removal of the caliper and installation of the new pads

leaks and ruptures of the piston boot. Overhaul or replace the caliper as necessary (see Section 4). Also inspect the brake disc carefully (see Section 5). If machining is necessary, follow the information in that Section to remove the disc.

6   Before installing the caliper guide pin bolts, clean them and check them for corrosion and damage. If they're significantly corroded or damaged, replace them. Be sure to tighten the caliper guide pin bolts to the torque listed in this Chapter's Specifications.

7   Install the brake pads on the opposite wheel, then install the

3.4a Wash down the disc and brake pads with brake cleaner to remove brake dust; DO NOT blow brake dust off with compressed air

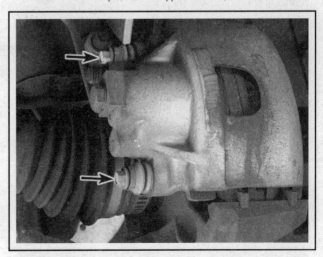

3.4b To remove the front caliper, remove these guide pin bolts (arrows)

3.4c To remove the rear caliper, remove these guide pin bolts (arrows)

3.4d Hang the caliper from the strut coil spring with a piece of wire - don't let it hang by the brake hose

wheels and lower the vehicle. Tighten the lug nuts to the torque listed in the Chapter 1 Specifications. Add brake fluid to the reservoir until it's full (see Chapter 1).

8   Pump the brakes several times to seat the pads against the disc,

then check the fluid level again.

9   Carefully test the operation of the brakes before placing the vehicle in normal operation. Try to avoid heavy brake applications until the brakes have been applied lightly several times to seat the pads.

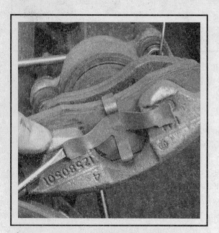

3.4e Pry the outer brake pad retaining spring from the caliper . . .

3.4f . . . and remove the outer brake pad

3.4g Pull the inner brake pad retaining spring loose from the piston and remove the pad

3.4h Remove the guide pin bushings

3.4i Remove the bushing boots, inspect them for tears and replace as necessary

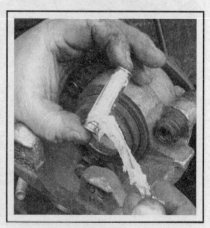

3.4j Lubricate the guide pin bushings with multi-purpose grease before installing them

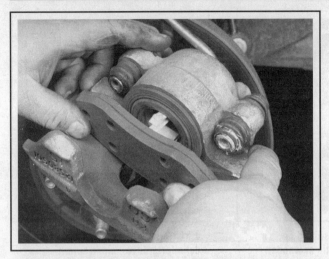

3.4k Install the inner brake pad - make sure the retaining spring is fully seated into its bore in the piston

3.4l Install the outer brake pad - make sure the retaining spring is properly engaged with the caliper body

4 Disc brake caliper - removal, overhaul and installation

❊❊ WARNING:

Dust created by the brake system may contain asbestos, which is harmful to your health. Never blow it out with compressed air and don't inhale any of it. An approved filtering mask should be worn when working on the brakes. Do not, under any circumstances, use petroleum-based solvents to clean brake parts. Use brake system cleaner only.

➡Note: If an overhaul is indicated (usually because of fluid leaks, a stuck piston or broken bleeder screw) explore all options before beginning this procedure. New and factory rebuilt calipers are available on an exchange basis, which makes this job quite easy. If you decide to rebuild the calipers, make sure rebuild kits are available before proceeding. Always rebuild or replace the calipers in pairs - never rebuild just one of them.

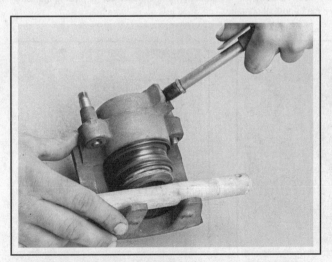

4.4 With a block of wood placed between the piston and the caliper frame, use compressed air to ease the piston out of the bore

REMOVAL

1   Loosen the front wheel lug nuts, raise the vehicle and support it securely on jackstands. Remove the front wheels (or the rear wheels, if you're working on a vehicle with rear discs).

2   Unscrew the banjo bolt from the caliper and detach the hose.

➡Note: If you're just removing the caliper for access to other components, don't disconnect the hose. Discard the sealing washers on each side of the fitting and use new ones during installation. Wrap a plastic bag around the end of the hose to prevent fluid loss and contamination.

3   Refer to the first few steps in Section 3 (caliper removal is the first part of the brake pad replacement procedure). Clean the caliper assembly with brake system cleaner.

❊❊ WARNING:

DO NOT, under any circumstances, use kerosene, gasoline or petroleum-based solvents to clean brake parts!

Be sure to check the pads as well and replace them if necessary (see Section 3).

OVERHAUL

▶ **Refer to illustrations 4.4, 4.5, 4.6, 4.10, 4.11a, 4.11b and 4.12**

4   Place several shop towels or a block of wood in the center of the caliper to act as a cushion, then use compressed air, directed into the fluid inlet, to remove the piston (see illustration). Use only enough air pressure to ease the piston out of the bore. If the piston is blown out, even with the cushion in place, it may be damaged.

❊❊ WARNING:

Never place your fingers in front of the piston in an attempt to catch or protect it when applying compressed air, as serious injury could occur.

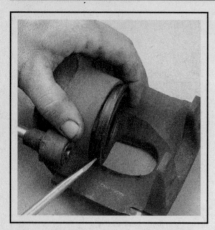

4.5 Carefully pry the dust boot out of the caliper

4.6 The piston seal should be removed with a plastic or wooden tool to avoid damage to the bore and the seal groove (a pencil will do the job)

4.10 Position the new seal in the cylinder groove - make sure it isn't twisted

5   Pry the dust boot from the caliper bore (see illustration).

6   Using a wood or plastic tool, remove the piston seal from the groove in the caliper bore (see illustration). Metal tools may cause bore damage.

7   Remove the bleeder screw, then remove and discard the guide pin bushings and sleeves.

8   Clean the remaining parts with brake system cleaner or clean brake fluid, then blow them dry with filtered, unlubricated compressed air.

9   Inspect the surfaces of the piston for nicks and burrs and loss of plating. If surface defects are present, the caliper must be replaced. Check the caliper bore in a similar way. Light polishing with crocus cloth is permissible to remove slight corrosion and stains. Discard the caliper pins if they're severely corroded or damaged.

10 Lubricate the new piston seal with clean brake fluid and position the seal in the cylinder groove using your fingers only (see illustration).

11 Install the new dust boot in the groove in the end of the piston (see illustration). Dip the piston in clean brake fluid and insert it squarely into the cylinder. Depress the piston to the bottom of the cylinder bore (see illustration).

12 Seat the boot in the caliper counterbore using a boot installation tool or a blunt punch (see illustration).

13 Install the new guide pin boots and bushings. Install the bleeder screw and tighten it securely.

14 Install the brake pads in the caliper.

INSTALLATION

15 Install the caliper assembly, tightening the caliper guide pin bolts to the torque listed in this Chapter's Specifications.

16 Connect the brake hose to the caliper using new sealing washers. Tighten the banjo bolt to the torque listed in this Chapter's Specifications.

17 Bleed the brake system (see Section 11).

18 Install the wheels and lug nuts. Lower the vehicle and tighten the lug nuts to the torque listed in the Chapter 1 Specifications.

19 After the job has been completed, firmly depress the brake pedal a few times to bring the pads into contact with the disc.

20 Carefully test the operation of the brakes before placing the vehicle in normal operation.

4.11a Slip the boot over the piston

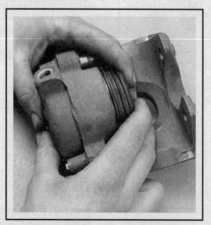

4.11b Push the piston straight into the cylinder - make sure it doesn't become cocked in the bore

4.12 If you don't have a boot installation tool, gently seat the boot with a drift punch

5   Brake disc - inspection, removal and installation

INSPECTION

▶ **Refer to illustrations 5.2, 5.3, 5.4a and 5.4b**

1   Loosen the wheel lug nuts, raise the front (or rear) of the vehicle and support it securely on jackstands. Apply the parking brake. Remove the front (or rear) wheels. Reinstall the lug nuts, flat side toward the disc, to hold the disc firmly against the hub.

2   Remove the brake caliper as described in Section 4. Visually inspect the disc surface for score marks and other damage (see illustration). Light scratches and shallow grooves are normal after use and won't affect brake operation. Deep grooves - over 0.015-inch deep - require disc removal and refinishing by an automotive machine shop. Be sure to check both sides of the disc.

3   To check disc runout, place a dial indicator at a point about 1/2-inch from the outer edge of the disc (see illustration). Set the indicator to zero and turn the disc. The indicator reading should not exceed the runout limit listed in this Chapter's Specifications. If it does, the disc should be refinished by an automotive machine shop.

➡**Note: Professionals recommend resurfacing the brake discs** regardless of the dial indicator reading (to produce a smooth, flat surface that will eliminate brake pedal pulsations and other undesirable symptoms related to questionable discs). At the very least, if you elect not to have the discs resurfaced, deglaze them with sandpaper or emery cloth.

4   The disc must not be machined to a thickness less than the specified minimum thickness. The minimum (or discard) thickness is cast into the disc (see illustration). The disc thickness can be checked with a micrometer (see illustration).

REMOVAL AND INSTALLATION

▶ **Refer to illustration 5.8**

5   Loosen the wheel lug nuts, raise the vehicle and place it securely on jackstands.

6   Remove the wheel. If you're removing a rear disc, block the front wheels and release the parking brake.

7   Remove the caliper (see Section 4).

8   Remove the retaining clips, if present, from the wheel studs (see illustration).

5.2 The brake pads on this vehicle were obviously neglected, as they wore down to the rivets, then cut deep grooves into the disc, which must be replaced

5.3 Use a dial indicator to check disc runout - if the reading exceeds the specified runout limit, the disc will have to be machined or replaced

5.4a On some models, the minimum thickness is cast into the inside of the disc - on others, it's located on the outside of the disc

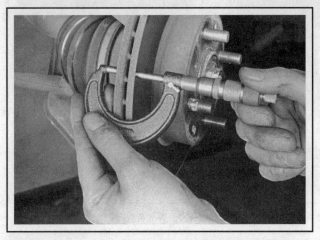

5.4b Use a micrometer to measure the thickness of the disc at several points

5.8 Cut off and discard the disc retaining washers, if present (it isn't necessary to reinstall them)

9 Pull the disc off the hub.

10 If you're removing a rear disc and the disc won't come off, remove the plug from the parking brake adjusting access hole. Turn the adjusting star wheel with a suitable tool (such as a screwdriver or brake adjusting tool) and retract the parking brake shoes (see Sec-

tion 13 for details on the parking brake shoes on models with rear disc brakes).

11 Installation is the reverse of removal. Make sure you tighten the caliper guide pin bolts to the torque listed in this Chapter's Specifications.

6 Drum brake shoes - replacement

▶ Refer to illustrations 6.2, 6.4a through 6.4x and 6.5

❋❋ WARNING:

Drum brake shoes must be replaced on both wheels at the same time - never replace the shoes on only one wheel. Also, the dust created by the brake system may contain asbestos, which is harmful to your health. Never blow it out with compressed air and don't inhale any of it. An approved filtering mask should be worn when working on the brakes. Do not, under any circumstances, use petroleum-based solvents to clean brake parts. Use brake system cleaner only!

❋❋ CAUTION:

Whenever the brake shoes are replaced, the hold-down springs should also be replaced. Due to the continuous heating/cooling cycle that the springs are subjected to, they lose their tension over a period of time and may allow the shoes to drag on the drum and wear at a much faster rate than normal.

1 Loosen the wheel lug nuts, raise the rear of the vehicle and support it securely on jackstands. Block the front wheels to keep the vehicle from rolling. Release the parking brake. Remove the wheel.

➡Note: All four front or rear brake shoes must be replaced at the same time, but to avoid mixing up parts, work on only one brake assembly at a time.

2 Remove the brake drum.

➡Note: If the drum won't come off, retract the brake shoes by inserting a screwdriver or brake adjusting tool through the hole

6.2 Before disassembling the brake shoe assembly, wash it thoroughly with brake system cleaner

in the backing plate and turning the adjuster screw star wheel. The drum should now come off.

Wash down the brake assembly with brake cleaner (see illustration).

3 Remove the hub and bearing assembly (see Chapter 10).

4 Follow illustrations 6.4a through 6.4x for the inspection and replacement of the brake shoes. Be sure to stay in order and read the caption under each illustration.

5 Before reinstalling the drum it should be checked for cracks, score marks, deep scratches and hard spots, which will appear as small discolored areas. If the hard spots cannot be removed with fine

6.4a Twist the hold-down spring pin 90 degrees

6.4b Remove the brake shoe hold-down spring

6.4c Withdraw one of the brake shoes from the wheel cylinder to relieve some of the return spring pressure

6.4d Unhook the self-adjuster
lever spring

6.4e Remove the self-adjuster lever

6.4f Unhook the upper return spring
from both brake shoes

emery cloth or if any of the other conditions listed above exist, the
drum must be taken to an automotive machine shop to have it turned.

➡Note: Professionals recommend resurfacing the drums whenever a brake job is done. Resurfacing will eliminate the possibility of out-of-round drums. If the drums are worn so much that

they can't be resurfaced without exceeding the maximum allowable diameter (stamped into the drum) (see illustration), then
new ones will be required. At the very least, if you elect not to
have the drums resurfaced, remove the glazing from the surface
with emery cloth or sandpaper using a swirling motion.

6.4g Remove the self-adjuster from
both brake shoes

6.4h Unhook the anchor plate spring
from both brake shoes

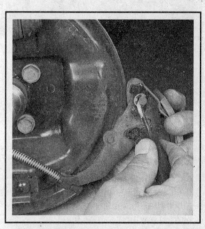

6.4i Remove the E-clip and separate
the parking brake arm from the rear
brake shoe assembly

6.4j Unhook and detach the parking
brake cable from the lever

6.4k Apply high-temperature grease
to all areas where the brake shoes
make contact with the backing plate

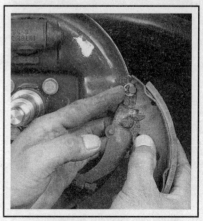

6.4l Transfer the parking brake lever
to the new rear shoe and install the
E-clip. Make sure the clip is
properly seated

6.4m Correctly position the rear shoe in place

6.4n Install the rear brake shoe hold-down spring

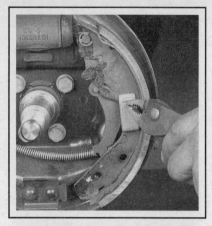

6.4o Twist the hold-down spring pin 90 degrees

6.4p Attach the anchor plate spring to both brake shoes, then move the front brake shoe up into position. Move the anchor spring into the correct position behind the anchor plate

6.4q Attach the return spring first to the rear brake shoe then to the front shoe

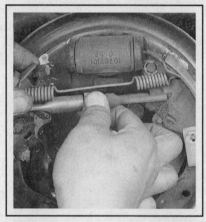

6.4r Engage the self-adjuster with the front shoe, then the rear shoe

6.4s Move the front shoe into position and push the brake shoe assembly against the brake backing plate; make sure both shoes are properly seated against the wheel cylinder pistons

6.4t Install the front brake shoe hold-down spring and turn the spring hold-down pin 90 degrees

6.4u Install the self-adjuster lever onto the front brake shoe and the self-adjuster

6.4v Hook the lower end of the self-adjuster lever spring into its hole in the front brake shoe

6.4w Hook the upper end of the spring onto the self-adjuster lever

6.4x Install the self-adjuster lever onto its pivot pin; make sure the small lever at the upper left is installed between the adjuster and the front brake shoe as shown

6.5 The maximum allowable diameter is stamped into the drum

6   Install the hub and bearing assembly (see Chapter 10).

7   Install the brake drum. Pump the brake pedal several times, then turn the adjuster star wheel using a screwdriver inserted through the hole in the backing plate until the brake shoes slightly drag on the drum as the drum is turned. Now, back off the adjuster until the shoes don't drag on the drum.

8   Mount the wheel, install the lug nuts, then lower the vehicle. Tighten the lug nuts to the torque listed in the Chapter 1 Specifications.

9   Make a number of forward and reverse stops to adjust the brakes until satisfactory pedal action is obtained.

10 Carefully test the operation of the brakes before placing the vehicle in normal operation.

7   Wheel cylinder - removal, overhaul and installation

➡Note: **If an overhaul is indicated (usually because of fluid leakage or sticky operation) explore all options before beginning the job. New wheel cylinders are available, which makes this job quite easy. If you decide to rebuild the wheel cylinder, make sure a rebuild kit is available before proceeding. Never overhaul only one wheel cylinder. Always rebuild both of them at the same time.**

REMOVAL

▶ **Refer to illustration 7.2**

1   Remove the brake shoes (see Section 6).

2   Unscrew the brake line fitting from the rear of the wheel cylinder (see illustration). If available, use a flare-nut wrench to avoid rounding off the corners on the fitting. Don't pull the metal line out of the wheel cylinder - it could bend, making installation difficult.

3   Remove the two bolts securing the wheel cylinder to the brake backing plate.

4   Remove the wheel cylinder.

5   Plug the end of the brake line to prevent the loss of brake fluid and the entry of dirt.

7.2 To detach the wheel cylinder from the brake backing plate, disconnect the brake line fitting and remove the wheel cylinder bolts

OVERHAUL

6   To disassemble the wheel cylinder, remove the rubber dust boot from each end of the cylinder, then push out the two pistons, the cups and the cup expanders and spring assembly. Discard the rubber parts and use new ones from the rebuild kit when reassembling the wheel cylinder.

7   Inspect the pistons for scoring and scuff marks. If present, the wheel cylinder should be replaced.

8   Examine the inside of the cylinder bore for score marks and corrosion. If these conditions exist, the cylinder can be honed slightly to restore it, but replacement is recommended.

9   If the cylinder is in good condition, clean it with brake system cleaner.

✳✳ WARNING:

DO NOT, under any circumstances, use gasoline or petroleum-based solvents to clean brake parts!

---

10   Remove the bleeder screw and make sure the hole is clean.

11   Lubricate the cylinder bore with clean brake fluid, then insert one of the new rubber cups into the bore. Make sure the lip on the rubber cup faces in.

12   Install the cup expander and spring assembly from the other side of the cylinder, then install the remaining cup in the cylinder bore.

13   Install the pistons.

14   Install the boots.

INSTALLATION

15   Installation is the reverse of removal. Attach the brake line to the wheel cylinder before installing the mounting bolts and tighten the line fitting after the wheel cylinder mountings bolts have been tightened. If available, use a flare-nut wrench to tighten the line fitting. Make sure you tighten the line fitting securely and the wheel cylinder mounting bolts to the torque listed in this Chapter's Specifications.

16   Install the brake shoes and brake drum (see Section 6).

17   Bleed the brake system (see Section 11). Don't drive the vehicle in traffic until brake operation has been thoroughly tested.

---

8   Master cylinder - removal and installation

REMOVAL

▶ **Refer to illustrations 8.3 and 8.4**

1   On ABS equipped models, the vacuum in the vacuum booster must be pumped down prior to removing the master cylinder, otherwise foreign matter may be sucked into the vacuum booster. With the ignition switch on the Off position, pump the brake pedal 4 to 5 times until a firm pedal is achieved without the aid of any vacuum assist.

2   Place rags under the brake line fittings and prepare caps or plastic bags to cover the ends of the lines once they're disconnected.

✳✳ CAUTION:

Brake fluid will damage paint. Cover all painted surfaces and avoid spilling fluid during this procedure.

---

3   Unhook the clip and unplug the electrical connector from the brake fluid level sensor (see illustration).

4   Loosen the tube nuts at the ends of the brake lines where they enter the master cylinder. To prevent rounding off the flats on these nuts, use a flare-nut wrench, which wraps around the nut (see illustration). Pull the brake lines away from the master cylinder slightly and plug the ends to prevent contamination. Also plug the openings in the master cylinder to prevent fluid spillage.

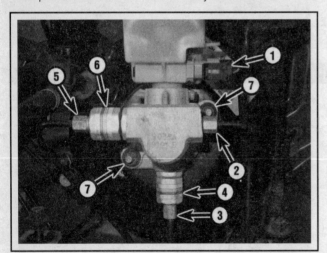

8.3   Master cylinder mounting details

1   *Electrical connector for fluid level warning switch*
2   *Left front brake line (right front brake line, not visible in this photo, is behind rear brake line and proportioning valve on right side of master cylinder)*
3   *Rear brake line*
4   *Proportioning valve*
5   *Rear brake line*
6   *Proportioning valve*
7   *Master cylinder mounting nuts*

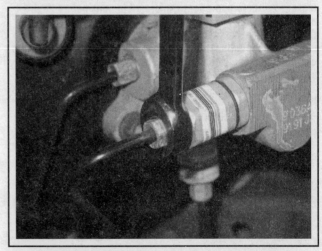

8.4   To prevent rounding off the corners on brake line tube nut fittings at the master cylinder (or the wheel cylinders), use a flare-nut wrench to loosen them

5   On ABS equipped models, clean the area where the master cylinder attaches to the vacuum booster with an aerosol brake cleaner.

6   Remove the two master cylinder mounting nuts (see illustration 8.3) and remove the master cylinder from the vehicle.

7   Remove the reservoir cap, then discard any fluid remaining in the reservoir.

INSTALLATION

▶ **Refer to illustration 8.9**

8   Bench bleed the master cylinder before installing it. Mount the master cylinder in a vise, with the jaws of the vise clamping on the mounting flange.

9   Attach a pair of master cylinder bleeder tubes to the outlet ports of the master cylinder (see illustration).

10 Fill the reservoir with brake fluid of the recommended type (see Chapter 1).

11 Using a large Phillips screwdriver, slowly push the pistons into the master cylinder. This expels air from the pressure chambers and into the reservoir. Because the tubes are submerged in fluid, air cannot be drawn back into the master cylinder when you release the pistons.

12 Repeat the procedure until no more air bubbles are present.

13 Remove the bleed tubes, one at a time, and install plugs in the open ports to prevent fluid leakage and air from entering. Install the reservoir cap.

14 Install the master cylinder over the studs on the power brake booster and tighten the nuts only finger-tight at this time.

15 Thread the front brake line fittings into the master cylinder and the rear lines into the proportioning valves. Since the master cylinder is still a bit loose, it can be moved slightly so the fittings thread in easily. Don't strip the threads as the fittings are tightened.

16 Tighten the master cylinder mounting nuts to the torque listed in

8.9 The best way to bleed the master cylinder before installing it on the vehicle is with a pair of bleeder tubes that direct fluid into the reservoir during bleeding

this Chapter's Specifications. Tighten the brake line tube nut fittings securely.

17 Fill the master cylinder reservoir with fluid, then bleed the master cylinder and the rest of the brake system (see Section 11). To bleed the master cylinder on the vehicle, have an assistant depress the brake pedal and hold it down. Loosen the fitting to allow air and fluid to escape. Tighten the fitting, then allow your assistant to return the pedal to its rest position. Repeat this procedure on the other fittings until the fluid is free of air bubbles.

18 Reinstall any components that were removed for access to the master cylinder.

19 Re-check the brake fluid level, then check the operation of the brake system carefully before driving the vehicle in traffic.

9   Proportioning valves - description, check and replacement

DESCRIPTION

1   All non-ABS models have two screw-in proportioning valves (see illustration 8.3) that balance front to rear braking by limiting the rise in hydraulic pressure in the rear brake lines when it exceeds the level established by the factory as a safe pressure. Under light pedal pressure, the valve allows full hydraulic pressure to the front and rear brakes. But above a certain pressure (known as the "split point"), the proportioning valves reduce the amount of pressure increase to the rear brakes in accordance with a predetermined ratio. This decreases the chance of rear wheel lock-up and skidding.

2   Models equipped with ABS do not use proportioning valves to balance front-to-rear braking pressure. Instead, they use Electronic Brake Distribution (EBD), which uses the ABS system to control the rear wheel slip.

CHECK

3   If either rear wheel skids prematurely under hard braking, it could indicate a defective proportioning valve. If this occurs, drive the vehicle to a dealer immediately and have the system checked out by a competent professional in the dealer service department. A pair of special pressure gauges is required for diagnosing the proportioning valve.

REPLACEMENT

❋❋ **CAUTION:**

Brake fluid will damage paint. Cover all painted surfaces and avoid spilling fluid during this procedure.

4   Loosen the brake hydraulic fluid lines from the proportioning valve with a flare-nut wrench to prevent rounding off the corners of the tube nut fittings (see illustration 8.4). Back off the fittings and disconnect the lines. Plug the ends of the lines to prevent loss of brake fluid and the entry of dirt.

5   Unscrew the valve from the master cylinder.

6   Install a new O-ring on the new proportioning valve. Even if you are going to re-use the old proportioning valve, discard the old O-ring and install a new one. Then lubricate the O-ring with a little clean brake fluid.

7   Screw in the proportioning valve and tighten it to the torque listed in this Chapter's Specifications.

8   Installation is otherwise the reverse of removal.

9   Bleed the brake system (see Section 11).

10 Brake hoses and lines - inspection and replacement

1 About every six months, with the vehicle raised and placed securely on jackstands, the flexible hoses which connect the steel brake lines with the front and rear brake assemblies should be inspected for cracks, chafing of the outer cover, leaks, blisters and other damage. These are important and vulnerable parts of the brake system and inspection should be complete. A light and mirror will be needed for a thorough check. If a hose exhibits any of the above defects, replace it with a new one.

FLEXIBLE HOSE REPLACEMENT

▶ **Refer to illustration 10.3**

2 Clean all dirt away from the ends of the hose.

3 Disconnect the brake line from the hose fitting (see illustration). Be careful not to bend the frame bracket or line. If the threaded fitting is corroded, soak it with penetrating oil and allow the penetrate time to loosen it up, then try again. If you try to break loose a brake nut that's frozen, you will kink the metal line, which will then have to be replaced.

4 Separate the metal line from the connection and pull the flexible hose through the bracket. Immediately plug the metal line to prevent excessive fluid loss or contamination.

5 Unscrew the banjo bolt at the caliper and disconnect the hose from the caliper, discarding the sealing washers on either side of the fitting.

6 Using new sealing washers, attach the new brake hose to the caliper. Tighten the banjo bolt to the torque listed this Chapter's Specifications.

7 Insert the other end of the new hose through the bracket. Make sure the hose isn't kinked or twisted, then attach metal line to the hose and tighten the brake line fitting nut securely.

8 Carefully check to make sure the suspension or steering components don't make contact with the hose. Have an assistant push down on the vehicle while you watch to see whether the hose interferes with suspension operation. If you're replacing a front hose, have your assistant turn the steering wheel lock-to-lock while you make sure the hose doesn't interfere with the steering linkage or the steering knuckle.

9 Bleed the brake system (see Section 11).

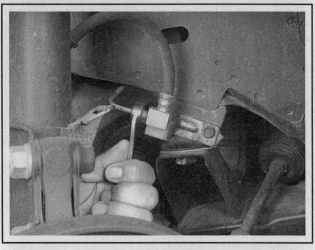

10.3 To disconnect a flexible brake hose from a metal brake line, simply unscrew the threaded fitting nut as shown, but make sure you don't kink the metal line; if the fitting nut is frozen, soak it with penetrating oil and try again

METAL BRAKE LINES

10 When replacing brake lines, be sure to use the correct parts. Don't use copper tubing for any brake system components. Purchase steel brake lines from a dealer parts department or auto parts store.

11 Prefabricated brake line, with the tube ends already flared and fittings installed, is available at auto parts stores and dealer parts departments. These lines are also sometimes bent to the proper shapes.

12 When installing the new line make sure it's well supported in the brackets and has plenty of clearance between moving or hot components. Make sure you tighten the fittings securely.

13 After installation, check the master cylinder fluid level and add fluid as necessary. Bleed the brake system as outlined in Section 11 and test the brakes carefully before placing the vehicle into normal operation.

11 Brake hydraulic system - bleeding

▶ **Refer to illustrations 11.7, 11.8 and 11.10**

> ※※ **WARNING:**
>
> Wear eye protection when bleeding the brake system. If the fluid comes in contact with your eyes, immediately rinse them with water and seek medical attention.

➡**Note: Bleeding the brake system is necessary to remove any air that's trapped in the system when it's opened during removal and installation of a hose, line, caliper, wheel cylinder or master cylinder.**

1 If a brake line was disconnected only at a wheel, then only that caliper or wheel cylinder must be bled.

2 On conventional (non-ABS) brake systems, if air has entered the system due to low fluid level, all four brakes must be bled.

> ※※ **WARNING:**
>
> If this has occurred on a model with an Anti-lock Brake System (ABS), or if the lines to the Hydraulic Control Unit (HCU) have been disconnected, the vehicle must be towed to a dealer service department or other repair shop equipped with a DRB II scan tool to have the system properly bled.

3 If a brake line is disconnected at a fitting located between the master cylinder and any of the brakes, that part of the system served by the disconnected line must be bled.

4 Remove any residual vacuum from the brake power booster (if equipped) by applying the brake several times with the engine off.

5 Remove the master cylinder reservoir cover and fill the reservoir with brake fluid. Reinstall the cover.

➡**Note: Check the fluid level often during the bleeding opera-**

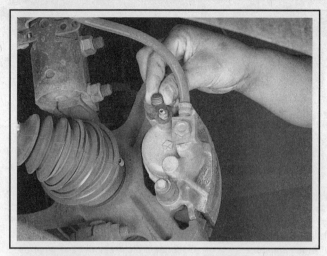

11.7 Remove the cap from each bleeder valve

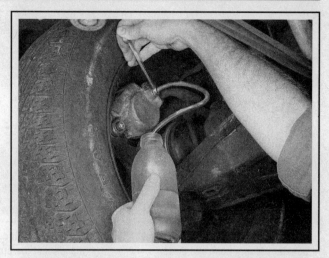

11.8 When bleeding the brakes, a hose is connected to the bleed screw at the caliper or wheel cylinder and then submerged in brake fluid - air will be seen as bubbles in the tube and container (all air must be expelled before moving to the next wheel)

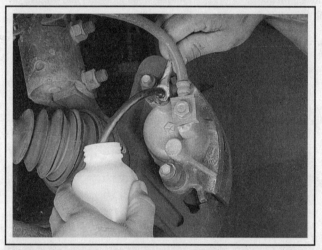

11.10 Open the bleeder screw just enough to allow a flow of fluid to leave the valve

tion and add fluid as necessary to prevent the fluid level from falling low enough to allow air bubbles into the master cylinder.

6   Have an assistant on hand, as well as a supply of new brake fluid, an empty clear container, a length of 3/16-inch clear plastic or vinyl tubing to fit over the bleeder valve and a wrench to open and close the bleeder valve.

7   Beginning at the right rear wheel, remove the bleeder cap (see illustration), loosen the bleeder screw slightly, then tighten it to a point where it's snug but can still be loosened quickly and easily.

8   Place one end of the tubing over the bleeder screw fitting and submerge the other end in brake fluid in the container (see illustration).

9   Have the assistant pump the brakes a few times to get pressure in the system, then hold the pedal firmly depressed.

10   While the pedal is held depressed, open the bleeder screw just enough to allow a flow of fluid to leave the valve (see illustration). Watch for air bubbles to exit the submerged end of the tube. When the fluid flow slows after a couple of seconds, tighten the screw and have your assistant release the pedal.

11   Repeat Steps 9 and 10 until no more air is seen leaving the tube, then tighten the bleeder screw and proceed to the left rear wheel, the right front wheel and the left front wheel, in that order, and perform the same procedure. Be sure to check the fluid in the master cylinder reservoir frequently.

12   Never use old brake fluid. It contains moisture, which will deteriorate the brake system components and can even boil if the temperature of the brake fluid rises high enough, which will render the brakes useless.

13   Refill the master cylinder with fluid at the end of the operation. Reinstall the bleeder caps.

14   Check the operation of the brakes. The pedal should feel solid when depressed, with no sponginess. If necessary, repeat the entire process.

✳✳ **WARNING:**

Do not operate the vehicle if the pedal feels low or spongy, if the ABS light on the dash won't go off or if you are in doubt about the effectiveness of the brake system.

12 Power brake booster - check, removal and installation

OPERATING CHECK

1   Depress the brake pedal several times with the engine off and make sure that there is no change in the pedal reserve distance.

2   Depress the pedal and start the engine. If the pedal goes down slightly, operation is normal.

AIRTIGHTNESS CHECK

3   Start the engine and turn it off after one or two minutes. Depress the brake pedal several times slowly. If the pedal goes down farther the first time but gradually rises after the second or third depression, the booster is airtight.

12.8 To disconnect the power brake booster vacuum hose and these accessory vacuum lines, simply pull straight forward; be sure to inspect and, if necessary, replace the grommet where the vacuum hose and lines insert into the booster

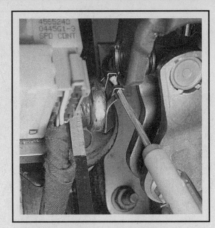

12.10 Pry off the retainer and disconnect the power brake booster pushrod from the top of the brake pedal. Discard the old retainer clip and buy a new clip for reassembly

12.11 Working up under the dash, disconnect the brake booster pushrod from the brake pedal arm, then remove these four nuts and remove the booster

4   Depress the brake pedal while the engine is running, then stop the engine with the pedal depressed. If there is no change in the pedal reserve travel after holding the pedal for 30 seconds, the booster is airtight.

5   The power brake booster unit requires no special maintenance apart from periodic inspection of the vacuum hoses and the case. The booster is not serviceable. If a problem develops, it must be replaced with a new one.

REMOVAL

▶ Refer to illustrations 12.8, 12.10 and 12.11

6   Remove the battery (see Chapter 5).

7   Remove the air filter housing (See Chapter 4).

8   Disconnect the vacuum hose from the power brake booster (see illustration).

9   Remove the master cylinder from the power brake booster (see Section 8).

10   Working inside the vehicle under the dash, disconnect the power brake pushrod from the top of the brake pedal by prying off the retaining clip (see illustration). For safety reasons, discard the old pushrod retaining clip and buy a new clip for reassembly.

11   Remove the nuts attaching the booster to the firewall (see illustration).

12   Working inside the engine compartment, carefully withdraw the power brake booster unit from the firewall and out of the engine compartment.

INSTALLATION

13   To install the booster, place it into position on the firewall and tighten the retaining nuts to the torque listed in this Chapter's Specifications. Connect the pushrod to the brake pedal.

✳✳ WARNING:

Use a new retainer clip. DO NOT reuse the old clip.

14   Install the master cylinder (see Section 8) and then bleed the brake hydraulic system (see Section 11).

15   The remaining installation steps are the reverse of removal.

16   Carefully test the operation of the brakes before placing the vehicle in normal operation.

13 Parking brake shoes (models with rear disc brakes) - replacement and adjustment

✳✳ WARNING:

Dust created by the brake system may contain asbestos, which is harmful to your health. Never blow it out with compressed air and don't inhale any of it. An approved filtering mask should be worn when working on the brakes. Do not, under any circumstances, use petroleum-based solvents to clean brake parts. Use brake system cleaner only.

➡Note: Parking brake shoes should be replaced on both wheels at the same time - never replace the shoes on only one wheel.

REMOVAL

▶ Refer to illustrations 13.6a through 13.6h

1   The parking brake system should be checked as a normal part of driving. With the vehicle parked on a hill, apply the brake, place the transmission in Neutral and verify that the parking brake alone will hold the vehicle (be sure to stay in the vehicle during this check). Additionally, every 24 months - and any time a fault is suspected - the assembly itself should be visually inspected.

13.6a Remove the front brake shoe hold-down clip by pushing in on the clip and turning the retaining pin 90-degrees

13.6b Remove the rear brake shoe hold-down clip

13.6c Pull the upper end of the rear brake shoe away from the parking brake actuator lever . . .

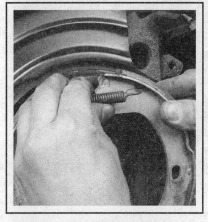

13.6d . . . then unhook the upper spring from the rear shoe

13.6e Disengage the lower end of the rear shoe from the adjuster . . .

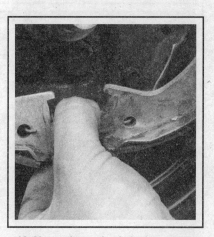

13.6f . . . then unhook the lower spring from the rear shoe and remove the rear shoe

2   Loosen the rear wheel lug nuts, raise the rear of the vehicle and support it securely on jackstands. Block the front wheels and remove the rear wheels. Release the parking brake.

3   Remove the rear calipers (see Section 4). Support the caliper assemblies with a Bungee cord or heavy wire and don't disconnect the brake line from the caliper.

4   Remove the rear discs (see Section 5). Remove the rear hub and bearing assemblies (see Chapter 10).

5   Clean the parking brake assembly with brake system cleaner.

6   Follow the accompanying sequence of photos to remove the parking brake shoes (see illustrations). Be sure to stay in order and read the caption under each illustration.

13.6g Unhook the lower spring from the front brake shoe

13.6h Unhook the upper spring from the front brake shoe

13.10a Lubricate the friction points on the brake backing plate with high-temperature grease

13.10b Insert the end of the parking brake cable into the parking brake actuator lever, if removed

13.10c Insert the pin for the hold-down clip through the brake backing plate

13.10d Install the front parking brake shoe

8   Clean the backing plate with a suitable solvent.
9   Inspect the drum (see Section 6).

INSPECTION

7   Inspect the lining contact pattern to determine whether the shoes are bent or have been improperly adjusted. The lining should show contact across the entire width, extending from head to toe. Shoes showing contact only on one side should be replaced.

INSTALLATION

▶ **Refer to illustrations 13.10a through 13.10l**

10 Follow the accompanying sequence of photos to install the new

13.10e Install the front brake shoe hold-down clip

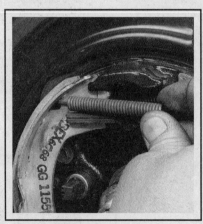

13.10f Hook the upper return spring to the front parking brake shoe

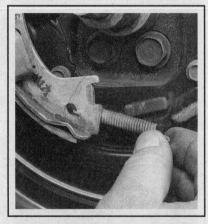

13.10g Hook the lower return spring to the front parking brake shoe

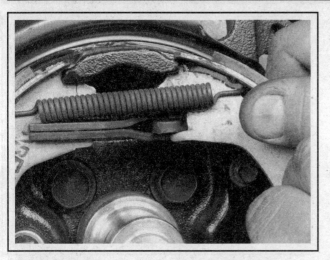

13.10h Hook the upper return spring to the rear parking brake shoe

13.10i Hook the lower return spring to the rear parking brake shoe

shoes (see illustrations), then install the hub and bearing assemblies (see Chapter 10).

11 Before installing the disc, rotate the star wheel on the adjuster until the distance across the friction surfaces of the parking brake shoes is 6-3/4 inches.

12 Install the disc (see Section 5). Using a screwdriver or brake adjusting tool, turn the star wheel on the parking brake shoe adjuster until the shoes slightly drag as the disc is turned, then back off the adjuster until the shoes don't drag.

13 Install the caliper (see Section 4).

14 Repeat this sequence for the other parking brake shoes at the other rear wheel.

13.10j Pull the rear parking brake shoe back and engage it with the actuator lever

13.10k Insert the pin for the rear hold-down clip through the backing plate and install the rear hold-down clip

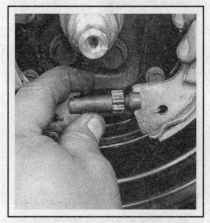

13.10l Install the adjuster between the parking brake shoes, then turn the adjuster star wheel until the distance across the friction surfaces of the shoes is 6-3/4 inches

14 Parking brake lever and automatic adjuster assembly - removal and installation

REMOVAL

▶ Refer to illustrations 14.3, 14.4, 14.5 and 14.6

1 Remove the center console (see Chapter 11).
2 Lower the parking brake lever.

※ WARNING:

The self-adjusting mechanism of the parking brake lever assembly contains a clock spring loaded to about twenty pounds. Use care in handling the parking brake lever assembly. Do not release the self-adjuster lockout device before installing the cables into the equalizer. Keep your hands away from the self-adjuster sector and pawl. Careless handling of the parking brake lever adjuster mechanism could cause serious injury.

3 Grasp the parking brake front output cable and pull it toward the rear. Continue to pull on the front cable until a 3/16-inch drill bit, Allen wrench or small screwdriver can be inserted into the brake handle and the brake sector gear to lock it into place (see illustration). Push the

drill bit or Allen wrench all the way through the mechanism. Locking the sector gear relieves the strain on the rear parking brake cables.

4   Disconnect both rear brake cables from the parking brake equalizer (see illustration).

5   Unplug the electrical connector ground from the brake warning light switch (see illustration).

6   Remove the nuts attaching the parking brake lever assembly to the center console bracket (see illustration).

7   Remove the parking brake lever, the automatic adjuster and the front output cable as an assembly.

INSTALLATION

8   Install the parking brake lever and automatic adjuster assembly onto the center console bracket, install the nuts and tighten securely.

9   Reconnect the electrical connector ground onto the brake warning light switch.

10   Insert each parking brake cable through the hole in the equalizer on the front output cable. Make sure the cables are properly installed and are aligned with the cable track in the lever.

11   Using a pair of pliers, pull out the drill bit or Allen wrench installed in the parking brake sector gear mechanism with a firm and quick motion. When the tool is removed from the parking brake pedal adjuster mechanism, the adjuster automatically adjusts the parking brake cables.

14.3 To lock the sector into place, insert a 3/16-inch drill bit, Allen wrench or small screwdriver into the parking brake lever mechanism; the tool must go all the way through both sides of the parking brake mechanism

12   Apply and release the parking brake lever several times. The rear wheels should rotate freely without the brakes dragging.

13   Install the center console (see Chapter 11).

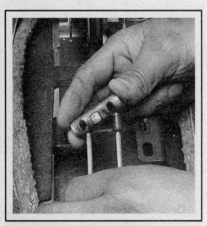

14.4 Hold onto the equalizer and disconnect both rear brake cables

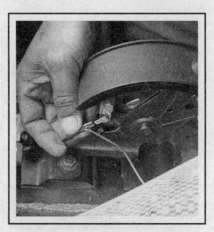

14.5 Unplug the electrical connector from the brake warning light switch

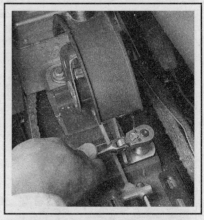

14.6 Remove the bolt attaching the parking brake lever assembly to the floor pan

15 Parking brake cables - replacement

FRONT CABLE

Replacement

➡Note: The front output cable is an integral part of the brake lever assembly and cannot be replaced separately. Do not attempt to remove the cable from the lever assembly since it is attached to the clock spring.

1   Remove the parking brake lever assembly (see Section 14).

2   Install the parking brake lever assembly (see Section 14).

REAR CABLES

Removal

▶ Refer to illustrations 15.8, 15.10, 15.12, 15.13, 15.14 and 15.15

➡Note: Disconnect only one rear parking brake cable from the rear brakes at a time. If you disconnect both cables simultaneously, it will be extremely difficult to connect both of them to the equalizer.

3   Remove the center console (see Chapter 11).

4   Lower the parking brake lever.

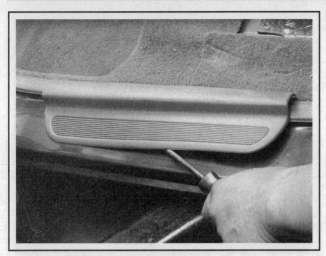

15.8 Remove the left (driver's side) sill molding and kick molding

5   Grasp the parking brake front output cable and pull it toward the rear. Continue to pull on the front cable until a 3/16-inch drill bit or Allen bolt can be inserted into the brake handle and the brake sector gear to lock it into place (see illustration 14.3). Push the drill bit or Allen wrench all the way through the mechanism. Locking the sector gear relieves the strain on the rear parking brake cables.

6   Disconnect one of the rear brake cables from the parking brake equalizer (see illustration 14.4).

7   Remove the rear seat assembly (see Chapter 11).

8   On four-door models, carefully pry on the retaining clips and remove the rear doorsill scuff plate on each side (see illustration).

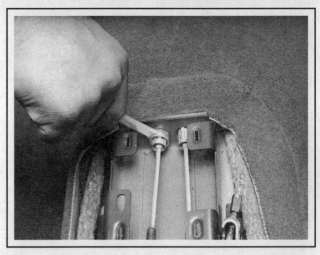

15.10 Use a 1/2-inch box wrench and compress the retaining tabs of the cable housing - then pull the cable through the opening in the center console bracket

9   Fold the rear carpeting forward to expose the parking brake cables.

10 Compress the retaining tabs of the cable housing with a 1/2-inch box wrench or a pair of pliers (see illustration). Pull the cable through the opening in the center console bracket.

11 Loosen the rear wheel lug nuts, raise the rear of the vehicle and place it securely on jackstands. Remove the rear wheels.

12 On models with rear drum brakes, remove the rear drum (see Section 6) and the rear hub and bearing assembly (see Chapter 10). Disassemble the brake shoe assembly (see Section 6). Disconnect the cable from the parking brake lever (see illustration). Using a small hose clamp, squeeze the cable housing retainer tabs and pull the cable through the brake backing plate. If you don't have a clamp handy, use a 1/2-inch box wrench or a pair of pliers (using pliers isn't as easy).

13 On models with rear disc brakes, remove the rear caliper (see Section 4) and disc (see Section 5). Disassemble the parking brake shoe assembly (see Section 13). Disconnect the cable from the parking brake actuator lever (see illustration), then squeeze the retainer tabs on the cable housing and pull the cable through the backing plate.

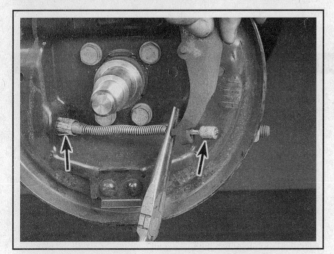

15.12 On models with rear drum brakes, disconnect the cable from the parking brake actuator lever (right arrow), squeeze the retainer tabs (left arrow) on the cable housing and pull the cable through the brake backing plate

15.13 On models with rear disc brakes, disconnect the cable from the parking brake actuator lever, then squeeze the retainer tabs on the cable housing and pull the cable through the brake backing plate

14 Remove the cable bracket from the frame rail (see illustration).

15 Remove the cable sealing grommet from the floor pan (see illustration).

16 Remove the cable assembly from the vehicle.

Installation

17 Insert the front end of the rear cable through the hole in the floor-pan. Make sure the sealing grommet is installed correctly to ensure a proper seal.

18 Insert the rear end of the new rear cable through the hole in the brake backing plate. Make sure the cable is pulled through the hole far enough to allow the retaining tabs to expand all the way around the cable, locking the cable to the backing plate.

19 Install the cable bracket onto the frame rail and tighten the bolt securely.

20 On models with rear drum brakes, connect the cable to the parking brake lever, reassemble the rear brake assembly (see Section 13), install the rear hub and bearing assembly (see Chapter 10) and install the rear drum.

21 On models with rear disc brakes, connect the cable to the parking brake actuator lever, reassemble the parking brake shoe assembly (see Section 13), install the disc (see Section 5) and install the rear caliper (see Section 4).

22 Install the rear wheels, hand tighten the wheel lug nuts, remove the jackstands and lower the vehicle. Tighten the rear wheel lug nuts to the torque listed in the Chapter 1 Specifications.

23 Grasp the parking brake cable grommet at the floor pan and pull hard to make sure the grommet is fully seated to the floor pan.

24 Route the cable under the carpeting and up to the cable retaining bracket on the floor pan.

25 Install the cable through the cable retaining bracket and make sure the cable retainers have expanded out to hold the cable in place in the bracket.

26 Grasp the equalizer firmly, pull it to the rear and connect the rear cable to the equalizer (see illustration 14.4).

27 Repeat this procedure for the other rear cable, if you're replacing both cables.

28 Fold back the carpet and on four-door models, install both rear door opening sill scuff plates. Install the rear seat assembly (see Chapter 11).

29 Install the center console (see Chapter 11).

15.14 Remove the cable retaining bolts attaching the rear cable to the floorpan and frame

15.15 Remove the cable sealing grommet from the floorpan and remove the cable assembly

16 Brake light switch - check, replacement and adjustment

▶ **Refer to illustration 16.7**

1 The brake light switch is a normally-open switch that controls the operation of the vehicle brake lights. The switch is located near the top of the brake pedal and is attached to the bracket. When the brake pedal is applied, a spring-loaded plunger closes the circuit to the left and right brake lights.

2 On models with cruise control, the brake light switch also deactivates the cruise control system when the brake pedal is depressed.

CHECK

3 If the brake lights don't come on when the brake pedal is applied, check the brake light fuse (see Chapter 12). If the fuse has blown, look for a short in the brake light circuit.

4 If the fuse is okay, use a test light or voltmeter to verify that there's voltage to the switch. If there's no voltage to the switch, look for an open or short in the power wire to the switch. Repair the power wire.

5 If the brake lights still don't come on when the brake pedal is applied, unplug the electrical connector from the brake light switch and, using an ohmmeter, verify that there's continuity between the switch terminals when the brake pedal is applied, i.e. the switch is closed. If there isn't, replace the switch.

6 If there is continuity between the switch terminals when the brake is applied, but the brake lights don't come on when the brake pedal is applied, check the wiring between the switch and the brake lights for an open circuit.

REPLACEMENT

7 Depress and hold the brake pedal, then rotate the brake light switch about 30-degrees in a counterclockwise direction (see illustration).

8 Pull the switch to the rear and remove it from its mounting bracket.

9 Unplug the electrical connector from the switch.

10 Hold onto the switch and pull the plunger outward until it has ratcheted to its fully extended position.

11 Install and adjust the new switch (see following).

ADJUSTMENT

12 Depress the brake pedal as far as it will go, then install the switch in the bracket by aligning the index key on the switch with the slot at the top of the square hole in the mounting bracket.

❋❋ CAUTION:

Don't use excessive force when pulling back on the brake pedal to adjust the switch. If you use too much force, you will damage the switch or the striker. When the switch is fully installed in the bracket, rotate the switch clockwise about 30-degrees to lock the switch into the bracket.

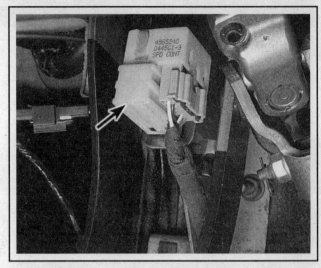

16.7 Hold the brake pedal down, then rotate the brake light switch about 30-degrees in a counterclockwise direction and remove the switch

13 Gently pull back on the brake pedal until the pedal stops moving. The switch plunger will ratchet backward to the correct position.

14 Plug the electrical connector into the switch.

Specifications

General

Brake fluid type	See Chapter 1

Disc brakes

Minimum pad lining thickness	See Chapter 1
Disc minimum thickness	Cast into disc
Disc runout limit (front and rear)	0.005 inch (0.13 mm)
Disc thickness variation (front and rear)	0.0005 inch (0.013 mm)

Drum brakes

Shoe lining minimum thickness	See Chapter 1
Maximum drum diameter	Cast into drum

Torque specifications

	Ft-lbs (unless otherwise indicated)	Nm
ABS wheel speed sensor mounting bolts	108 in-lbs	12
Brake hose banjo bolts		
2000 through 2002 models	35	48
2003 and later models	216 in-lbs	24
Caliper guide pin bolts (front and rear)	192 in-lbs	22
Master cylinder-to-brake booster mounting nuts		
2000 through 2002 models	250 in-lbs	28
2003 and later models	160 in-lbs	18
Power brake booster mounting nuts	25	34
Proportioning valves (non-ABS models)		
2000 and 2001	30	40
2002 and later models	155 in-lbs	17.5
Rear caliper mounting bracket bolts	55	75

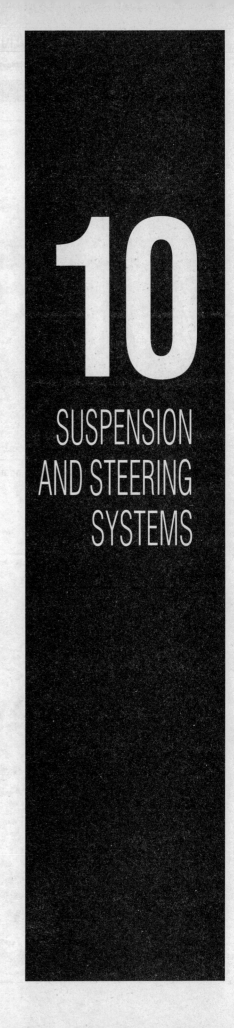

10

SUSPENSION AND STEERING SYSTEMS

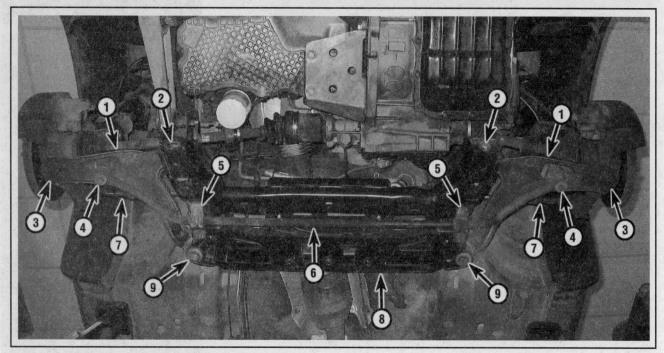

1.1a Front steering and suspension components

1	Control arms	4	Stabilizer bar link bolts	7	Tie-rod ends
2	Control arm front pivot bolts	5	Stabilizer bar bushing clamps	8	Crossmember
3	Control arm-to-steering knuckle balljoints	6	Stabilizer bar	9	Control arm rear bolts

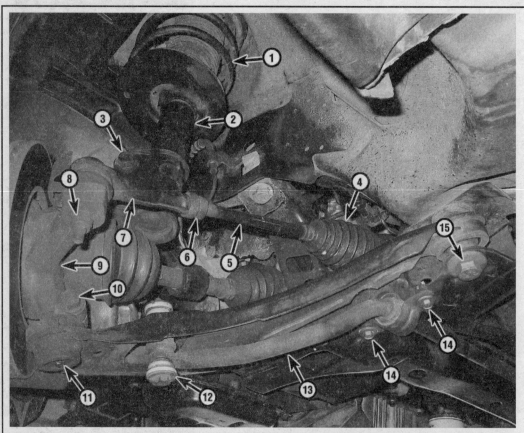

1.1b Front steering and suspension components

1   Strut coil spring
2   Strut housing
3   Strut-to-steering knuckle bolts (other bolt not visible)
4   Steering gear boot
5   Tie-rod
6   Tie-rod end jam nut
7   Tie-rod end
8   Tie-rod end ballstud nut
9   Steering knuckle
10 Steering knuckle-to-tie-rod end ballstud nut
11 Control arm balljoint
12 Stabilizer bar link bolt and nut
13 Stabilizer bar
14 Stabilizer bar bushing clamp bolts
15 Control arm rear bolt

▶ **Refer to illustrations 1.1a, 1.1b and 1.2**

The front suspension is a MacPherson strut design. The upper end of each strut assembly is attached to the vehicle body by three nuts. The lower ends of the struts are bolted to the upper ends of the steering knuckles. The lower ends of the steering knuckles are attached to balljoints mounted in the outer ends of the control arms (see illustrations).

The rear suspension also utilizes strut/coil spring assemblies. The upper end of each strut is attached to the vehicle body by a strut support. The lower end of the strut is attached to the rear knuckle. The knuckle is located by a pair of suspension arms on each side, and a longitudinally mounted trailing arm between the body and the knuckle (see illustration).

The power-assisted rack-and-pinion steering gear is attached to the front crossmember. The steering gear actuates the tie-rods, the outer ends of which are attached to the steering knuckles. The lower end of the steering column is attached to the steering gear by a pinch bolt.

Frequently, when working on the suspension or steering system components, you will encounter fasteners that seem impossible to loosen. That's because fasteners on the underside of the vehicle are continually subjected to water, road grime, mud, etc., and can become rusted or "frozen" in place, making them extremely difficult to remove. In order to unscrew these stubborn fasteners without damaging them or other components, be sure to use lots of penetrant (penetrating oil) and allow it to soak in for a while. Using a wire brush to clean exposed threads will also ease removal of the nut or bolt and prevent damage to the threads. Sometimes a sharp blow with a hammer and punch will break the bond between a nut and bolt threads, but care must be taken to prevent the punch from slipping off the fastener and ruining the threads. Heating the stuck fastener and surrounding area with a torch sometimes helps too, but isn't recommended because of the obvious dangers associated with fire. Long breaker bars and extension, or "cheater," pipes will increase leverage, but never use an extension pipe on a ratchet - the ratcheting mechanism could be damaged. Sometimes tightening the nut or bolt first will help to break it loose. Fasteners that require drastic measures to remove should always be replaced with new ones.

Since most of the procedures in this Chapter involve jacking up the vehicle and working underneath it, a good pair of jackstands will be needed. A hydraulic floor jack is the preferred type of jack to lift the vehicle, and it can also be used to support certain components during various operations.

✲✲ WARNING:

Never, under any circumstances, rely on a jack to support the vehicle while working on it. Whenever any of the suspension or steering fasteners are loosened or removed they must be inspected and, if necessary, replaced with new ones of the same part number or of original equipment quality and design. Torque specifications must be followed for proper reassembly and component retention. Never attempt to heat or straighten any suspension or steering components. Instead, replace any bent or damaged part with a new one.

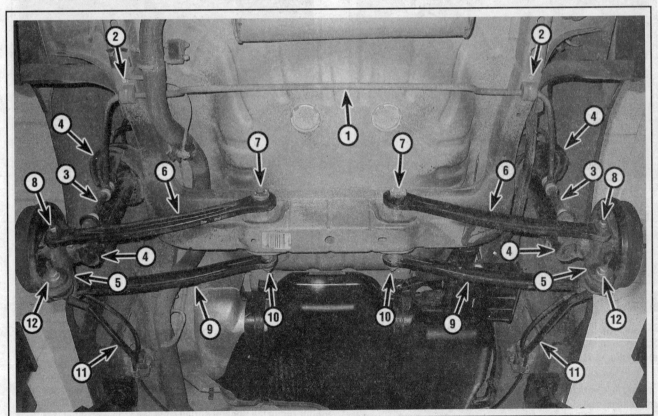

1.2 Rear suspension components

1	Stabilizer bar	5	Rear spindles
2	Stabilizer bar bushing clamps	6	Rear lateral arms
3	Stabilizer bar links	7	Rear lateral arm pivot bolts
4	Strut assemblies	8	Rear lateral arm-to-spindle bolts/nuts

9	Front lateral arms
10	Front lateral arm pivot bolts
11	Trailing arms
12	Trailing arm-to-spindle nuts

2   Strut assembly (front) - removal and installation

REMOVAL

▶ **Refer to illustrations 2.4 and 2.6**

1   Loosen the front wheel lug nuts.

2   Raise the vehicle and support it securely on jackstands. Remove the front wheels.

➡**Note: If both strut assemblies are going to be removed, mark the assemblies right and left so they will be reinstalled on the correct side.**

3   Mark the position of the strut to the steering knuckle. This is only necessary if special camber adjusting bolts have been installed in place of the regular strut-to-knuckle bolts.

4   Disconnect the ground wire from the strut (see illustration). On ABS equipped models, the wheel speed sensor is also attached to the strut.

5   Remove the strut-to-steering knuckle nuts and bolts (see illustration 2.4).

6   Remove the upper mounting nuts (see illustration), disengage the strut from the steering knuckle and detach it from the vehicle.

7   Inspect the strut and coil spring assembly for leaking fluid, dents, damage and corrosion. If the strut is damaged, see Section 3.

INSTALLATION

8   To install the strut, place it in position with the studs extending up through the shock tower. Install the nuts and tighten them to the torque listed in this Chapter's Specifications.

9   Attach the strut to the steering knuckle, then insert the strut-to-steering knuckle bolts. Install the nuts, but don't tighten them yet.

10   Attach the ground wire (and ABS speed sensor, if equipped) bracket to the strut and tighten the bolt.

11   Tighten the steering knuckle-to-strut nuts to the torque listed in this Chapter's Specifications.

☀ CAUTION:

Do NOT turn the bolts if they are serrated. Instead, hold the bolt stationary and tighten the nut onto the bolt.

12   Install the wheels and lower the vehicle. Tighten the lug nuts to the torque listed in the Chapter 1 Specifications.

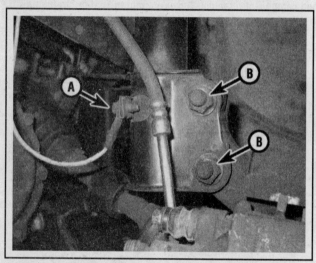

2.4   Remove this bolt (A) and detach the ground wire from the strut, then remove the strut-to-knuckle nuts/bolts (B)

2.6   Remove these three upper mounting nuts - DO NOT remove the center nut!

3   Strut /coil spring - replacement

➡**Note: You'll need a spring compressor for this procedure. Spring compressors are available on a daily rental basis at most auto parts stores or equipment yards.**

1   If the struts or coil springs exhibit the telltale signs of wear (leaking fluid, loss of damping capability, chipped, sagging or cracked coil springs) explore all options before beginning any work. The strut/shock absorber assemblies are not serviceable and must be replaced if a problem develops. However, strut assemblies complete with springs may be available on an exchange basis, which eliminates much time and work. Whichever route you choose to take, check on the cost and availability of parts before disassembling your vehicle.

☀ WARNING:

Disassembling a strut is potentially dangerous and utmost attention must be directed to the job, or serious injury may result. Use only a high-quality spring compressor and carefully follow the manufacturer's instructions furnished with the tool. After removing the coil spring from the strut assembly, set it aside in a safe, isolated area.

DISASSEMBLY

▶ **Refer to illustrations 3.3, 3.4, 3.5, 3.6, 3.7a, 3.7b, 3.8 and 3.9**

2   Remove the strut and spring assembly (see Section 2).
3   Mount the strut clevis bracket portion of the strut assembly (see illustration).

✳✳ CAUTION:

Do not clamp any other part of the strut assembly in the vise, because it might be damaged. Line the vise jaws with wood or rags to prevent damage to the unit and don't tighten the vise excessively.

3.3 Mount the strut clevis bracket portion of the strut assembly in a vise

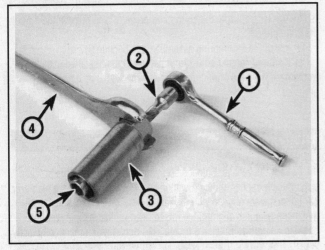

3.5 Here's the setup that can be used to unscrew the damper shaft nut

1   *1/4-inch drive ratchet*
2   *Extension*
3   *13/16-inch spark plug socket*
4   *Wrench to turn socket*
5   *10 mm socket (to hold damper shaft)*

4   Following the tool manufacturer's instructions, install the spring compressor (which can be obtained at most auto parts stores or equipment yards on a daily rental basis) on the spring and compress it sufficiently to relieve all pressure from the upper spring seat (see illustration). This can be verified by wiggling the spring.
5   While holding the damper shaft from turning, loosen the shaft nut with a socket. A special tool is available to do this, but a substitute can be made from a 13/16-inch spark plug socket (with a hex surface at the top), a ratchet, a 1/4-inch drive extension inserted through the hole in the spark plug socket, and a 10 mm socket attached to the extension (see illustration).
6   Remove the nut and suspension support (see illustration). Inspect the bearing in the suspension support for smooth operation. If it doesn't turn smoothly, replace the suspension support. Check the rubber portion of the suspension support for cracking and general deterioration. If there is any separation of the rubber, replace it.
7   Remove the upper spring seat and dust boot from the damper shaft (see illustrations). Check the spring seat for cracking and hardness; replace it if necessary.
8   Slide the rubber bumper off the damper shaft (see illustration).
9   Carefully lift the compressed spring from the assembly (see illustration) and set it in a safe place.

3.4 Install the spring compressor in accordance with the tool manufacturer's instructions and compress the spring until all pressure is removed from the upper suspension support

3.6 Remove the damper shaft nut and suspension support

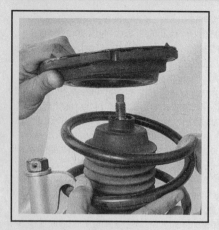

3.7a Remove the upper spring seat . . .

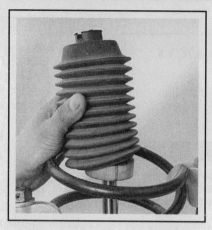

3.7b . . . and the dust boot from the damper shaft

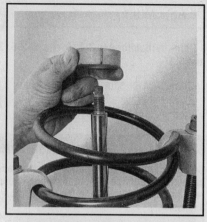

3.8 Remove the rubber bumper from the damper shaft

❊❊ **WARNING:**

When removing the compressed spring, lift it off very carefully and set it in a safe place. Keep the ends of the spring away from your body.

➡**Note:** Mark the spring so it can be reinstalled on the same side of the vehicle from which it was removed.

REASSEMBLY

10 Extend the damper rod to its full length and install the rubber bumper.

11 Position the coil spring with the smaller coils going on first and carefully place the coil onto the damper shaft.

12 Install the rubber bumper onto the damper shaft and push it all the way down until it bottoms out.

13 Install the upper spring seat and dust boot onto the damper shaft.

14 Install the suspension support and align the support notch with the clevis bracket on the strut support. Install the suspension support to the damper shaft.

15 Install the nut damper shaft and tighten it to the torque listed in this Chapter's Specifications.

16 Equally loosen the coil spring compressors until the top coil is properly seated against the upper spring seat and suspension support.

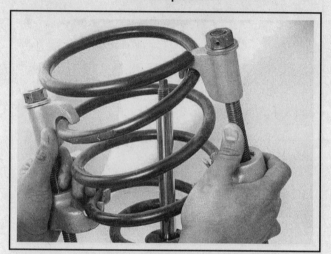

3.9 Carefully lift the compressed spring from the assembly and set it in a safe place - keep the ends of the spring pointed away from your body

Relieve all tension from the spring compressors and remove them from the coil spring.

17 Install the strut/spring assembly (see Section 2).

18 If special camber adjusting bolts have been installed sometime in the past, have the front end alignment checked and, if necessary, adjusted.

---

4   Stabilizer bar and bushings (front) - removal and installation

▶ **Refer to illustrations 4.2 and 4.3**

1   Loosen the front wheel lug nuts, raise the front of the vehicle, support it securely on jackstands and remove the front wheels.

2   Remove the stabilizer bar link bolts, nuts and insulators from the control arms (see illustration).

3   Support the stabilizer bar and remove the stabilizer bar clamp bolts and clamps from the front suspension crossmember (see illustration). Remove the stabilizer bar from the vehicle.

4   Check the bar for damage, corrosion and signs of twisting.

5   Check the clamps, bushings and retainers for distortion, damage and wear. Replace the bushings by prying them open at the split and removing them. Install the new bushings with the split facing toward

the front of the vehicle. Silicone spray lubricant will ease this process.

6   Attach the bar to the crossmember so the cutouts in the stabilizer bar bushings are aligned with the raised bead on the crossmember. Install the retainers aligning the raised bead on the retainer with the bushing cutouts and install the bolts, but don't tighten them completely yet.

7   Install the link bolts, insulators and nuts, tightening the bolts to the torque listed in this Chapter's Specifications.

8   Raise the control arms to normal ride height and tighten the bolts to the torque listed in this Chapter's Specifications.

9   Install the wheels and lower the vehicle. Tighten the lug nuts to the torque listed in the Chapter 1 Specifications.

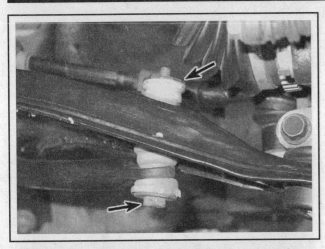

4.2 Hold the nut (upper arrow) with a wrench, unscrew the bolt (lower arrow) and detach the link from the control arm (note the order of the insulators as you remove the bolt) - repeat on the other side

4.3 Support the stabilizer bar and remove both stabilizer bar clamp bolts and clamps from the front suspension crossmember

5 Control arm - removal, inspection and installation

REMOVAL

▶ **Refer to illustrations 5.2, 5.3, 5.4 and 5.5**

1 Loosen the wheel lug nuts on the side to be dismantled, raise the front of the vehicle, support it securely on jackstands and remove the wheel.

2 Remove the balljoint clamping bolt and nut (see illustration).

3 Disconnect the stabilizer bar link from the control arm (see Section 4). If you're removing the right side control arm, remove the engine torque strut (see illustration).

4 Use a prybar to disconnect the control arm from the steering knuckle (see illustration). Pull the balljoint stud from the steering knuckle.

✷✷ CAUTION:

Do not move the steering knuckle/strut assembly out or you may separate the inner CV joint.

5.2 Remove the nut from this pinch bolt, then remove the bolt

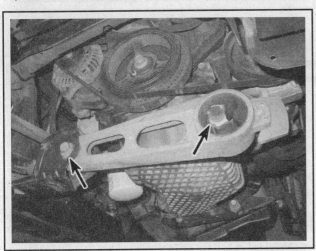

5.3 If you're removing the right control arm you'll have to remove the engine torque strut in order to remove the front pivot bolt

5.4 Separate the control arm from the steering knuckle with a prybar

5   Remove the bolt and nut that attach the front of the control arm to the crossmember (see illustration).

➡ **Note: Turn the bolt, not the nut (the nut is equipped with an anti-rotation tang).**

6   Remove the bolt that attaches the rear of the control arm to the crossmember (see illustration 5.5).

7   Remove the control arm.

INSPECTION

8   Make sure the control arm is straight. If it's bent, replace it. Do not attempt to straighten a bent control arm.

9   Inspect the bushings. If they're cracked, torn or worn out, replace the control arm.

INSTALLATION

10 Installation is the reverse of removal. Tighten the control arm's rear bolt first, then the front bolt and nut. Be sure to tighten all fasteners to the torque listed in this Chapter's Specifications.

11 Install the wheel and lug nuts, lower the vehicle and tighten the lug nuts to the torque listed in the Chapter 1 Specifications.

12 It's a good idea to have the front wheel alignment checked, and if necessary, adjusted after this job has been performed.

5.5 To detach the control arm from the crossmember, remove the pivot bolt (upper arrow) and nut (center arrow), then remove the vertical rear mounting bolt (lower arrow)

6   Balljoints - check and replacement

▶ **Refer to illustration 6.2**

1   The suspension balljoints are designed to operate without freeplay.

2   With the weight of the vehicle resting on its wheels, try to wiggle the snapped-off grease fitting in the center of the balljoint, using only your fingers (see illustration). If there is any movement, the balljoint is worn and must be replaced with a new one. Remove the control arm (see Section 5) and take it to an automotive machine shop to have the old balljoint pressed out and a new one pressed in.

6.2 If you can feel any play when wiggling the snapped-off grease fitting, the balljoint is worn out and should be replaced

7   Steering knuckle, hub and bearing - removal, inspection and installation

REMOVAL

1   With the vehicle weight resting on the front suspension, remove the hub cap, cotter pin, nut lock and spring washer. Loosen, but do not remove, the front hub (driveaxle) nut and wheel lug nuts.

2   Raise the front of the vehicle, support it securely on jackstands and remove the front wheels.

3   Remove the caliper and brake pads (see Chapter 9). Taking care not to twist the brake hose, hang the caliper out of the way in the wheel well with a piece of wire. If the vehicle is equipped with ABS, remove the wheel speed sensor from the steering knuckle.

4   Disconnect the tie-rod end from the steering knuckle (see Section 16).

5   Move the tie-rod out of the way and secure it with a piece of wire.

6   Detach the lower control arm balljoint from the steering knuckle (see Section 5).

7   With the knuckle and hub assembly in the straight-ahead position, grasp it securely and pull it directly out and off the driveaxle splines.

❋❋ **CAUTION:**

Be careful not to pull the driveaxle out or you may disengage the inner CV joint. It may be necessary to tap on the axle end with a soft-face hammer to dislodge the driveaxle from the hub. Secure the end of the driveaxle with a piece of wire.

8   Mark the position of the strut to the steering knuckle.

➡**Note: This is only necessary if special camber adjusting bolts have been installed in place of the regular strut-to-knuckle bolts.**

9   Remove the strut-to-steering knuckle nuts and bolts (see illustration 2.4). Remove the steering knuckle.

INSPECTION

10   Place the assembly on a clean work surface and wipe it off with a lint-free cloth. Inspect the knuckle for rust, damage and cracks. Check the bearings by rotating them to make sure they move freely, without excessive noise or looseness. The bearings should be packed with an adequate supply of clean grease.

➡**Note: Further disassembly will have to be left to your dealer service department or a repair shop because of the special tools required.**

INSTALLATION

11   Prior to installation, clean the CV joint seal and the hub grease seal with solvent (don't get any solvent on the CV joint boot). Lubricate the entire circumference of the CV joint wear sleeve and seal contact surface with multi-purpose grease (see Chapter 8).

12   Carefully place the knuckle and hub assembly in position. Align the splines of the axle and the hub and slide the hub into place.

13   Install the knuckle-to-strut bolts and nuts, followed by the balljoint pinch bolt and nut.

❋❋ CAUTION:

Do not turn the bolts if they are serrated. Instead, hold the bolt stationary and tighten the nut onto the bolt.

Tighten all fasteners to the torque listed in this Chapter's Specifications.

14   Reattach the tie-rod end to the steering knuckle and tighten the nut (see Section 16).

15   Install the brake disc, pads and caliper (see Chapter 9).

16   Push the CV joint completely into the hub to make sure it is seated and install the washer and hub nut finger tight.

17   Install the wheels and lug nuts, then lower the vehicle and tighten the wheel lug nuts to the torque specified in Chapter 1.

18   With an assistant applying the brakes, tighten the hub nut to the torque specified in Chapter 8. Install the spring washer, nut lock and a new cotter pin.

19   If special camber adjusting bolts have been installed sometime in the past, have the front end alignment checked and, if necessary, adjusted.

8   Stabilizer bar and bushings (rear) - removal and installation

▶ **Refer to illustrations 8.2 and 8.3**

1   Loosen the rear wheel lug nuts. Raise the rear of the vehicle and place it securely on jackstands. Remove the rear wheels.

2   Detach the stabilizer bar links from the bar (see illustration). Rotate the stabilizer bar down to clear the bar links.

3   Unbolt the stabilizer bar bushing clamps from the body (see illustration).

4   The stabilizer bar can now be removed from the vehicle. Pull the clamps off the stabilizer bar (if they haven't fallen off already) using a rocking motion.

5   Check the bushings for wear, hardness, distortion, cracking and other signs of deterioration, replacing them if necessary. Also check the link bushings for these signs.

6   Using a wire brush, clean the areas of the bar where the bushings ride.

7   Installation is the reverse of the removal procedure. Install the new bushings with the split facing toward the front of the vehicle. If necessary, use a light coat of vegetable oil to ease bushing and clamp installation (don't use petroleum-based products or brake fluid, as these will damage the rubber).

8   Installation is the reverse of removal.

8.2   Before disassembling the link bolts, mark the rubber bushings so you'll know where to put them when you reassemble the links. To disconnect the links from the stabilizer bar, remove the nut at the lower end of each link bolt, then pull the bolt straight up. When you pull up the link bolt, the sleeve in the middle and the rubber bushings will fall off, so be ready to catch them!

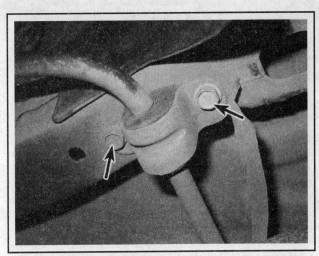

8.3   To detach the stabilizer bar from the pan, remove the two bolts from each bushing clamp

9 Strut assembly (rear) - removal and installation

REMOVAL

▶ **Refer to illustrations 9.3, 9.5 and 9.6**

1   Open the trunk and peel back the carpet and insulation.
2   Loosen the rear wheel lug nuts, raise the rear of the vehicle and support it securely on jackstands. Remove the wheel.
3   Detach the brake hose bracket from the strut (see illustration). If the vehicle is equipped with ABS, detach the ABS sensor wire from the strut.
4   Detach the stabilizer bar link from the strut (see Section 8).
5   Remove the strut-to-knuckle bolts and nuts (see illustration).
6   Have an assistant support the strut, then remove the three strut upper mounting nuts (see illustration). Remove the strut from the fenderwell.
7   Inspect the strut and coil spring assembly for leaking fluid, dents, damage and corrosion. If the strut or coil spring is damaged, replace it (see Section 3).

INSTALLATION

8   Maneuver the assembly up into the fenderwell and insert the mounting studs through the holes in the body. Install the nuts, but don't tighten them yet.
9   Connect the strut to the knuckle and install the bolts and nuts. Tighten the nuts to the torque listed in this Chapter's Specifications.

❋❋ CAUTION:

Do not turn the bolts if they are serrated. Instead, hold the bolt stationary and tighten the nut onto the bolt.

10 Attach the brake hose (and ABS speed sensor, if equipped) bracket to the strut and tighten the bolt.
11 Install the wheel and lug nuts, lower the vehicle and tighten the lug nuts to the torque listed in the Chapter 1 Specifications.
12 Tighten the strut upper mounting nuts to the torque listed in this Chapter's Specifications.
13 Push the trunk carpet and insulation back into place.

9.3 Remove the bolt and nut and detach the brake hose bracket from the strut. If the vehicle is equipped with ABS, detach the ABS sensor wire from the strut, too

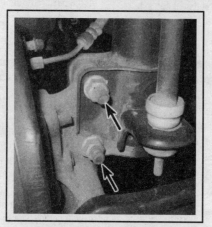

9.5 Remove the strut-to-knuckle mounting bolts and nuts

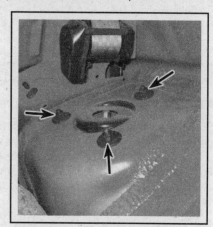

9.6 Remove the three upper strut mounting nuts while an assistant supports the strut

10 Trailing arms - removal and installation

REMOVAL

▶ **Refer to illustrations 10.3 and 10.4**

1   Loosen the rear wheel lug nuts, raise the rear of the vehicle and support it securely on jackstands. Remove the wheel.
2   Remove the rear brake drum or caliper assembly (see Chapter 9). Taking care not to twist the brake hose, hang the brake assembly or disc/caliper assembly out of the way in the wheel well with a piece of wire.
3   Place a wrench on the flat on the rear of the trailing arm to keep it from turning. At the knuckle, remove the rear nut, retainer and bushing (see illustration).
4   To detach the forward end of the trailing arm from the vehicle, remove the parking brake cable bracket nuts (see illustration), lower the parking cable and bracket and move them out of the way, then remove the two nuts above the parking cable bracket and remove the trailing arm.

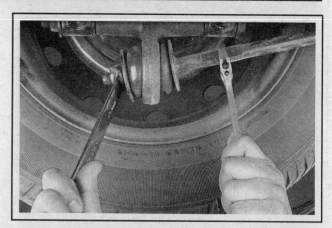

10.3 To keep the trailing arm from turning, secure it with a large adjustable wrench on the flat on the rear portion, then remove the rear nut

10.4 To detach the forward end of the trailing arm, remove these two parking brake cable bracket nuts, lower the parking cable and bracket and move them out of the way, then remove the two nuts above the parking cable bracket and pull the front end of the trailing arm down

5   If you're removing both trailing arms, make sure that you mark the two arms "left" and "right" to avoid switching them accidentally during installation.

6   Slide off the bushing and retainer from the end of the trailing arm.

INSPECTION

7   Make sure the trailing arm is straight. If it's bent, replace it. Do not attempt to straighten a bent trailing arm.

8   Inspect the bushings. If they're cracked, torn or worn out, replace all of them as a set.

INSTALLATION

9   Installation is the reverse of removal. Tighten all fasteners to the torque listed in this Chapter's Specifications.

10   Install the wheel and lug nuts, lower the vehicle and tighten the lug nuts to the torque listed in the Chapter 1 Specifications.

11 Lateral arms - removal and installation

REMOVAL

▶ **Refer to illustrations 11.2, 11.3, 11.4 and 11.5**

➡**Note: This procedure applies to either pair of lateral arms.**

1   Raise the rear of the vehicle and support it securely on jack-stands. Block the front wheels.

2   Mark the position of the toe adjustment cam (see illustration).

3   The front lateral arms have the word "FORWARD" stamped on their front sides (see illustration). The front arms are not interchangeable with the rear arms. Make sure that you can make out the FORWARD stamping on the front lateral arm you're going to remove. If the forward side of the front lateral arm is not indicated, make your own mark before removing the arm. Also, mark the arms LEFT and RIGHT so they can be returned to their original locations (if they're not being replaced with new ones).

11.2 To ensure that the correct rear wheel toe is preserved, mark the position of the toe adjustment cam

11.3 There should be a stamping that says "FORWARD" on each front lateral arm to ensure that the front arms are installed facing in the correct direction; if you don't see this word on the front of the lateral arm(s) you're planning to remove, mark it before disassembly

11.4 Remove this nut, then pull out the through-bolt to detach the lateral arms from the knuckle

4   Remove the nut and washer from the through-bolt that attaches the lateral arms to the rear knuckle (see illustration), then pull out the through-bolt.

5   Remove the lateral arm pivot bolt nut (see illustration) and washer, then pull out the pivot bolt and toe adjustment cam. Note that the lateral arm pivot bolt is longer than the through-bolt that connects the lateral arms to the spindle.

6   Remove the front and rear lateral arms.

INSPECTION

7   Inspect the arms for damage. If an arm is bent, replace it with a new one - don't try to straighten it.

8   Inspect all rubber bushings. If any bushing is cracked, torn or deteriorated, replace the arm.

INSTALLATION

9   If you're installing the original arms, install them in their original locations. If you're installing new arms, note that the word FORWARD on the front arms must be facing the front of the vehicle, and the rear arms must be installed with the larger diameter bushings at the center of the vehicle. Finally, you might have noticed that the left rear arm looks as if it was installed incorrectly because its edges are facing forward, whereas the edges on the other three arms all face to the rear. But that's the way it's supposed to be! Do NOT try to reverse the left rear lateral arm. When installed, the bow in each arm must be pointing downward.

10   Installation is otherwise the reverse of removal. After you have

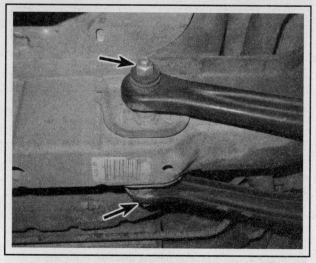

11.5 To disconnect the lateral arms from the pan, remove the nut and knock out the pivot bolt

bolted everything together, snug all fasteners but don't torque them yet.

11   Raise the knuckle with a floorjack to simulate normal ride height. Now tighten the lateral arm bolts and nuts to the torque listed in this Chapter's Specifications.

12   Install the wheel and lug nuts, then lower the vehicle. Tighten the lug nuts to the torque listed in the Chapter 1 Specifications.

13   Have the rear wheel alignment checked and, if necessary, adjusted.

12   Hub and bearing assembly (rear) - removal and installation

▶ Refer to illustrations 12.3a, 12.3b and 12.3c

✳✳ WARNING:

Dust created by the brake system is harmful to your health. Never blow it out with compressed air and don't inhale any of it. Do not, under any circumstances, use petroleum-based solvents to clean brake parts. Use brake system cleaner only.

1   Loosen the wheel lug nuts, raise the vehicle and support it securely on jackstands. Remove the wheel.

2   On models with rear drum brakes, remove the drum; on models with rear disc brakes, remove the caliper and the brake disc (see Chapter 9).

3   Follow the accompanying photos to remove the hub and bearing assembly (see illustrations). Installation is the reverse of removal. Be sure to lubricate the spindle with a thin layer of wheel bearing grease and tighten the hub-to-spindle nut to the torque listed in this Chapter's Specifications.

4   Install the brake drum, or disc and caliper (see Chapter 9).

5   Install the wheel and wheel lug nuts. Lower the vehicle and tighten the lug nuts to the torque listed in the Chapter 1 Specifications.

12.3a Remove the dust cap

12.3b Remove the mounting nut

12.3c Remove the hub and bearing assembly

13 Knuckle (rear) - removal and installation

▶ Refer to illustration 13.5

1   Loosen the rear wheel lug nuts, raise the vehicle and support it on jackstands. Block the front wheels and remove the rear wheel.

2   On models with ABS, remove the wheel speed sensor from the brake backing plate.

3   On models with rear drum brakes, remove the rear brake drum and brake shoe assembly, disconnect the parking brake cable from the parking brake lever and disconnect the brake hose from the wheel cylinder (see Chapter 9). On models with rear disc brakes, remove the caliper and brake disc, remove the parking brake shoes and disconnect the parking brake cable from the actuator lever (see Chapter 9).

4   Remove the rear hub and bearing assembly (see Section 12).

5   On models with rear drum brakes, unbolt the brake backing plate (see illustration) and remove it. On models with rear disc brakes unbolt the adapter mounting plate and remove it.

6   Loosen, but do not remove, the rear strut-to-knuckle mounting nuts.

7   Remove the nut and bolt that attach the lateral links to the knuckle (see illustration 11.4).

8   Remove the nut and washer that attach the trailing arm to the knuckle (see illustration 10.3).

13.5 To remove the brake backing plate, remove these four bolts

9   Remove the rear strut-to-knuckle mounting nuts and bolts and slide the spindle down and out of the strut clevis bracket.

10   Installation is the reverse of removal.

➡**Note: Refer to Section 11 for the correct lateral arm bolt and nut tightening procedure. Be sure to tighten all suspension fasteners to the torque listed in this Chapter's Specifications.**

11   Install the wheel and lug nuts. Lower the vehicle and tighten the lug nuts to the torque listed in the Chapter 1 Specifications.

14 Steering system - general information

All models are equipped with rack-and-pinion steering. The steering gear - which is located behind the engine, above the transaxle, in front of the firewall - operates the steering knuckles via tie rods connected to steering arms on the strut assemblies. The tie-rod ends can be replaced by unscrewing them from the inner tie rods. Adjustment sleeves between the inner tie rods and the tie-rod ends are used to adjust front wheel toe.

The power assist system consists of a belt-driven pump and associated lines and hoses. The fluid level in the power steering pump reservoir should be checked periodically (see Chapter 1).

The steering wheel operates the steering shaft, which actuates the steering gear through a short steering column and a couple of universal joints (referred to by Chrysler as the upper and lower intermediate column couplers). Looseness in the steering can be caused by wear in these universal joints, the steering gear, the tie-rod ends and loose retaining bolts.

15 Steering wheel - removal and installation

REMOVAL

▶ Refer to illustrations 15.3, 15.4, 15.5a, 15.5b, 15.6, 15.7 and 15.8

1   Make sure the front wheels are in the straight-ahead position.

2   Disconnect the cable from the negative terminal of the battery. Wait at least two minutes before proceeding.

3   On models with cruise control, remove the two cruise control switches from the steering wheel to gain access to the airbag module retaining screws (see Chapter 12). If the vehicle isn't equipped with cruise control, remove the screws and the covers on each side of the steering wheel for access (see illustration).

4   Remove the airbag module retaining bolt on each side (see illustration).

5   Lift the airbag module off the steering wheel (see illustration) and unplug the electrical connectors for the airbag and the horn (see illustration).

15.3 Remove the cover screws and pull off the cover on each side

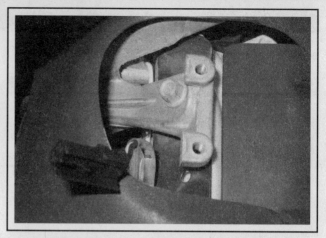

15.4 To detach the airbag module from the steering wheel, remove the two module retaining bolts (left bolt shown, right bolt identical)

✳✳ WARNING:

Carry the airbag with the trim cover side facing away from your body to minimize injury if the airbag module accidentally deploys. Set the airbag module aside in a safe, isolated location and set it down with the trim cover side facing up.

6   Remove the steering wheel retaining nut or bolt (see illustration).

7   Mark the relationship of the steering wheel to the steering shaft (see illustration).

8   Use a puller to disconnect the steering wheel from the shaft (see illustration).

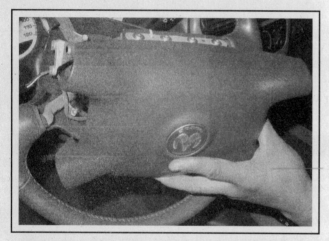

15.5a Lift the airbag module off the steering wheel . . .

15.5b . . . unplug the module electrical connector and the horn switch connector

15.6 Remove the sueering wheel retaining nut (2001 and earlier models) or bolt (2002 and later models)

15.7 Mark the relationship of the steering wheel to the steering shaft

➥**Note: On 2002 and later models, first reinstall the steering wheel bolt until about 1/2-inch of threads are showing between the bolt head and the steering wheel; this will give the puller screw something to bear down on.**

9   If it's necessary to remove the clockspring, temporarily reinstall the steering wheel onto the steering shaft, then turn the wheel 1/2-turn clockwise. Turn the ignition key to LOCK to lock the steering shaft in this position. Now remove the steering wheel from the shaft.

10 Remove the steering column covers (see Chapter 11) and the multi-function switch (see Chapter 12).

11 Unclip the latches, disconnect the electrical connector and remove the clockspring from the steering column.

12 Rotate the clockspring rotor 1/2-turn counterclockwise and lock it in this position by inserting a straightened-out paper clip through the hole in the 10 o'clock position.

INSTALLATION

13 If it was removed, reinstall the clockspring:

a) *Remove the paper clip from the clockspring, then turn the clock-spring rotor 1/2-turn (180-degrees) clockwise.*

b) *Install the clockspring on the steering column, making sure the latch hooks engage properly. Also make sure the electrical connector is plugged in completely.*

c) *If a new clockspring is being installed, it will already be in the centered position and secured with a "grenade pin;" because of this, you'll have to turn the front wheels to the straight-ahead position before installing it. Be sure to remove the pin.*

14 If you're not sure if the clockspring is centered and locked into position (see Step 12) or if the steering shaft has accidentally turned since removal, center the clockspring properly. Rotate the clockspring rotor by hand in a clockwise direction until the end of its travel (don't apply too much force). Now rotate the clockspring rotor counterclockwise three full turns; the wires should align at the bottom of the clockspring. If they don't, continue turning the rotor counterclockwise a little more until they do, but don't turn it more than 180-degrees (1/2-turn). Make sure the locking mechanism engages securely.

15 Install the wheel on the steering shaft, aligning the marks. Make

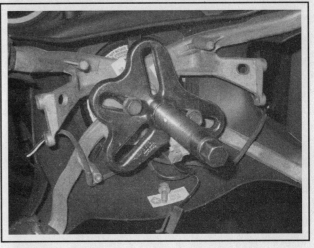

15.8 Use a steering wheel puller to remove the wheel from the shaft

sure the flats on the steering wheel hub align with the flats on the clockspring rotor. Install the steering wheel nut or bolt, tightening it to the torque listed in this Chapter's Specifications.

16 Plug in the horn and airbag module electrical connectors.

17 Tighten the airbag module retaining bolts to the torque listed in this Chapter's Specifications.

18 Plug in and install the cruise control switches, if equipped, and tighten the screws securely (see Chapter 12, if necessary). On vehicles without cruise control, install the covers.

19 Connect the negative battery cable.

20 From the *passenger side* of the vehicle, turn the ignition switch to the Off position, then turn it to the On position. Check that the instrument cluster AIRBAG lamp is illuminated for six to eight seconds and then goes out indicating the airbag system is functioning properly. If the lamp fails to light, blinks on and off or stays on, there is a malfunction in the airbag system. If any of these conditions exist, a dealer service department or other qualified repair shop should diagnose the vehicle.

16 Tie-rod ends - removal and installation

REMOVAL

▶ **Refer to illustrations 16.2, 16.3a, 16.3b and 16.4**

1   Loosen the wheel lug nuts. Apply the parking brake, raise the front of the vehicle and support it securely on jackstands. Remove the front wheel.

2   Loosen the nut on the tie-rod end stud. If the stud turns, hold it with a wrench (see illustration).

3   Hold the tie-rod with a pair of locking pliers or wrench and loosen the jam nut enough to mark the position of the tie-rod end in relation to the threads (see illustrations).

16.2 If the tie-rod end stud spins when loosening the nut, hold the stud with a wrench or socket

4   Disconnect the tie-rod from the steering knuckle arm with a puller (see illustration). Remove the nut and detach the tie-rod. Note how the heat shield is installed.

5   Unscrew the tie-rod end from the tie-rod.

INSTALLATION

6   Thread the tie-rod end on to the marked position and insert the tie-rod stud into the steering knuckle arm. Make sure the heat shield is positioned properly. Tighten the jam nut securely.

7   Install the nut on the stud and tighten it to the torque listed in this Chapter's Specifications.

8   Install the wheel and lug nuts. Lower the vehicle and tighten the lug nuts to the torque listed in the Chapter 1 Specifications.

9   Have the alignment checked and, if necessary, adjusted.

16.3a Hold the tie-rod with a wrench or a pair of pliers while loosening the jam nut

16.3b Back off the jam nut and mark the exposed threads

16.4 Use a two-jaw puller to push the tie-rod end stud out of the steering knuckle

17 Steering gear - removal and installation

REMOVAL

▶ **Refer to illustrations 17.3, 17.7a, 17.7b, 17.8, 17.9, 17.10 and 17.13**

✳ WARNING:

These models are equipped with airbags. Make sure the steering shaft is not turned while the steering gear is removed or you could damage the airbag system. To prevent the shaft from turning, turn the ignition key to the lock position before beginning work or run the seat belt through the steering wheel and clip the seat belt into place. Due to the possible damage to the airbag system, we recommend only experienced mechanics attempt this procedure.

➡**Note: These models are designed and assembled using net build front suspension alignment settings. The front suspension alignment settings are determined as the vehicle is being built by accurately locating the front suspension crossmember to the** master gauge holes located in the underbody of the vehicle. Therefore whenever the front crossmember is removed from the vehicle's body it must be reinstalled in the exact same location on the body to maintain correct front end alignment.

1   Disconnect the negative battery cable.

2   Turn the front wheels of the vehicle to the full-left position, then turn them back until the retaining pin in the coupler is accessible. Turn the ignition key to the Lock position to keep the steering column from rotating after the coupler is disengaged from the steering gear.

✳ CAUTION:

Failure to lock the steering shaft could allow it to rotate beyond its normal number of turns in either direction, which will damage the airbag module clockspring. Mark the relationship of the steering coupler to the steering gear shaft to ensure proper reassembly.

3   Working in the passenger compartment, disconnect the steering gear coupler from the steering column shaft coupler (see illustration).

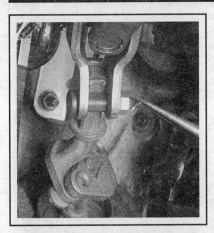

17.3 Mark the relationship of the steering coupler to the steering gear shaft to ensure proper reassembly, then remove the clip and disconnect the steering gear coupler from the steering column shaft coupler

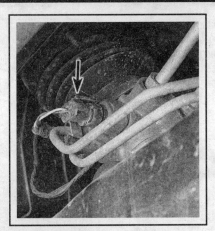

17.7a Unhook the clip and disconnect the electrical connector from the power steering fluid pressure switch . . .

17.7b . . . then remove the power steering fluid hose bracket - leave the bracket attached to the hoses

4   Loosen the front wheel lug nuts, raise the front of the vehicle and support it securely on jackstands. Remove both wheels.

5   Disconnect the tie-rod ends from the steering knuckle (see Section 16).

6   Unbolt the engine torque strut from the crossmember (see illustration 5.3).

7   Disconnect the electrical connector from the power steering fluid pressure switch (see illustration). Remove the power steering fluid hose bracket - leave the bracket attached to the hoses (see illustration).

8   Place a drain pan under the steering gear. Detach the power steering pressure and return lines (see illustration) and cap the ends to prevent excessive fluid loss and contamination. Detach the bracket from the top of the steering gear assembly.

9   Using an awl or scribe, accurately scribe a line marking the location of the front crossmember-to-underbody (see illustration).

10 Position a transmission jack, or equivalent, under the center of the front crossmember (see illustration).

11 Remove the two front mounting bolts which attach the front crossmember to the frame rails of underbody, then loosen the two rear mounting bolts which attach the front crossmember and control arms to the underbody.

12 Continue to loosen the rear bolts while lowering the transmission jack and front suspension crossmember - don't remove the rear bolts, as they keep the crossmember attached to the control arms. Lower the crossmember sufficiently to remove the bolts securing the steering gear to the crossmember.

✳✳ CAUTION:

Before removing the front crossmember, the exact location of the crossmember must be accurately scribed on the underbody of the vehicle.

✳✳ CAUTION:

Do not allow the crossmember to hang from the lower control arms as they will be damaged - it must be supported by the jack at all times.

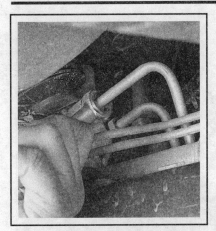

17.8 Place a drain pan under the steering gear and detach the power steering pressure and return lines - cap the ends to prevent excessive fluid loss and contamination

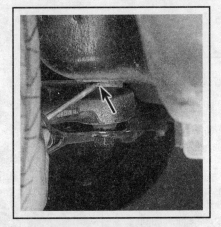

17.9 Using an awl, accurately scribe a line marking the location of front crossmember-to-underbody

17.10 Position a transmission jack under the center of the front crossmember to support it

13 Remove the four mounting bolts which attach the steering gear to the crossmember (see illustration), slide the steering gear forward in the vehicle to disengage the coupler from the steering gear shaft. Once the steering gear is disengaged, do NOT turn the steering gear shaft or the steering column shaft (see the **Warning** at the beginning of this Section). Remove the steering gear assembly from the vehicle.

INSTALLATION

14 Install the steering gear onto the crossmember and install the four mounting bolts (see illustration 17.13). Tighten the bolts to the torque listed in this Chapter's Specifications.

15 Slowly raise the steering gear assembly up and engage the coupler onto the steering gear shaft. If you're installing the old steering gear, align the mark on the steering coupler with the mark on the steering gear shaft and insert the shaft into the coupler. If you're installing a new steering gear assembly, rotate the steering gear shaft back from its full-left position until the master spline on the steering gear shaft is aligned with the master spline on the coupler, then insert the shaft into the coupler.

16 Continue to raise the crossmember and align the four mounting holes in the crossmember. Install the rear two mounting bolts then the front two mounting bolts. Tighten the four bolts in a crisscross pattern until the crossmember is up against the underbody.

17 Using a soft face hammer, tap the crossmember into correct alignment with the scribe marks made on the underbody in Step 9 (see illustration 17.9).

✳✳ CAUTION:

This alignment is necessary to maintain the net build front suspension alignment settings.

When the alignment is correct, first tighten the rear two bolts, then the front two bolts to the torque listed in this Chapter's Specifications

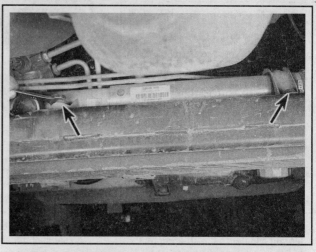

17.13 Remove the four mounting bolts attaching the steering gear to the crossmember

18 Attach the power steering fluid pressure and return lines to the correct ports on the steering gear (see illustration 17.8) and tighten the tube fittings securely. Connect the electrical connector onto the power steering fluid pressure switch (see illustration 17.7a). Install the power steering fluid hose bracket and tighten the bolt securely (see illustration 17.7b).

19 Connect the tie-rod ends to the steering knuckle (see Section 16).

20 Install the wheel and lug nuts. Lower the vehicle and tighten the lug nuts to the torque listed in the Chapter 1 Specifications.

21 Working in the passenger compartment, connect the steering gear coupler onto the steering column shaft coupler (see illustration 17.3). Tighten the bolt and nut to the torque listed in this Chapter's Specifications. Install the retaining clip onto the bolt.

22 Have the front end alignment checked and, if necessary, adjusted.

18 Power steering pump - removal and installation

▶ **Refer to illustrations 18.8a and 18.8b**

REMOVAL

1 Using a suction gun, remove as much fluid from the power steering fluid reservoir as you can.

2 Remove the drivebelt from the power steering pump and the air conditioning compressor (see Chapter 1).

3 Unscrew the pressure line fitting from the bottom of the power steering pump. To do this, you'll probably need a ratchet, long extension and a crowfoot attachment.

4 Detach the fluid supply hose from the fitting on the pump.

5 Remove the bolt securing the rear of the pump to the bracket, then remove the two bolts securing the bracket to the engine block.

6 Working through the holes in the pulley, remove the three pump mounting bolts.

7 Remove the pump from the engine compartment.

8 If you're installing a new pump, you'll need a special puller to remove the pulley from the old pump and another special tool to install it on the new pump (see illustrations). These tools are available at most

18.8a If you're installing a new pump, you'll need a special puller to remove the pulley from the old pump . . .

18.8b . . . and another special tool to install it on the new pump

auto parts stores. When installing the pulley, press it onto the pump shaft until the hub of the pulley is flush with the end of the shaft.

INSTALLATION

9   Installation is the reverse of removal, noting the following points:

a) *Before installing the pump, check the condition of the O-ring on the pressure line fitting, replacing it if necessary.*

b) *Tighten the pump mounting bracket bolts and the pump-to-mounting bracket bolts to the torque listed in this Chapter's Specifications.*

c) *Refill the power steering fluid reservoir with the recommended fluid (see Chapter 1).*

d) *Bleed the power steering system as described in the next Section.*

19 Power steering system - bleeding

1   Following any operation in which the power steering fluid lines have been disconnected, the power steering system must be bled to remove all air and obtain proper steering performance.

2   With the front wheels in the straight ahead position, check the power steering fluid level and, if low, add fluid until it reaches the Cold mark on the dipstick.

3   Start the engine and allow it to run at fast idle. Recheck the fluid level and add more if necessary to reach the Cold mark on the dipstick.

4   Bleed the system by turning the wheels from side to side, without hitting the stops. This will work the air out of the system. Keep the

reservoir full of fluid as this is done.

5   When the air is worked out of the system, return the wheels to the straight-ahead position and leave the vehicle running for several more minutes before shutting it off.

6   Road test the vehicle to be sure the steering system is functioning normally and noise free.

7   Recheck the fluid level to be sure it is up to the Hot mark on the dipstick while the engine is at normal operating temperature. Add fluid if necessary (see Chapter 1).

20 Wheels and tires - general information

▶ **Refer to illustration 20.1**

1   All vehicles covered by this manual are equipped with metric-sized fiberglass or steel belted radial tires (see illustration). Use of other size or type of tires may affect the ride and handling of the vehicle. Don't mix different types of tires, such as radials and bias belted, on the same vehicle as handling may be seriously affected. It's recommended that tires be replaced in pairs on the same axle, but if only one tire is being replaced, be sure it's the same size, structure and tread design as the other.

2   Because tire pressure has a substantial effect on handling and wear, the pressure on all tires should be checked at least once a month or before any extended trips (see Chapter 1).

3   Wheels must be replaced if they are bent, dented, leak air, have elongated bolt holes, are heavily rusted, out of vertical symmetry or if the lug nuts won't stay tight. Wheel repairs that use welding or peening are not recommended.

4   Tire and wheel balance is important in the overall handling, braking and performance of the vehicle. Unbalanced wheels can adversely affect handling and ride characteristics as well as tire life. Whenever a tire is installed on a wheel, a shop with the proper equipment should balance the tire and wheel.

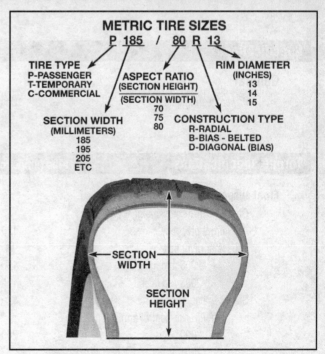

20.1 Metric tire size code

21 Wheel alignment - general information

♦ **Refer to illustration 21.1**

A wheel alignment refers to the adjustments made to the wheels so they are in proper angular relationship to the suspension and the ground. Wheels that are out of proper alignment not only affect vehicle control, but also increase tire wear. The alignment angles normally measured are camber, caster and toe-in (see illustration). Toe-in is the only routine adjustment made; camber is adjustable, but only after installing special undersized strut-to-knuckle bolts. The other angles should be measured to check for bent or worn suspension parts.

Getting the proper wheel alignment is a very exacting process, one in which complicated and expensive machines are necessary to perform the job properly. Because of this, you should have a technician with the proper equipment perform these tasks. We will, however, use this space to give you a basic idea of what is involved with a wheel alignment so you can better understand the process and deal intelligently with the shop that does the work.

Toe-in is the turning in of the wheels. The purpose of a toe specification is to ensure parallel rolling of the wheels. In a vehicle with zero toe-in, the distance between the front edges of the wheels will be the same as the distance between the rear edges of the wheels. The actual amount of toe-in is normally only a fraction of an inch. On the front end, toe-in is controlled by the tie-rod end position on the tie-rod. On the rear end, it's controlled by an adjuster cam on the rear lateral link. Incorrect toe-in will cause the tires to wear improperly by making them scrub against the road surface.

Camber is the tilting of the wheels from vertical when viewed from one end of the vehicle. When the wheels tilt out at the top, the camber is said to be positive (+). When the wheels tilt in at the top the camber is negative (-). The amount of tilt is measured in degrees from vertical and this measurement is called the camber angle. This angle affects the amount of tire tread which contacts the road and compensates for changes in the suspension geometry when the vehicle is cornering or

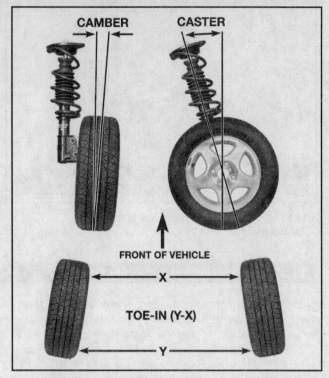

21.1 Camber, caster and toe-in angles

traveling over an undulating surface.

Caster is the tilting of the front steering axis from the vertical. A tilt toward the rear is positive caster and a tilt toward the front is negative caster.

Specifications

General

Power steering fluid	Chapter 1

Torque specifications	Ft-lbs (unless otherwise indicated)	Nm
Front suspension		
Control arm-to-crossmember pivot bolts		
Front pivot bolt	120	163
Rear pivot bolt		
2000 through 2002	150	20a
2003 and later	175	237
Stabilizer bar		
Stabilizer-to-control arm link nut		
2000	200 in-lbs	23
2001 on	23	31
Stabilizer bushing clamp bolts	21	28

Torque specifications	Ft-lbs (unless otherwise indicated)	Nm
Steering knuckle-to-balljoint stud pinch clamp bolt and nut	70	95
Strut/coil spring assembly		
Upper mounting nuts	25	34
Strut-to-knuckle bolt and nut	40 plus 90 degrees	54 plus 90 degrees
Damper shaft nut	55	75

Rear suspension

Brake support plate-to-knuckle bolts	55	75
Lateral arms		
Lateral arm-to-spindle bolt and nut	70	95
Lateral arm-to-rear crossmember bolt and nut	65	88
Rear hub retaining nut	160	217
Stabilizer bar		
Stabilizer-to-link bolt and nut	200 in-lbs	23
Stabilizer bushing clamp bolts	25	34
Strut/coil spring assembly		
Upper mounting nuts	25	34
Strut-to-knuckle bolt and nut	65	88
Damper shaft nut	55	75
Trailing arms		
Trailing arm-to-frame bracket nuts	70	95
Trailing arm-to-spindle nut	70	95

Steering system

Airbag module retaining bolts	90 in-lbs	10
Crossmember-to-underbody bolts		
Front bolts	105	142
Rear bolts	150	203
Power steering pump		
Power steering pump bracket mounting bolts	40	54
Power steering pump mounting bolts	250 in-lbs	28
Steering gear assembly		
Steering gear coupler pinch bolt	250 in-lbs	28
Steering gear-to-crossmember mounting bolts	45	61
Steering wheel nut		
2000 models	45	61
2001 models	40	54
Steering wheel bolt (2002 and later models)	40	54
Tie-rods		
Tie-rod end adjusting nut	55	75
Steering knuckle-to-tie-rod end ballstud nut	40	54

Notes

Section

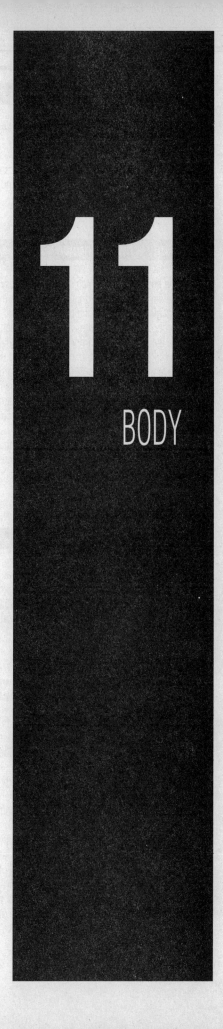

11

BODY

1 General information

The models covered by this manual feature a "unibody" layout, using a floor pan with front and rear frame side rails which support the body components, front and rear suspension systems and other mechanical components. Certain components are particularly vulnerable to accident damage and can be unbolted and repaired or replaced.

Among these parts are the body moldings, bumpers, front fenders, the hood and trunk lids and all glass. Only general body maintenance practices and body panel repair procedures within the scope of the do-it-yourselfer are included in this Chapter.

2 Body - maintenance

1 The condition of the vehicle's body is very important, because the resale value depends a great deal on it. It's much more difficult to repair a neglected or damaged body than it is to repair mechanical components. The hidden areas of the body, such as the wheel wells, the frame and the engine compartment, are equally important, although they don't require as frequent attention as the rest of the body.

2 Once a year, or every 12,000 miles, it's a good idea to have the underside of the body steam cleaned. All traces of dirt and oil will be removed and the area can then be inspected carefully for rust, damaged brake lines, frayed electrical wires, damaged cables and other problems.

3 At the same time, clean the engine and the engine compartment with a steam cleaner or water-soluble degreaser.

4 The wheel wells should be given close attention, since under-

coating can peel away and stones and dirt thrown up by the tires can cause the paint to chip and flake, allowing rust to set in. If rust is found, clean down to the bare metal and apply an anti-rust paint.

5 The body should be washed about once a week. Wet the vehicle thoroughly to soften the dirt, then wash it down with a soft sponge and plenty of clean soapy water. If the surplus dirt is not washed off very carefully, it can wear down the paint.

6 Spots of tar or asphalt thrown up from the road should be removed with a cloth soaked in solvent.

7 Once every six months, wax the body and chrome trim. If a chrome cleaner is used to remove rust from any of the vehicle's plated parts, remember that the cleaner also removes part of the chrome, so use it sparingly.

3 Vinyl trim - maintenance

Don't clean vinyl trim with detergents, caustic soap or petroleum-based cleaners. Plain soap and water works just fine, with a soft brush to clean dirt that may be ingrained. Wash the vinyl as frequently as the rest of the vehicle. After cleaning, application of a high-quality

rubber and vinyl protectant will help prevent oxidation and cracks. The protectant can also be applied to weatherstripping, vacuum lines and rubber hoses, which often fail as a result of chemical degradation, and to the tires.

4 Upholstery and carpets - maintenance

1 Every three months remove the floor mats and clean the interior of the vehicle (more frequently if necessary). Use a stiff whiskbroom to brush the carpeting and loosen dirt and dust, then vacuum the upholstery and carpets thoroughly, especially along seams and crevices.

2 Dirt and stains can be removed from carpeting with basic household or automotive carpet shampoos available in spray cans. Follow the directions and vacuum again, then use a stiff brush to bring back the "nap" of the carpet.

3 Most interiors have cloth or vinyl upholstery, either of which can be cleaned and maintained with a number of material-specific cleaners or shampoos available in auto supply stores. Follow the directions on the product for usage, and always spot-test any upholstery cleaner on an inconspicuous area (like the bottom edge of a back seat cushion) to ensure that it doesn't cause a color shift in the material.

4 After cleaning, vinyl upholstery should be treated with a protectant.

➡ **Note: Make sure the protectant container indicates the product can be used on seats - some products may make a seat too slippery.**

✳ CAUTION:

Do not use a protectant on vinyl-covered steering wheels.

5 Leather upholstery requires special care. It should be cleaned regularly with saddle soap or leather cleaner. Never use alcohol, gasoline, nail polish remover or thinner to clean leather upholstery.

6 After cleaning, regularly treat leather upholstery with a leather conditioner, rubbed in with a soft cotton cloth. Never use car wax on leather upholstery.

7 In areas where the interior of the vehicle is subject to bright sunlight, cover leather seating areas of the seats with a sheet if the vehicle is to be left out for any length of time.

MINOR DAMAGE

Repair of scratches

1   If the scratch is superficial and does not penetrate to the metal of the body, repair is very simple. Lightly rub the scratched area with a fine rubbing compound to remove loose paint and built up wax. Rinse the area with clean water.

2   Apply touch-up paint to the scratch, using a small brush. Continue to apply thin layers of paint until the surface of the paint in the scratch is level with the surrounding paint. Allow the new paint at least two weeks to harden, then blend it into the surrounding paint by rubbing with a very fine rubbing compound. Finally, apply a coat of wax to the scratch area.

3   If the scratch has penetrated the paint and exposed the metal of the body, causing the metal to rust, a different repair technique is required. Remove all loose rust from the bottom of the scratch with a pocket knife, then apply rust inhibiting paint to prevent the formation of rust in the future. Using a rubber or nylon applicator, coat the scratched area with glaze-type filler. If required, the filler can be mixed with thinner to provide a very thin paste, which is ideal for filling narrow scratches. Before the glaze filler in the scratch hardens, wrap a piece of smooth cotton cloth around the tip of a finger. Dip the cloth in thinner and then quickly wipe it along the surface of the scratch. This will ensure that the surface of the filler is slightly hollow. The scratch can now be painted over as described earlier in this Section.

Repair of dents

See photo sequence

4   When repairing dents, the first job is to pull the dent out until the affected area is as close as possible to its original shape. There is no point in trying to restore the original shape completely as the metal in the damaged area will have stretched on impact and cannot be restored to its original contours. It is better to bring the level of the dent up to a point which is about 1/8-inch below the level of the surrounding metal. In cases where the dent is very shallow, it is not worth trying to pull it out at all.

5   If the back side of the dent is accessible, it can be hammered out gently from behind using a soft-face hammer. While doing this, hold a block of wood firmly against the opposite side of the metal to absorb the hammer blows and prevent the metal from being stretched.

6   If the dent is in a section of the body which has double layers, or some other factor makes it inaccessible from behind, a different technique is required. Drill several small holes through the metal inside the damaged area, particularly in the deeper sections. Screw long, self tapping screws into the holes just enough for them to get a good grip in the metal. Now pulling on the protruding heads of the screws with locking pliers can pull out the dent.

7   The next stage of repair is the removal of paint from the damaged area and from an inch or so of the surrounding metal. This is easily done with a wire brush or sanding disk in a drill motor, although it can be done just as effectively by hand with sandpaper. To complete the preparation for filling, score the surface of the bare metal with a screwdriver or the tang of a file or drill small holes in the affected area. This will provide a good grip for the filler material. To complete the repair, see the subsection on *filling and painting*.

Repair of rust holes or gashes

8   Remove all paint from the affected area and from an inch or so of the surrounding metal using a sanding disk or wire brush mounted in a drill motor. If these are not available, a few sheets of sandpaper will do the job just as effectively.

9   With the paint removed, you will be able to determine the severity of the corrosion and decide whether to replace the whole panel, if possible, or repair the affected area. New body panels are not as expensive as most people think and it is often quicker to install a new panel than to repair large areas of rust.

10 Remove all trim pieces from the affected area except those which will act as a guide to the original shape of the damaged body, such as headlight shells, etc. Using metal snips or a hacksaw blade, remove all loose metal and any other metal that is badly affected by rust. Hammer the edges of the hole on the inside to create a slight depression for the filler material.

11 Wire brush the affected area to remove the powdery rust from the surface of the metal. If the back of the rusted area is accessible, treat it with rust inhibiting paint.

12 Before filling is done, block the hole in some way. This can be done with sheet metal riveted or screwed into place, or by stuffing the hole with wire mesh.

13 Once the hole is blocked off, the affected area can be filled and painted. See the following subsection on *filling and painting*.

Filling and painting

14 Many types of body fillers are available, but generally speaking, body repair kits which contain filler paste and a tube of resin hardener are best for this type of repair work. A wide, flexible plastic or nylon applicator will be necessary for imparting a smooth and contoured finish to the surface of the filler material. Mix up a small amount of filler on a clean piece of wood or cardboard (use the hardener sparingly). Follow the manufacturer's instructions on the package, otherwise the filler will set incorrectly.

15 Using the applicator, apply the filler paste to the prepared area. Draw the applicator across the surface of the filler to achieve the desired contour and to level the filler surface. As soon as a contour that approximates the original one is achieved, stop working the paste. If you continue, the paste will begin to stick to the applicator. Continue to add thin layers of paste at 20-minute intervals until the level of the filler is just above the surrounding metal.

16 Once the filler has hardened, the excess can be removed with a body file. From then on, progressively finer grades of sandpaper should be used, starting with a 180-grit paper and finishing with 600-grit wet-or-dry paper. Always wrap the sandpaper around a flat rubber or wooden block, otherwise the surface of the filler will not be completely flat. During the sanding of the filler surface, the wet-or-dry paper should be periodically rinsed in water. This will ensure that a very smooth finish is produced in the final stage.

17 At this point, the repair area should be surrounded by a ring of bare metal, which in turn should be encircled by the finely feathered edge of good paint. Rinse the repair area with clean water until all of the dust produced by the sanding operation is gone.

18 Spray the entire area with a light coat of primer. This will reveal any imperfections in the surface of the filler. Repair the imperfections with fresh filler paste or glaze filler and once more smooth the surface with sandpaper. Repeat this spray-and-repair procedure until you are satisfied that the surface of the filler and the feathered edge of the paint are perfect.

These photos illustrate a method of repairing simple dents. They are intended to supplement Body repair - minor damage in this Chapter and should not be used as the sole instructions for body repair on these vehicles.

1   If you can't access the backside of the body panel to hammer out the dent, pull it out with a slide-hammer-type dent puller. In the deepest portion of the dent or along the crease line, drill or punch hole(s) at least one inch apart . . .

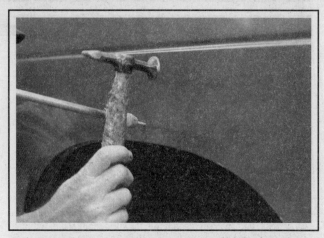

2   . . . then screw the slide-hammer into the hole and operate it. Tap with a hammer near the edge of the dent to help 'pop' the metal back to its original shape. When you're finished, the dent area should be close to its original contour and about 1/8-inch below the surface of the surrounding metal

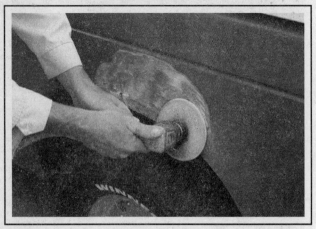

3   Using coarse-grit sandpaper, remove the paint down to the bare metal. Hand sanding works fine, but the disc sander shown here makes the job faster. Use finer (about 320-grit) sandpaper to feather-edge the paint at least one inch around the dent area

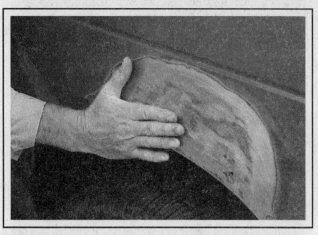

4   When the paint is removed, touch will probably be more helpful than sight for telling if the metal is straight. Hammer down the high spots or raise the low spots as necessary. Clean the repair area with wax/silicone remover

5   Following label instructions, mix up a batch of plastic filler and hardener. The ratio of filler to hardener is critical, and, if you mix it incorrectly, it will either not cure properly or cure too quickly (you won't have time to file and sand it into shape)

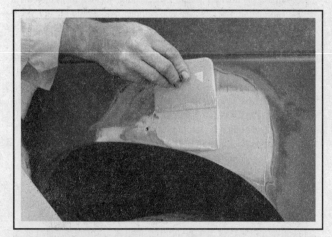

6   Working quickly so the filler doesn't harden, use a plastic applicator to press the body filler firmly into the metal, assuring it bonds completely. Work the filler until it matches the original contour and is slightly above the surrounding metal

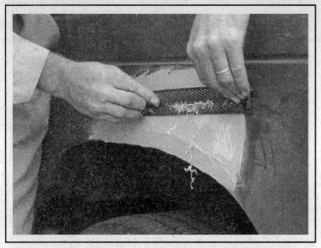

7   Let the filler harden until you can just dent it with your fingernail. Use a body file or Surform tool (shown here) to rough-shape the filler

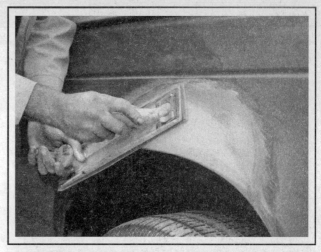

8   Use coarse-grit sandpaper and a sanding board or block to work the filler down until it's smooth and even. Work down to finer grits of sandpaper - always using a board or block - ending up with 360 or 400 grit

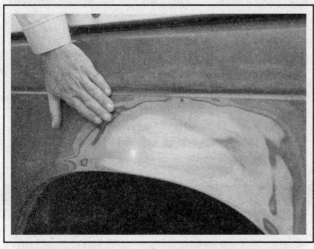

9   You shouldn't be able to feel any ridge at the transition from the filler to the bare metal or from the bare metal to the old paint. As soon as the repair is flat and uniform, remove the dust and mask off the adjacent panels or trim pieces

10   Apply several layers of primer to the area. Don't spray the primer on too heavy, so it sags or runs, and make sure each coat is dry before you spray on the next one. A professional-type spray gun is being used here, but aerosol spray primer is available inexpensively from auto parts stores

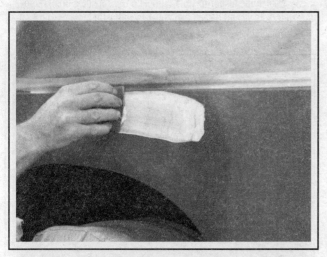

11   The primer will help reveal imperfections or scratches. Fill these with glazing compound. Follow the label instructions and sand it with 360 or 400-grit sandpaper until it's smooth. Repeat the glazing, sanding and respraying until the primer reveals a perfectly smooth surface

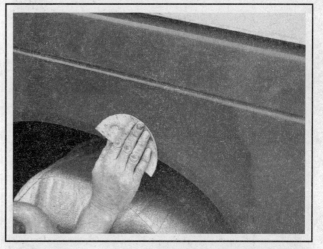

12   Finish sand the primer with very fine sandpaper (400 or 600-grit) to remove the primer overspray. Clean the area with water and allow it to dry. Use a tack rag to remove any dust, then apply the finish coat. Don't attempt to rub out or wax the repair area until the paint has dried completely (at least two weeks)

Rinse the area with clean water and allow it to dry completely.

19 The repair area is now ready for painting. Spray painting must be carried out in a warm, dry, windless and dust free atmosphere. These conditions can be created if you have access to a large indoor work area, but if you are forced to work in the open, you will have to pick the day very carefully. If you are working indoors, dousing the floor in the work area with water will help settle the dust, which would otherwise be in the air. If the repair area is confined to one body panel, mask off the surrounding panels. This will help minimize the effects of a slight mismatch in paint color. Trim pieces such as chrome strips, door handles, etc., will also need to be masked off or removed. Use masking tape and several thickness of newspaper for the masking operations.

20 Before spraying, shake the paint can thoroughly, then spray a test area until the spray painting technique is mastered. Cover the repair area with a thick coat of primer. The thickness should be built up using several thin layers of primer rather than one thick one. Using 600-grit wet-or-dry sandpaper, rub down the surface of the primer until it is very smooth. While doing this, the work area should be thoroughly rinsed with water and the wet-or-dry sandpaper periodically rinsed as well. Allow the primer to dry before spraying additional coats.

21 Spray on the top coat, again building up the thickness by using several thin layers of paint. Begin spraying in the center of the repair

area and then, using a circular motion, work out until the whole repair area and about two inches of the surrounding original paint is covered. Remove all masking material 10 to 15 minutes after spraying on the final coat of paint. Allow the new paint at least two weeks to harden, then use a very fine rubbing compound to blend the edges of the new paint into the existing paint. Finally, apply a coat of wax.

MAJOR DAMAGE

22 Major damage must be repaired by an auto body shop specifically equipped to perform unibody repairs. These shops have the specialized equipment required to do the job properly.

23 If the damage is extensive, the body must be checked for proper alignment or the vehicle's handling characteristics may be adversely affected and other components may wear at an accelerated rate.

24 Due to the fact that all of the major body components (hood, fenders, etc.) are separate and replaceable units, any seriously damaged components should be replaced rather than repaired. Sometimes the components can be found in a auto salvage or wrecking yard that specializes in used vehicle components, often at considerable savings over the cost of new parts.

6 Hinges and locks - maintenance

Once every 3000 miles, or every three months, the hinges and latch assemblies on the doors, hood and trunk should be given a few drops of light oil or lock lubricant. The door latch strikers should also be

lubricated with a thin coat of grease to reduce wear and ensure free movement. Lubricate the door and trunk locks with spray-on graphite lubricant.

7 Windshield and fixed glass - replacement

Replacement of the windshield and fixed glass requires the use of special fast-setting adhesive/caulk materials and some specialized

tools and techniques. These operations should be left to a dealer service department or a shop specializing in glass work.

8 Front fender - replacement

▶ **Refer to illustrations 8.2, 8.3a, 8.3b and 8.3c**

1 Remove the headlight assembly (Chapter 12).
2 Remove the fasteners and detach the fender splash shield (see illustration).

3 Detach the fender-to-front bumper cover retainers, then remove the bolts/screws and detach the fender (see illustrations).
4 Installation of the fender is the reverse of removal.

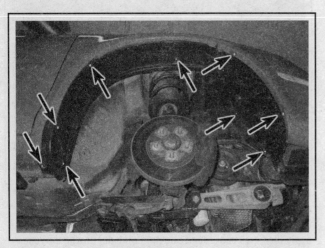

8.2 Fender splash shield retainer locations

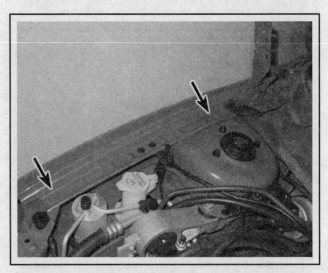

8.3a Remove the bolts along the top of the front fender

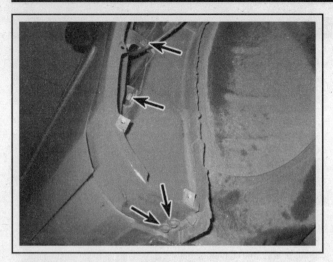

8.3b Remove the fender-to-body bolts at the rear of the fender . . .

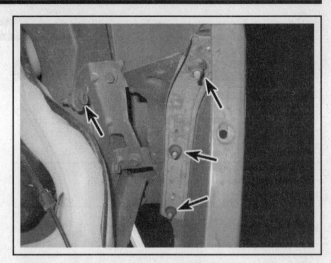

8.3c . . . and these fasteners under the front end of the fender

9   Hood - removal, installation and adjustment

▶ **Refer to illustrations 9.3, 9.4 and 9.11**

➡**Note: The hood is heavy and somewhat awkward to remove and install - at least two people should perform this procedure.**

REMOVAL AND INSTALLATION

1   Use blankets or pads to cover the cowl area of the body and both fenders. This will protect the body and paint as the hood is lifted off.

2   Open the hood and support it on the prop rod.

3   Scribe alignment marks around the bolt heads to insure proper alignment during installation (a permanent-type felt-tip marker also will work for this) (see illustration).

4   Disconnect any hoses or electrical wire harnesses which will interfere with removal (see illustration).

5   Have an assistant support one side of the hood while you support the other. Simultaneously remove the hood-to-hinge bolts.

6   Lift off the hood.

7   Installation is the reverse of removal.

ADJUSTMENT

8   Fore-and-aft and side-to-side adjustment of the hood is done by moving the hood in relation to the hinge plate after loosening the bolts. The hood must be aligned so there is a 5/32-inch gap (approximate) to the front fenders and flush with the top surface.

9   Scribe or trace a line around the entire hinge plate so you can judge the amount of movement.

10   Loosen the bolts and move the hood into correct alignment. Move it only a little at a time. Tighten the hinge bolts and carefully lower the hood to check the alignment.

11   Adjust the hood bumpers on the radiator support so the hood is flush with the fenders when closed (see illustration).

12   The latch assembly can also be adjusted up-and-down and side-to-side after loosening the bolts (see Section 10).

13   The hood latch assembly, as well as the hinges, should be periodically lubricated with white lithium-base grease to prevent sticking and wear.

9.3 Use a marking pen to outline the hinge plate and bolt heads

9.4 Detach the washer hose before removing the hood

9.11 To adjust the hood until its flush with the fenders, screw the hood bumpers in-or-out

10 Hood latch and cable - removal and installation

✳✳ WARNING:

These models have airbags. Always disconnect the negative battery cable and wait two minutes before working in the vicinity of the impact sensors, steering column or instrument panel to avoid the possibility of accidental deployment of the airbag, which could cause personal injury (see Chapter 12).

LATCH

▶ **Refer to illustration 10.3**

1   Open the hood and support it on the prop rod.
2   Remove the radiator grille (see Section 11).
3   Remove the nuts and detach the latch assembly, then disconnect the release cable from the hood latch. (see illustration).

10.3   Remove the two hood latch nuts

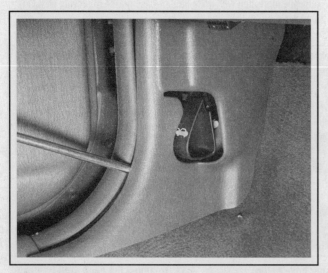

10.5a   Carefully pry the kick panel free and remove it

CABLE

▶ **Refer to illustrations 10.4, 10.5a and 10.5b**

4   Disconnect the release cable from the hood latch (see illustration).
5   In the passenger compartment, remove the left front kick panel (see illustration). Remove the bolts securing the hood release handle and move the handle away from the cowl panel (see illustration).
6   Under the dash, remove the cable grommet from the firewall.
7   Connect a piece of heavy string or flexible wire to the engine compartment end of the cable, then pull the release cable and wire through the firewall into the passenger compartment. Disconnect the string or wire from the old cable.
8   Connect the string or wire to the new cable and pull it through the firewall into the engine compartment.
9   The remainder of the installation is the reverse of removal. Make sure the grommet seats properly in the firewall.

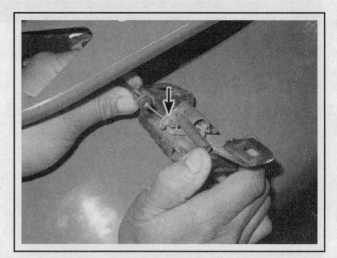

10.4   Detach the cable from the latch

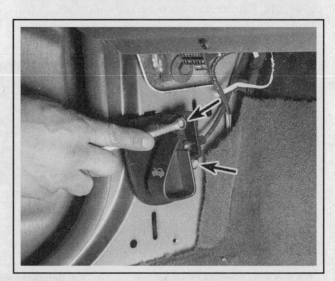

10.5b   Remove the handle mounting bolts

11 Radiator grille - removal and installation

▶ **Refer to illustration 11.2**

1   Open the hood and support it on the prop rod.
2   Remove the screws and detach the grille (see illustration).
3   Installation is the reverse of removal.

11.2 Remove the screws and detach the grille

12 Bumpers - removal and installation

▶ **Refer to illustrations 12.4, 12.5, 12.6a and 12.6b**

1   Remove the radiator grille (see Section 11).
2   Disconnect any wiring or other components that would interfere with front bumper cover removal.
3   Have an assistant support the bumper cover as the bolts and nuts are removed.
4   Remove the front bolts and side screws and partially remove the front bumper cover (see illustration).
5   Remove the clips and disconnect the side marker lights from the bumper cover (see illustration).
6   Remove the nuts on the back and the bolts on each side and detach the front bumper reinforcement. The bumper can now be removed (see illustrations).
7   Installation is the reverse of removal.

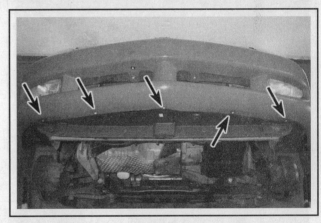

12.4 Remove the lower front bumper cover bolts

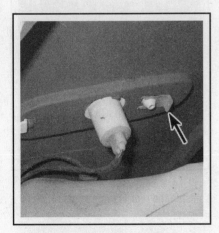

12.5 Remove this clip to detach the side marker light from the bumper cover

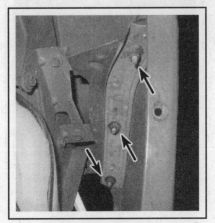

12.6a Remove the nuts attaching the bumper cover to the fender

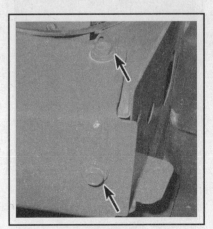

12.6b Remove the bolts from the ends of the front bumper reinforcement (there's one more on the underside of each end)

REAR BUMPER

▶ Refer to illustrations 12.10, 12.13a, 12.13b, 12.13c, 12.13d, and 12.14

8 Open the trunk lid.

9 Disconnect any wiring or other components that would interfere with bumper cover removal.

10 Remove the rear fenderwell splash shields (see illustration).

11 Remove the rear license plate light assembly (see Chapter 12).

12 Have an assistant support the bumper cover as the fasteners are removed.

13 Remove the fasteners and detach the rear bumper cover (see illustrations).

14 Remove the nuts and bolts and detach the rear bumper reinforcement (see illustration).

15 Installation is the reverse of removal.

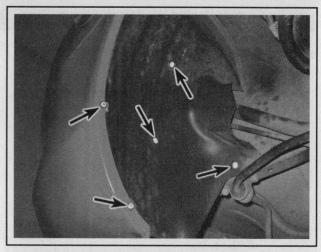

12.10 Remove the retainers and screws and detach the rear fenderwell splash shields

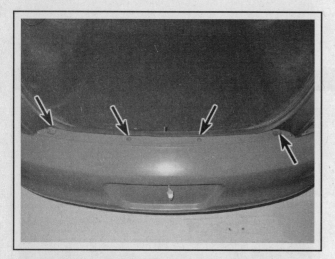

12.13a Remove the fasteners along the top of the rear bumper cover

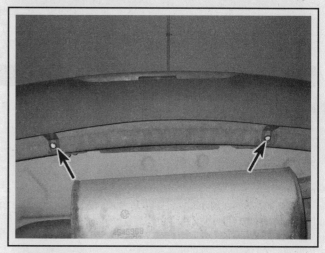

12.13b Remove the bumper cover lower fasteners . . .

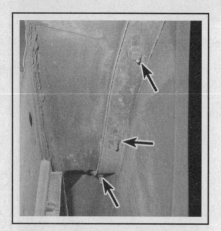

12.13c . . . the nuts along each side of the bumper cover . . .

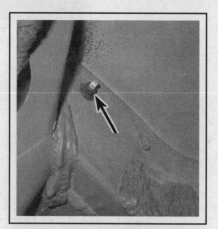

12.13d . . . and the nuts accessible from inside the trunk on each side (you'll have to pull back the carpet)

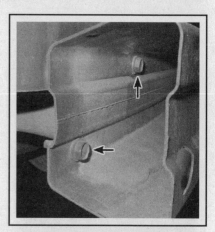

12.14 Remove the bolts located inside the rear bumper reinforcement

13 Door trim panel - removal and installation

▶ Refer to illustrations 13.3, 13.4a, 13.4b, 13.14c, 13.5, 13.7, 13.8 and 13.9

➡Note: This procedure applies to both the front and rear doors.

1   Disconnect the negative cable from the battery.

2   Open the door and completely lower the window glass.

3   On manual window models, remove the window crank retaining clip by working a cloth back-and-forth behind the handle to dislodge the retainer (see illustration).

4   Remove the retaining screws holding the trim panel to the door (see illustrations).

5   Disengage the push-in fasteners around the perimeter of the trim panel (see illustration). Tilt the trim panel outward to clear the locator pins on the backside of the trim panel.

6   Grasp the trim panel and pull up sharply to detach it from the retainer channel in the inner belt weatherstrip on top of the door.

7   Move the trim panel away from the door and disengage the latch linkage from the backside of the door handle (see illustration).

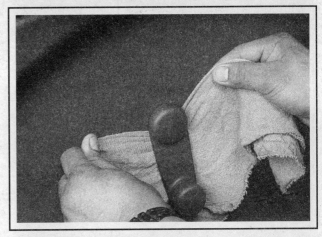

13.3 Working a cloth back-and-forth behind the window crank will dislodge the retaining clip

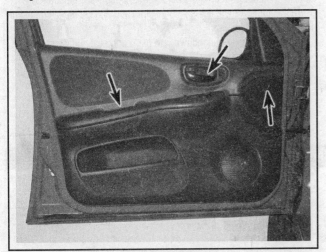

13.4a Door trim panel screw locations

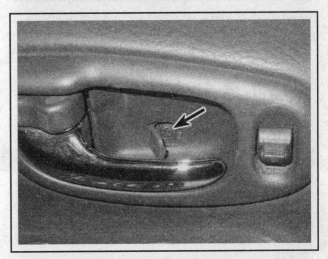

13.4b Pull the interior door release handle out and remove the trim panel-to-door mounting screw

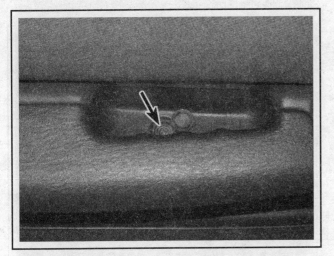

13.4c Remove the cover for access to the retaining screw in the pull cup

13.5 Use a flat-blade screwdriver to disengage the push-in fasteners around the perimeter of the trim panel, tilt the trim panel outward to clear the locator pins, then detach the door trim panel

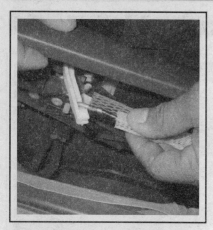

13.7 Pull the trim panel partially away from the door and disengage the door handle latch linkage

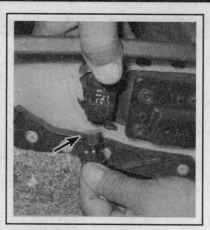

13.8 Disconnect the electrical connectors

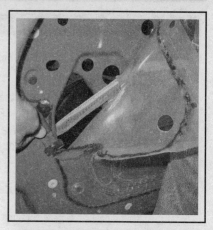

13.9 Carefully peel the watershield from the door, taking care not to tear it

✳✳ CAUTION:

Do not allow the trim panel to hang from the electrical wires.

8   On models so equipped, disconnect the electrical connectors and remove the trim panel (see illustration).

9   If it is necessary to remove the watershield, remove the mounting bracket then detach the watershield from the door (see illustration).

10 Reconnect any electrical connectors and engage the latch linkage to the door handle clip.

11 Press the trim panel down into place and seat all fasteners.

12 Install the door handle screws and window crank.

14 Door latch, outside handle and lock cylinder - removal and installation

➡Note: This procedure applies to both the front and rear doors.

1   Remove the door trim panel and watershield, then close the window completely (see Section 13).

LATCH

▶ **Refer to illustrations 14.4 and 14.5**

2   Disconnect the link rods from the latch.

3   On power door lock models, disconnect the electrical connector from the door lock motor.

4   Remove the three mounting screws from the end of the door (it may be necessary to use an impact-type screwdriver to loosen them) (see illustration).

5   Detach the clips from the control rods and detach the latch from the door (see illustration).

6   Place the latch in position, install the screws and tighten them securely.

7   Connect the link rods to the latch.

LOCK CYLINDER

▶ **Refer to illustration 14.8**

8   Disconnect the link and electrical connector, remove the lock cylinder clip retainer and withdraw the lock cylinder from the door handle assembly (see illustration)

9   Installation is the reverse of removal.

OUTSIDE HANDLE

▶ **Refer to illustration 14.10**

10 Disconnect the link, remove the mounting bolts and detach the handle from the door (see illustration).

11 Place the handle in position, attach the links and install the bolts. Tighten the bolt securely.

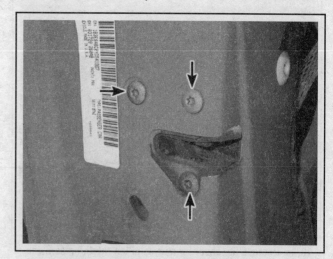

14.4 Use a Torx-head driver and remove the three door latch screws

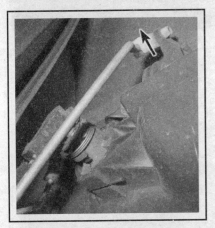

14.5 To detach any of the control rods, unsnap the clip from the rod, then pull the rod out of the lever

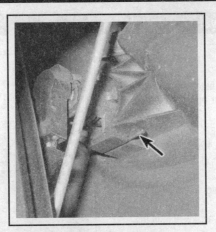

14.8 Remove the lock cylinder by pushing off the clip at its base

14.10 The door handle is retained by two bolts

15 Door window glass - removal and installation

REMOVAL

▶ Refer to illustrations 15.4, 15.6 and 15.7

1   Remove the door trim panel (see Section 13).
2   Remove the door speaker (if equipped) and the door trim panel mounting bracket (see Section 13). Disengage the linkage from the lock button bell crank.
3   Remove the watershield (see Section 13).
4   Carefully pull up and remove the inner belt weatherstrip (see illustration).
5   If it's not already down, lower the window to gain access to the glass attachment fasteners.
6   Carefully support the glass, then remove the two bolts holding the regulator lift channel to the door glass (see illustration).
7   Carefully lift the door glass up and out of opening at the top of the door (see illustration).

INSTALLATION

8   Installation is the reverse of removal. Tighten all bolts and nuts

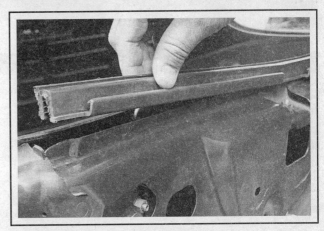

15.4 Carefully pull up and remove the inner belt weatherstrip

securely. Make sure the watershield is sealed properly; replace the adhesive if necessary.

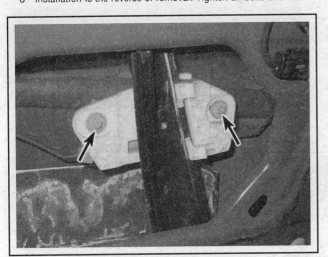

15.6 The window glass is held to the regulator by these two bolts

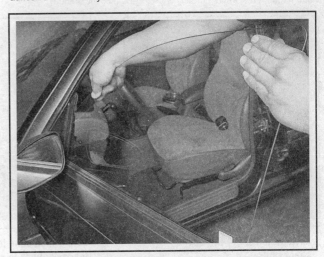

15.7 Carefully pull the door glass up, then lift it outward

16 Door window regulator - removal and installation

♦ **Refer to illustration 16.4**

➠**Note: Door window regulator removal and installation procedures are identical for both manual and power operated windows.**

1   Remove the door trim panel (see Section 13).
2   Remove the door window glass (see Section 15).
3   On power window models, unplug the electrical connector from the motor.
4   Remove the bolts holding the window regulator to the inner door panel (see illustration).
5   Remove the window regulator from the large access hole in the door.
6   Installation is the reverse of removal.

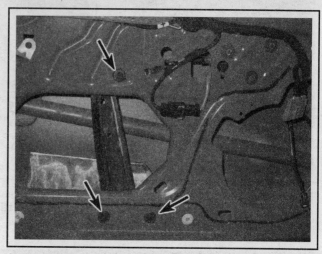

16.4 Remove the window regulator-to-inner door panel mounting bolts

17 Door - removal and installation

➠**Note 1: This procedure applies to both the front and rear doors.**

➠**Note 2: The door is heavy and somewhat awkward to remove and install - at least two people should perform this procedure.**

REMOVAL

♦ **Refer to illustrations 17.2, 17.3 and 17.4**

1   Open the door, peel back the rubber boot, and disconnect the wiring harness connector at the door pillar (see illustration 23.16).
2   Remove the bolts and detach the check strap from the door pillar (see illustration).
3   Place a jack or jackstands under the door or have an assistant on hand to support it when the bolts are removed (see illustration).

➠**Note: If a jack or jackstand is used, place a rag and a piece of wood between it and the door to protect the door's painted surfaces.**

4   Mark around the hinges, then remove the bolts securing first the lower then the upper hinge to the door and remove the door (see illustration).

INSTALLATION

5   Installation is the reverse of removal. Install the bolts and tighten them securely. Check door fit and adjust if necessary by loosening the hinge bolts.

➠**Note: Place scribe marks around the hinges before loosening the bolts to maintain a reference point.**

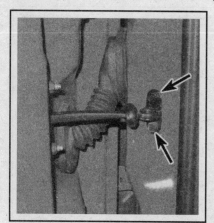

17.2 Remove the two door check bolts

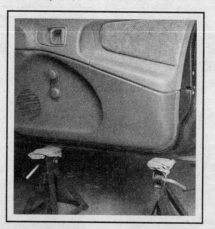

17.3 Place shop towels on top of the jackstands to prevent damage to the door

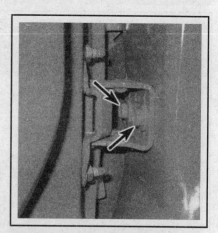

17.4 Remove the bolts from the lower hinge, then the upper hinge and lift the door off

18 Trunk lid and latch - removal, installation and adjustment

TRUNK LID

▶ **Refer to illustration 18.3 and 18.4**

1 Open the trunk lid and cover the edges of the trunk compartment with pads or cloths to protect the painted surfaces when the lid is removed.

2 Disconnect any cables or electrical connectors attached to the trunk lid that would interfere with removal.

3 Use a marking pen to make alignment marks around the hinge (see illustration).

4 While an assistant supports its weight, remove the hinge bolts from both sides and lift the trunk lid off (see illustration).

5 Installation is the reverse of removal.

➡**Note: Before tightening the bolts, align the marks made in Step 3.**

6 After installation, close the lid and see if it's in proper alignment with the surrounding panels. Fore-and-aft adjustment of the lid is con-trolled by the position of the hinge bolts in the holes. To adjust it, loosen the hinge bolts, reposition the lid and retighten the bolts.

7 The height of the lid in relation to the surrounding body panels when closed can be adjusted by loosening the lock striker bolts, repositioning the striker and re-tightening the bolts.

LATCH

▶ **Refer to illustration 18.9**

8 Open the trunk and detach the trim cover.

9 Remove the latch bolts and detach the latch from the trunk (see illustration).

10 Detach the remote cable and disconnect the trunk electrical connectors from the latch and remove it. The lock cylinder link will stay attached to the trunk.

11 Installation is the reverse of removal, making sure the lock cylinder link meshes properly with the latch assembly.

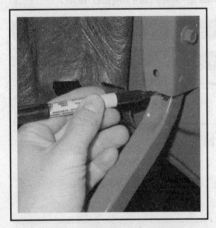

18.3 Use a marking pen to outline the position of the hinge

18.4 Remove the hinge bolts and lift the trunk lid off

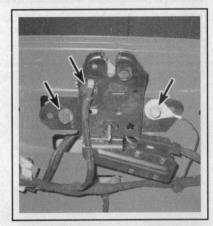

18.9 Remove the latch bolts, lower the latch and detach the electrical connectors and cables

19 Instrument panel top cover - removal and installation

▶ **Refer to illustrations 19.2 and 19.3**

※ WARNING:

These models have airbags. Always disconnect the negative battery cable and wait two minutes before working in the vicinity of the impact sensors, steering column or instrument panel to avoid the possibility of accidental deployment of the airbag, which could cause personal injury (see Chapter 12).

※ CAUTION:

This cover can be easily scratched; take care in removing the top cover to avoid damage.

1 Disconnect the negative cable from the battery.

2 Carefully pry out the A-pillar trim pieces (see illustration).

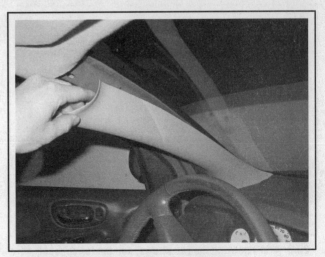

19.2 Carefully detach the A-pillar trim pieces

3   Grasp the rearward edge of the panel top cover, lift up at the instrument cluster and pull back to disengage the mounting clips (see illustration).

4   Pull the panel top cover assembly toward the rear until the forward pins disengage from the instrument panel.

5   Installation is the reverse of removal. Position the spring clips on the panel top cover and push them on until correctly seated.

19.3 Grasp the top cover securely and pull back sharply to detach it from the instrument panel

20 Dashboard trim panels - removal and installation

✳✳ WARNING:

These models have airbags. Always disconnect the negative battery cable and wait two minutes before working in the vicinity of the impact sensors, steering column or instrument panel to avoid the possibility of accidental deployment of the airbag, which could cause personal injury (see Chapter 12).

✳✳ CAUTION:

The following trim covers can be easily scratched; take care in removing them to avoid damage.

END COVERS

▶ **Refer to illustration 20.1**

1   Grasp the cover handle securely and rotate the cover out of the dashboard (see illustration). Installation is the reverse of removal.

LOWER LEFT PANEL COVER

▶ **Refer to illustration 20.3**

2   Remove the instrument cluster bezel (see Section 21).

3   Remove the screws, grasp the lower edge of the cover and pull back to detach it (see illustration).

4   Installation is the reverse of removal.

20.1 Rotate the left end cover out of the end of the dashboard

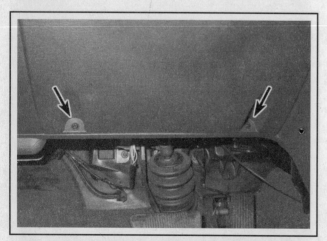

20.3 After removing the two screws, pull the lower panel straight back to remove it

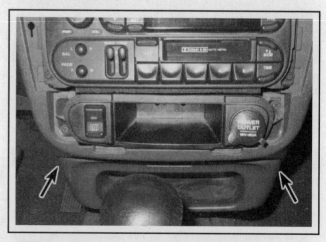

20.5 Pry at the corners of the lower storage bin, then pull it out

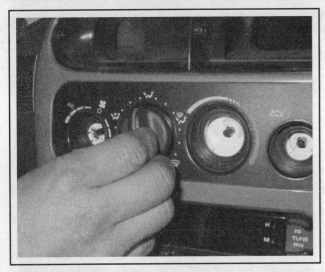

20.6 Pull off the center bezel knobs

20.7 Rotate the center vents down and out of the openings to remove them

CENTER BEZEL

▶ **Refer to illustrations 20.5, 20.6, 20.7, 20.8a and 20.8b**

5   Remove the lower storage bin (see illustration).

6   Remove the center bezel control knobs (see illustration).

7   Remove the two center vents by rotating them forward and pulling straight out (see illustration).

8   Remove the screws from the vent openings and rotate the bezel rearward and lower it from the dashboard (see illustrations).

9   Installation is the reverse of removal.

GLOVEBOX

▶ **Refer to illustration 20.10**

10 Remove the three screws along the bottom edge, then open the glovebox, squeeze the sides in and lower it from the dashboard (see illustration).

11 Installation is the reverse of removal.

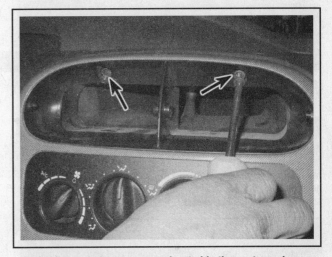

20.8a Remove the two screws located in the vent openings

20.8b Lower the center bezel until the tabs are clear of the dashboard, then pull it straight back to remove it

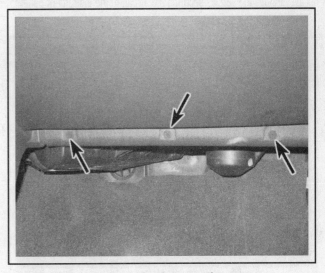

20.10 Remove the three glovebox mounting screws

21 Instrument cluster bezel - removal and installation

✳✳ WARNING:

These models have airbags. Always disable the airbag system before working in the vicinity of the impact sensors, steering column or instrument panel to avoid the possibility of accidental deployment of the airbag, which could cause personal injury (see Chapter 12).

REMOVAL

▶ **Refer to illustration 21.3**

1 Disconnect the cable from the negative battery ground terminal.
2 Remove the instrument panel top cover (see Section 19).
3 Pull the instrument cluster bezel straight out and remove it from the dashboard (see illustration).
4 Installation is the reverse of removal.

21.3 Grasp the cluster bezel securely and pull back to detach it

22 Passenger airbag - removal and installation

▶ **Refer to illustrations 22.4a, 22.4b, 22.5, 22.6a and 22.6b**

✳✳ WARNING:

Always disconnect the negative battery cable and wait two minutes before working in the vicinity of the impact sensors, steering column or instrument panel to avoid the possibility of accidental deployment of the airbag, which could cause personal injury (see Chapter 12).

1 Disconnect the negative cable from the battery and isolate it so there is no chance it can accidentally come in contact with the terminal.
2 Remove the instrument panel top cover (see Section 19).
3 Remove the glovebox.
4 Remove the screws securing the airbag to the instrument panel (see illustrations).

5 Remove the three securing nuts holding the back of the airbag assembly to the instrument panel support structure (see illustration).
6 Lift the airbag module up, squeeze the red locking tab, compress the lock and unplug the 4-pin module electrical connector, then remove the airbag module (see illustrations).

✳✳ WARNING:

Carry the airbag with the trim cover side FACING AWAY from your body to minimize injury if the airbag module accidentally deploys. Set the airbag module aside in a safe, isolated location and set it down with the trim cover side facing up.

7 Installation is the reverse of removal. Ensure that the red locking tab is in the locked position after attaching the connector (see illustration 22.6b). Tighten the screws and nuts securely.

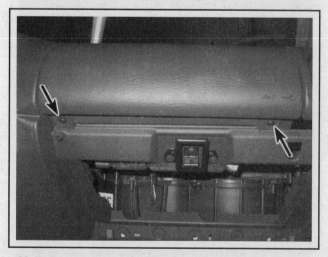

22.4a Remove the screws along the top of the passenger airbag cover

22.4b Remove the two lower airbag cover screws

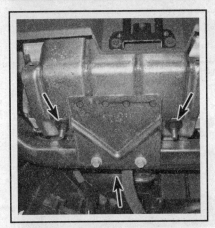

22.5 Remove the three airbag mounting nuts (the nut indicated by the lower arrow, which isn't visible in this photo, faces the front of the vehicle)

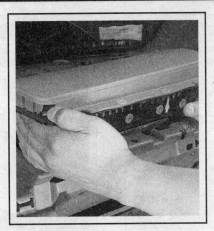

22.6a Lift the airbag module up and out of the instrument panel

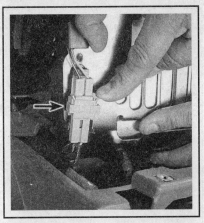

22.6b Unplug the four-pin electrical connector and remove the airbag module. Refer to Warning in text regarding how to safely carry the airbag assembly

8   From the driver's side of the vehicle turn the ignition switch to the Off position, then turn it to the On position. Check that the instrument cluster AIRBAG lamp illuminates for six to eight seconds and then goes out, indicating the airbag system is functioning properly. If the lamp fails to light, blinks on and off or stays on, there is a malfunction in the airbag system. If any of these conditions exist, a dealer service department or other qualified repair shop should diagnose the vehicle.

23 Instrument panel - removal and installation

✳✳ WARNING:

These models have airbags. Always disable the airbag system before working in the vicinity of the impact sensors, steering column or instrument panel to avoid the possibility of accidental deployment of the airbag, which could cause personal injury (see Chapter 12).

➡Note: It is a good idea to remove both front seats (see Section 28) to allow additional working space and lessen the chance of damage to the seats during this procedure.

REMOVAL

▸ Refer to illustrations 23.9, 23.10, 23.16 and 23.17

1   Disconnect the negative cable from battery ground terminal. Make sure the cable can't accidentally come into contact with the terminal.

2   Remove the A-pillar trim pieces (see illustration 19.2).

3   Remove the instrument panel top cover (see Section 19).

4   Remove the instrument cluster bezel (see Section 21).

5   Refer to Section 20 and remove the dashboard trim panels and glovebox.

6   Remove the steering column covers (Section 24)

7   Lock the steering wheel in the straight ahead position.

8   Disconnect the electrical connectors for the ignition switch, turn signal switch, immobilizer module (if equipped) and airbag clockspring.

9   Mark the relationship of the universal joint to the steering gear input shaft, then remove the pin from the steering shaft joint (see illustration).

10 Remove nuts attaching the steering column to the instrument panel (see illustration).

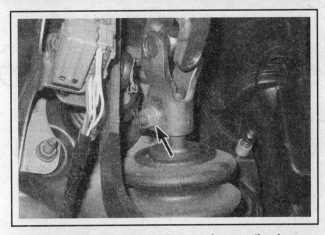

23.9 Pull off the clip, unscrew the nut and remove the pin from the steering shaft joint

23.10 Remove the steering column nuts

23.16 Disconnect the wire harness connectors located at the right and left A-pillars

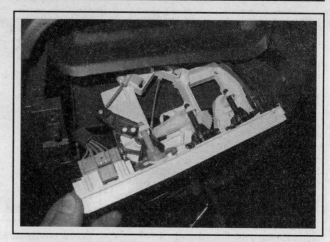

23.17 Remove the screws and rotate the air conditioning control out of the instrument panel

11 Remove the floor console (see Section 25).

12 Disconnect the Data Link Connector from the instrument panel bracket.

13 Remove the instrument panel center support mounting bolts.

14 Remove the A-pillar-to-instrument panel mounting bolts (there are two on each side).

15 Disconnect the antenna right side connector.

16 Disconnect the instrument panel harness connectors located at the A-pillars (see illustration).

17 Remove the two screws, disconnect the electrical and vacuum connectors and remove the heater/air conditioning unit by rotating

it 90-degrees as you lift it from the opening (see illustration).

18 Remove the two bolts at the top of the brake pedal/instrument panel support.

19 With the help of an assistant, lift up the instrument panel, pull rearward and withdraw it from the vehicle.

INSTALLATION

20 Installation is the reverse of removal, taking care to make sure the instrument panel will be vertical by first inserting two half-inch bolts or pins through the pilot holes in the pillars.

24 Steering column covers - removal and installation

▶ Refer to illustrations 24.2a and 24.2b

❋❋ WARNING:

These models have airbags. Always disable the airbag system before working in the vicinity of the impact sensors, steering column or instrument panel to avoid the possibility of accidental deployment of the airbag, which could cause personal injury (see Chapter 12).

❋❋ CAUTION:

These covers can be easily scratched; take care when removing them to avoid damage.

1 Disconnect the negative cable from battery ground terminal. Make sure the cable can't accidentally come into contact with the terminal.

2 Remove the screws from the lower steering column cover and separate the covers (see illustrations).

3 Installation is the reverse of removal.

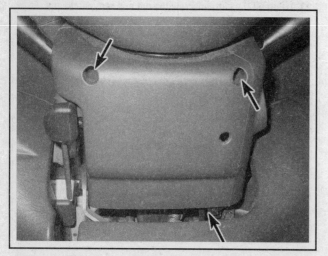

24.2a Remove the steering column cover screws . . .

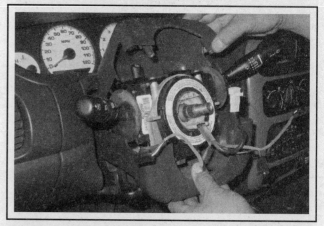

24.2b . . . then separate the lower and upper covers and remove them

25 Center console - removal and installation

▶ Refer to illustrations 25.3a, 25.3b and 25.5

⁂ WARNING:

These models have airbags. Always disconnect the negative battery cable and wait two minutes before working in the vicinity of the impact sensors, steering column or instrument panel to avoid the possibility of accidental deployment of the airbag, which could cause personal injury (see Chapter 12).

⁂ CAUTION:

The center console can be easily scratched; take care in removing the assembly to avoid damage.

1 Disconnect the negative cable from the battery.
2 Completely raise the parking brake lever.
3 Remove the front and rear plugs and center console-retaining screws (see illustrations).
4 On manual transaxle models, remove the shift lever knob.
5 Lift the rear portion of the center console up and over the shift lever and parking brake handle and remove from vehicle (see illustration).
6 Installation is the reverse of removal.

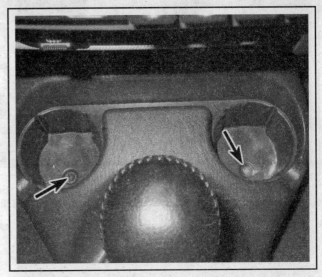

25.3a Remove the front center console retaining screws (arrows) . . .

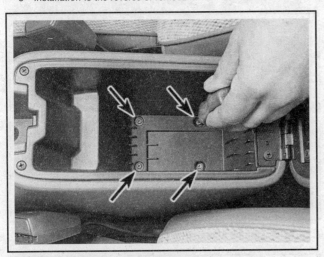

25.3b . . . followed by the rear screws

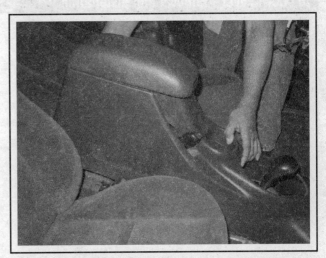

25.5 Lift the rear portion up and over the shift lever and parking brake handle and remove it from the vehicle

26 Mirrors - removal and installation

INTERIOR

▶ Refer to illustrations 26.1a and 26.1b

1 Use a Phillips head screwdriver to remove the set screw, disconnect the electrical connector, then slide the mirror up off the button on the windshield (see illustrations).
2 Installation is the reverse of removal.

EXTERIOR

▶ Refer to illustration 26.5

3 Detach the mirror bezel or remove the door trim panel (see Section 13), as necessary.
4 On power mirrors, remove the watershield (see Section 13), then unplug the electrical connector.

26.1a The interior mirror fits over a button bonded to the windshield and is held in place by a Phillips head set screw - loosen this screw . . .

26.1b . . . and slide the mirror assembly up off the button

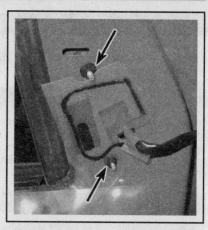

26.5 Remove the mounting nuts and detach the mirror

5   Remove the nuts and detach the mirror from the mirror stanchion on the door (see illustration).

6   Installation is the reverse of removal.

27 Cowl cover - removal and installation

▶ Refer to illustrations 27.1a, 27.1b and 27.3

1   Pry off the plastic trim cap on the windshield wiper arms, then detach the wiper arm retaining nuts and remove the wiper arms (see illustrations).
2   Carefully remove the rubber hood-sealing strip.
3   Remove the retaining screws on each side securing the cowl cover, then remove the cover (see illustration).
4   Installation is the reverse of removal.

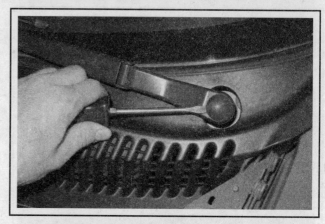

27.1a Pry off the plastic trim cap . . .

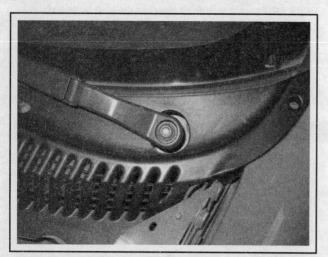

27.1b . . . remove the wiper arm retaining nuts, then mark the shafts and remove both wiper arms

27.3 Remove the retaining screws on each side and remove the cowl cover

28 Seats - removal and installation

✳✳ WARNING:

Some models are equipped with side-impact airbags located in the exterior sides of the front seats. On these models, be sure to disable the airbag system before removing a front seat. Also, do not disassemble these seats.

FRONT

▶ **Refer to illustrations 28.1 and 28.2**

1   Move the seat rearward and remove the seat track front bolts (see illustration).

2   Move the seat forward and remove the seat track rear bolts (see illustration).

3   Unplug any electrical connectors attached to the seat and lift the seat from the vehicle.

4   Installation is the reverse of removal.

REAR

▶ **Refer to illustrations 28.5, 28.6a and 28.6b**

5   Remove the seat cushion by grasping the front edge securely, then pulling up sharply to detach the cushion (see illustration).

6   After removing the seat cushion, remove the seat back and seat and seat belt retaining bolts, lift up at the rear to disengage the hooks from the slots and remove the seat back from the vehicle (see illustration).

7   Installation is the reverse of removal.

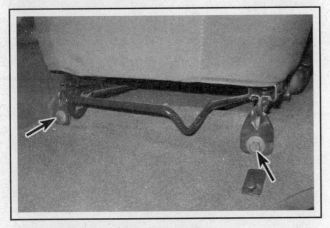

28.1 Remove the seat track front bolts

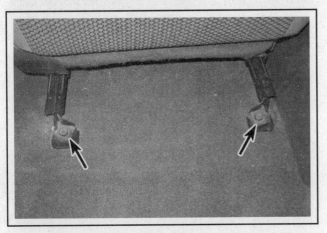

28.2 Remove the seat track rear bolts

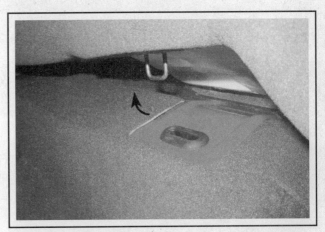

28.5 Grasp the front edge of the rear cushion securely and pull up to detach the hooks

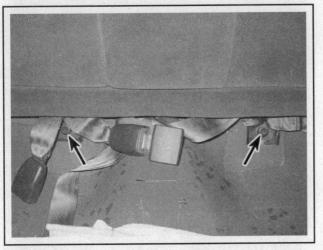

28.6a Remove the seat back and seat belt retaining bolts, then lift up . . .

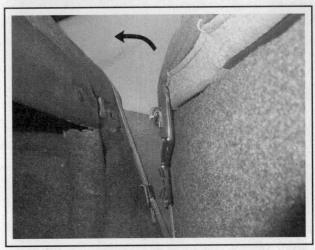

28.6b . . . at the rear to disengage the hooks at the top of the seat back

Notes

Section

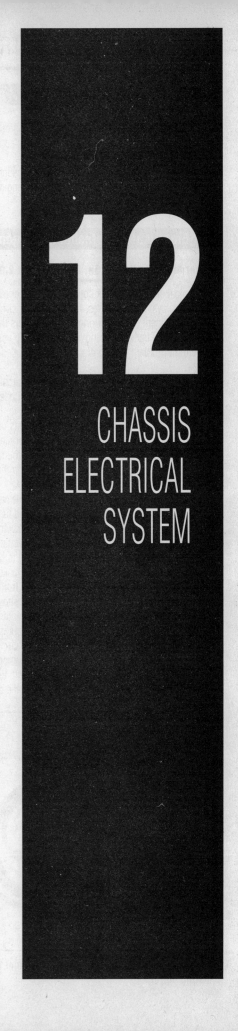

12

CHASSIS ELECTRICAL SYSTEM

1 General information

The electrical system is a 12-volt, negative ground type. Power for the lights and all electrical accessories is supplied by a lead/acid-type battery, which is charged by the alternator.

This Chapter covers repair and service procedures for the various electrical components not associated with the engine. Information on the battery, alternator, distributor and starter motor can be found in Chapter 5.

It should be noted that when portions of the electrical system are serviced, the cable should be disconnected from the negative battery terminal (see Chapter 5) to prevent electrical shorts and/or fires.

2 Electrical troubleshooting - general information

▶ **Refer to illustrations 2.5a, 2.5b, 2.6 and 2.9**

A typical electrical circuit consists of an electrical component, any switches, relays, motors, fuses, fusible links or circuit breakers related to that component and the wiring and connectors that link the component to both the battery and the chassis. To help you pinpoint an electrical circuit problem, wiring diagrams are included at the end of this Chapter.

Before tackling any troublesome electrical circuit, first study the appropriate wiring diagrams to get a complete understanding of what makes up that individual circuit. Trouble spots, for instance, can often be narrowed down by noting if other components related to the circuit are operating properly. If several components or circuits fail at one time, chances are the problem is in a fuse or ground connection, because several circuits are often routed through the same fuse and ground connections.

Electrical problems usually stem from simple causes, such as loose or corroded connections, a blown fuse, a melted fusible link or a failed relay. Visually inspect the condition of all fuses, wires and connections in a problem circuit before troubleshooting the circuit.

If test equipment and instruments are going to be utilized, use the diagrams to plan ahead of time where you will make the necessary connections in order to accurately pinpoint the trouble spot.

The basic tools needed for electrical troubleshooting include a circuit tester or voltmeter (a 12-volt bulb with a set of test leads can also be used), a continuity tester, which includes a bulb, battery and set of test leads, and a jumper wire, preferably with a circuit breaker incorporated, which can be used to bypass electrical components (see illustrations). Before attempting to locate a problem with test instruments, use the wiring diagram(s) to decide where to make the connections.

VOLTAGE CHECKS

Voltage checks should be performed if a circuit is not functioning properly. Connect one lead of a circuit tester to either the negative battery terminal or a known good ground. Connect the other lead to a connector in the circuit being tested, preferably nearest to the battery or fuse (see illustration). If the bulb of the tester lights, voltage is present,

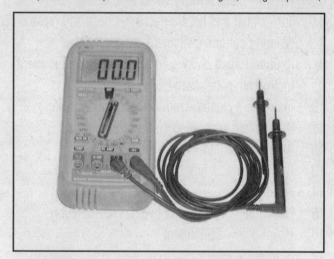

2.5a The most useful tool for electrical troubleshooting is a digital multimeter that can check volts, amps, and test continuity

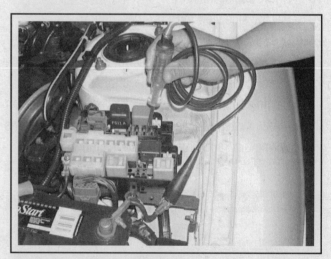

2.6 In use, a basic test light's lead is clipped to a known good ground, then the pointed probe can test connectors, wires or electrical sockets - if the bulb lights, the circuit being tested has battery voltage

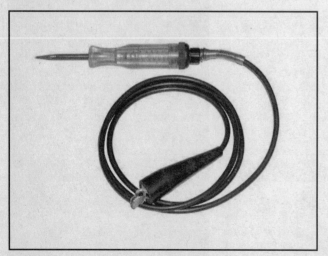

2.5b A simple test light is a very handy tool for testing voltage

which means that the part of the circuit between the connector and the battery is problem free. Continue checking the rest of the circuit in the same fashion. When you reach a point at which no voltage is present, the problem lies between that point and the last test point with voltage. Most of the time the problem can be traced to a loose connection.

→Note: Keep in mind that some circuits receive voltage only when the ignition key is in the Accessory or Run position.

FINDING A SHORT

One method of finding shorts in a live circuit is to remove the fuse and connect a test light in place of the fuse terminals (fabricate two jumper wires with small spade terminals, plug the jumper wires into the fuse box and connect the test light). There should be voltage present in the circuit. Move the suspected wiring harness from side-to-side while watching the test light. If the bulb goes off, there is a short to ground somewhere in that area, probably where the insulation has rubbed through.

GROUND CHECK

Perform a ground test to check whether a component is properly grounded. Disconnect the battery and connect one lead of a continuity tester or multimeter (set to the ohms scale), to a known good ground. Connect the other lead to the wire or ground connection being tested. If the resistance is low (less than 5 ohms), the ground is good. If the bulb on a self-powered test light does not go on, the ground is not good.

CONTINUITY CHECK

A continuity check is done to determine if there are any breaks in a circuit - if it is passing electricity properly. With the circuit off (no power in the circuit), a self-powered continuity tester or multimeter can be used to check the circuit. Connect the test leads to both ends of the circuit (or to the "power" end and a good ground), and if the test light comes on the circuit is passing current properly (see illustration). If the resistance is low (less than 5 ohms), there is continuity; if the reading is 10,000 ohms or higher, there is a break somewhere in the circuit. The same procedure can be used to test a switch, by connecting the continuity tester to the switch terminals. With the switch turned On, the test light should come on (or low resistance should be indicated on a meter).

FINDING AN OPEN CIRCUIT

When diagnosing for possible open circuits, it is often difficult to locate them by sight because the connectors hide oxidation or terminal misalignment. Merely wiggling a connector on a sensor or in the wiring harness may correct the open circuit condition. Remember this when an open circuit is indicated when troubleshooting a circuit. Inter-

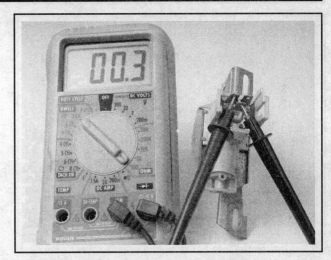

2.9 With a multimeter set to the ohm scale, resistance can be checked across two terminals - when checking for continuity, a low reading indicates continuity, a high reading or infinity indicates lack of continuity

mittent problems may also be caused by oxidized or loose connections.

Electrical troubleshooting is simple if you keep in mind that all electrical circuits are basically electricity running from the battery, through the wires, switches, relays, fuses and fusible links to each electrical component (light bulb, motor, etc.) and to ground, from which it is passed back to the battery. Any electrical problem is an interruption in the flow of electricity to and from the battery.

CONNECTORS

Most electrical connections on these vehicles are made with multi-wire plastic connectors. The mating halves of many connectors are secured with locking clips molded into the plastic connector shells. The mating halves of large connectors, such as some of those under the instrument panel, are held together by a bolt through the center of the connector.

To separate a connector with locking clips, use a small screwdriver to pry the clips apart carefully, then separate the connector halves. Pull only on the shell, never pull on the wiring harness as you may damage the individual wires and terminals inside the connectors. Look at the connector closely before trying to separate the halves. Often the locking clips are engaged in a way that is not immediately clear. Additionally, many connectors have more than one set of clips.

Each pair of connector terminals has a male half and a female half. When you look at the end view of a connector in a diagram, be sure to understand whether the view shows the harness side or the component side of the connector. Connector halves are mirror images of each other, and a terminal shown on the right side end-view of one half will be on the left side end view of the other half.

---

3   Fuses and fusible links - general information

FUSES

▶ **Refer to illustrations 3.1a, 3.1b, 3.1c and 3.3**

The electrical circuits of the vehicle are protected by a combination of fuses, circuit breakers and fusible links. Fuse blocks are located

under the left end of the instrument panel and in the engine compartment (see illustrations).

Each of the fuses is designed to protect a specific circuit, and the various circuits are identified on the fuse panel cover.

Miniaturized fuses are employed in the fuse blocks. These compact fuses, with blade terminal design, allow fingertip removal and replace-

3.1a The interior fuse box is located under a cover on the left end of the instrument panel

3.1b The main engine compartment fuse/relay box is located behind the air cleaner housing next to the battery

ment. If an electrical component fails, always check the fuse first. The best way to check a fuse is with a test light. Check for power at the exposed terminal tips of each fuse. If power is present on one side of the fuse but not the other, the fuse is blown. A blown fuse can also be confirmed by visually inspecting it (see illustration).

Be sure to replace blown fuses with the correct type. Fuses of different ratings are physically interchangeable, but only fuses of the proper rating should be used. Replacing a fuse with one of a higher or lower value than specified is not recommended. Each electrical circuit needs a specific amount of protection. The amperage value of each fuse is molded into the fuse body.

If the replacement fuse immediately fails, don't replace it again until

the cause of the problem is isolated and corrected. In most cases, this will be a short circuit in the wiring caused by a broken or deteriorated wire.

FUSIBLE LINKS

Some circuits are protected by fusible links. The links are used in circuits which are not ordinarily fused, or which carry high current.

Cartridge type fusible links are located in the engine compartment fusible link box and are similar to a large fuse. After disconnecting the negative battery cable or remote ground terminal, simply unplug and replace a fusible link of the same amperage.

3.1c Information on the fuse block components is found on the back of the cover

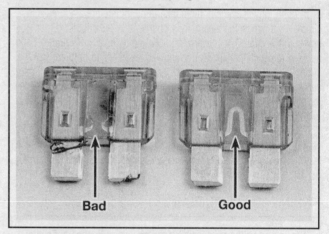

Bad **Good**

3.3 When a fuse blows, the element between the terminals melts

4 Circuit breakers - general information

Circuit breakers protect certain circuits, such as the power windows or heated seats. Depending on the vehicle's accessories, there may be one or two circuit breakers, located in the fuse/relay box in the engine compartment (see illustration 3.1c).

Because the circuit breakers reset automatically, an electrical overload in a circuit-breaker-protected system will cause the circuit to fail momentarily, then come back on. If the circuit does not come back on, check it immediately.

For a basic check, pull the circuit breaker up out of its socket on the fuse panel, but just far enough to probe with a voltmeter. The breaker should still contact the sockets.

With the voltmeter negative lead on a good chassis ground, touch each end prong of the circuit breaker with the positive meter probe. There should be battery voltage at each end. If there is battery voltage only at one end, the circuit breaker must be replaced.

Some circuit breakers must be reset manually.

5   Relays - general information and testing

GENERAL INFORMATION

1   Several electrical accessories in the vehicle, such as the fuel injection system, horns, starter, and fog lamps use relays to transmit the electrical signal to the component. Relays use a low-current circuit (the control circuit) to open and close a high-current circuit (the power circuit). If the relay is defective, that component will not operate properly. Most relays are mounted in the engine compartment fuse/relay box, with some specialized relays located above the interior fuse box in the dash (see illustrations 3.1a and 3.1b). If a faulty relay is suspected, it can be removed and tested using the procedure below or by a dealer service department or a repair shop. Defective relays must be replaced as a unit.

TESTING

▶ **Refer to illustrations 5.2a and 5.2b**

2   Most of the relays used in these vehicles are of a type often called "ISO" relays, which refers to the International Standards Organization. The terminals of ISO relays are numbered to indicate their usual circuit connections and functions. There are two basic layouts of terminals on the relays used in the covered vehicles (see illustrations).

3   Refer to the wiring diagram for the circuit to determine the proper connections for the relay you're testing. If you can't determine the correct connection from the wiring diagrams, however, you may be able to determine the test connections from the information that follows.

4   Two of the terminals are the relay control circuit and connect to the relay coil. The other relay terminals are the power circuit. When the

relay is energized, the coil creates a magnetic field that closes the larger contacts of the power circuit to provide power to the circuit loads.

5   Terminals 85 and 86 are normally the control circuit. If the relay contains a diode, terminal 86 must be connected to battery positive (B+) voltage and terminal 85 to ground. If the relay contains a resistor, terminals 85 and 86 can be connected in either direction with respect to B+ and ground.

6   Terminal 30 is normally connected to the battery voltage (B+) source for the circuit loads. Terminal 87 is connected to the ground side of the circuit, either directly or through a load. If the relay has several alternate terminals for load or ground connections, they usually are numbered 87A, 87B, 87C, and so on.

7   Use an ohmmeter to check continuity through the relay control coil.

   a) *Connect the meter according to the polarity shown in the illustration for one check; then reverse the ohmmeter leads and check continuity in the other direction.*

   b) *If the relay contains a resistor, resistance will be indicated on the meter, and should be the same value with the ohmmeter in either direction.*

   c) *If the relay contains a diode, resistance should be higher with the ohmmeter in the forward polarity direction than with the meter leads reversed.*

   d) *If the ohmmeter shows infinite resistance in both directions, replace the relay.*

8   Remove the relay from the vehicle and use the ohmmeter to check for continuity between the relay power circuit terminals. There should be no continuity between terminal 30 and 87 with the relay de-energized.

9   Connect a fused jumper wire to terminal 86 and the positive battery terminal. Connect another jumper wire between terminal 85 and ground. When the connections are made, the relay should click.

10   With the jumper wires connected, check for continuity between the power circuit terminals. Now, there should be continuity between terminals 30 and 87.

11   If the relay fails any of the above tests, replace it.

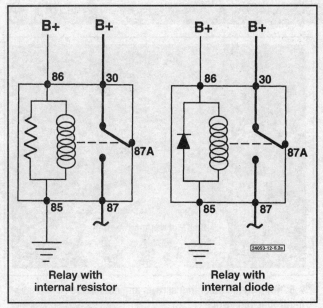

5.2a Typical ISO relay designs, terminal numbering and circuit connections

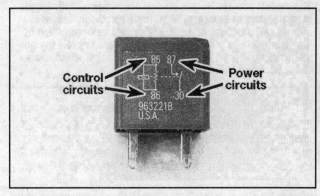

5.2b Most relays are marked on the outside to easily identify the control circuits and the power circuits - four terminal type shown

6 Turn signal and hazard flasher - check and replacement

▶ **Refer to illustration 6.4**

1   The turn signal and hazard flasher is a single combination unit.

2   When the flasher unit is functioning properly, an audible click can be heard during its operation. If the turn signals fail on one side or the other and the flasher unit does not make its characteristic clicking sound, or if a bulb on one side of the vehicle flashes much faster than normal but the bulb at the other end of the vehicle (on the same side) doesn't light at all, a faulty turn signal bulb may be indicated.

3   If both turn signals fail to blink, the problem may be due to a blown fuse, a faulty flasher unit, a defective switch or a loose or open connection. If a quick check of the fuse box indicates that the turn signal fuse has blown, check the wiring for a short before installing a new fuse.

4   To replace the flasher remove the upper steering column cover (see Chapter 11). Disconnect the electrical connector and remove the flasher unit from its mounting bracket. The flasher is located on top of

6.4 The turn signal hazard flasher unit is located in the steering column and mounted behind the multi-function switch

the steering column behind the multi function switch (see illustration).

5   Make sure that the replacement unit is identical to the original. Compare the old one to the new one before installing it.

6   Installation is the reverse of removal.

7 Steering column switches - replacement

▶ **Refer to illustration 7.5**

1   The multi-function switch is located on the steering column. It incorporates the turn signals, the hazard warning, the headlights, the headlight beam select (Hi/Lo), the headlight flasher, the instrument panel dimmer switch, the windshield wiper and windshield washer functions. There are two levers on the multi-function switch. The left side lever controls the signaling and the lighting while the right side controls the wipers and washer system. The cruise control switches are located on either side of the steering wheel hub.

2   Disconnect and isolate the negative cable so that it can't accidentally come in contact with the battery negative terminal.

MULTI-FUNCTION SWITCH

3   Remove the steering column covers (see Chapter 11).

4   Disconnect the electrical connectors from the backside of the switch.

7.5 Remove the mounting screws and detach the multi-function switch

5   Remove the multi-function switch retaining screws (see illustration).

6   Remove the multi-function switch by sliding the switch up off the column and lifting the assembly away.

7   Insert the connectors into the new multi-function switch, pushing in until they are securely locked in place.

8   The remainder of installation is the reverse of removal.

7.11 Remove the cruise control switch and disconnect the electrical connector

Make sure the clockspring is centered (see Chapter 10, Section 15).

Cruise control switches

▶ Refer to illustrations 7.11

9 Disconnect the battery negative cable and isolate it from the negative battery terminal.

10 Remove the two switch retaining screws from the back of steering wheel.

11 Withdraw the switch from the steering wheel and disconnect the electrical connector (see illustration).

12 Installation is the reverse of removal.

8 Ignition switch and key lock cylinder - replacement

IGNITION SWITCH

▶ Refer to illustrations 8.5 and 8.6

1 Disconnect the battery negative cable and isolate it from the negative battery terminal.

2 Remove the steering column covers (see Chapter 11).

3 Remove the key lock cylinder (see Steps 11 through 14).

4 Disconnect the electrical connector from the ignition switch.

5 Remove the Torx head screw securing the switch to the steering column (see illustration).

6 Depress the retaining tabs, then detach the switch and lower it from the steering column (see illustration).

7 When installing the switch, turn it to the ON position and make sure the actuator shaft in the housing is also in the ON position, then snap the switch into place.

8 The remainder of installation is the reverse of removal.

LOCK CYLINDER

▶ Refer to illustration 8.14

9 Disconnect and isolate the negative cable so that it can't accidentally come in contact with the battery negative terminal.

10 Insert the ignition key and turn the switch to the ON position.

11 Insert a small screwdriver through the access hole in the lower steering column cover and depress the retaining tab, then withdraw the lock cylinder from the housing (see illustration).

12 Before installing the lock cylinder, make sure the slot in the ignition switch is in the ON position. Insert the lock cylinder into the housing until the retaining tab locates the housing. Check the key operation.

8.5 Ignition switch retaining screw location

8.6 Use a screwdriver to depress the retaining tabs and withdraw the ignition switch

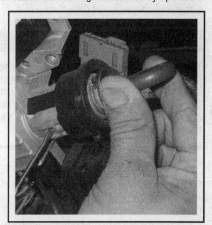

8.11 With the lock cylinder in the ON position, insert a small screwdriver and depress the retaining pin, then pull the lock cylinder straight out (steering column cover removed for clarity)

9   Instrument panel switches - replacement

⁕⁕ WARNING:

The models covered by this manual are equipped with Supplemental Restraint Systems (SRS), more commonly known as airbags. Always disable the airbag system before working in the vicinity of any airbag system components to avoid the possibility of accidental deployment of the airbag(s), which could cause personal injury (see Section 25).

ACCESSORY BEZEL SWITCHES

▶ **Refer to illustrations 9.4 and 9.5**

1   The rear window defogger switch and the cigar lighter/power

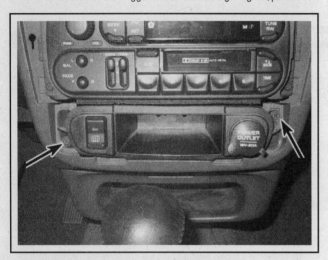

9.4 Remove the switch bezel retaining screws

outlet unit are located in the switch bezel in the dashboard. The defogger switch is part of the switch bezel and if it fails, the bezel and switch must be replaced as a unit. The cigar lighter/power outlet unit, however, can be replaced separately.

2   Disconnect and isolate the negative cable so that it can't accidentally come in contact with the battery negative terminal.

3   Remove the center bezel (see Chapter 11).

4   Remove the screws and detach the switch bezel from the dashboard (see illustration).

5   Disconnect the electrical connectors and remove the switch bezel housing (see illustration).

6   Detach the cigar lighter/power outlet unit from the bezel and transfer it to the replacement unit.

7   Installation is the reverse of the removal procedure.

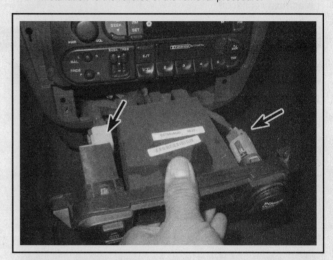

9.5 Detach the bezel unit and disconnect the electrical connectors

10   Instrument cluster - removal and installation

▶ **Refer to illustration 10.3**

⁕⁕ WARNING:

The models covered by this manual are equipped with Supplemental Restraint Systems (SRS), more commonly known as airbags. Always disable the airbag system before working in the vicinity of any airbag system components to avoid the possibility of accidental deployment of the airbag(s), which could cause personal injury (see Section 25).

1   Disconnect and isolate the negative cable so that it can't accidentally come in contact with the battery negative terminal.

2   Remove the instrument panel top cover and the instrument cluster bezel (see Chapter 11).

3   Remove the four cluster mounting screws and separate the instrument cluster from the instrument panel (see illustration).

4   Disconnect any electrical connectors that would interfere with removal.

5   Installation is the reverse of removal.

10.3 Remove the instrument cluster mounting screws - 2000 model shown, other models similar

11 Windshield wiper motor - replacement

▶ Refer to illustrations 11.2a, 11.2b, 11.4, 11.5 and 11.6

1   Disconnect and isolate the negative cable so that it can't accidentally come in contact with the battery negative terminal.

2   Pry out the plastic caps and remove the wiper arm nuts, mark their positions for ease of installation, then remove the wiper arms (see illustrations).

3   Remove the windshield cowl cover grille (see Chapter 11).

4   Disconnect the wiper motor harness connector (see illustration).

5   Remove the windshield wiper module mounting bolts/nuts and detach the wiper module assembly from the vehicle (see illustration).

6   Remove the wiper motor mounting screws and wiper arm nut, then separate the arm from the motor and the motor from the assembly (see illustration).

7   Activate the wiper motor before installation to make sure it is in the Park position. Installation is the reverse of removal.

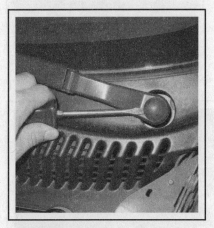

11.2a Use a screwdriver to pry out the plastic caps

11.2b Draw or paint a line across the stud and wiper arm to mark its position so it can be reinstalled in the same relationship to the windshield

11.4 Disconnect the wiper motor electrical connector

11.5 Remove the wiper motor and linkage assembly mounting bolts/nuts

11.6 Remove the wiper arm nut and the three retaining screws and detach the motor from the assembly

12 Antenna - replacement

▶ Refer to illustrations 12.2, 12.5a, 12.5b, 12.5c, 12.6a, 12.6b and 12.6c

✳✳ WARNING:

The models covered by this manual are equipped with Supplemental Restraint Systems (SRS), more commonly known as airbags. Always disable the airbag system before working in the vicinity of any airbag system components to avoid the possibility of accidental deployment of the airbag(s), which could cause personal injury (see Section 25).

1   Disconnect and isolate the negative cable so that it can't accidentally come in contact with the battery negative terminal.

12.2 Use a small wrench to unscrew the antenna mast

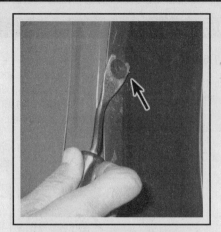

12.5a Pry out the plastic retaining clips from the fenderwell

12.5b Remove the screws retaining the lower end of the fender panel

2   Use a small open-end wrench and unscrew the antenna mast (see illustration).

3   Working in the right side interior kick panel area under the instrument panel, disconnect the antenna cable from the cable lead.

4   Raise the vehicle and place it securely on jackstands and remove the front right wheel.

5   Remove the right fenderwell inner panels to gain access to the antenna base (see illustrations).

6   Working in the right fenderwell, detach the grommet and carefully pull the antenna lead and grommet from the fender panel access hole (see illustration). Remove the antenna mounting nut and remove the antenna assembly from the fender (see illustrations). Don't lose the antenna adapter in the top of fender opening.

7   Installation is the reverse of removal. If removed, align the antenna adapter tongue with the fender hole and push it into place.

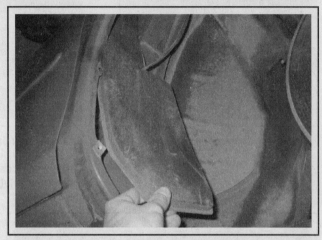

12.5c Remove the panel from the fenderwell

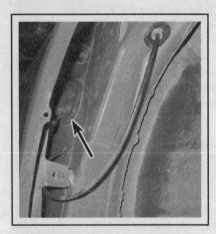

12.6a Detach the grommet and pull the antenna lead from the body

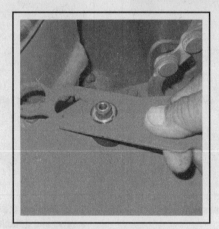

12.6b Remove the antenna retaining nut

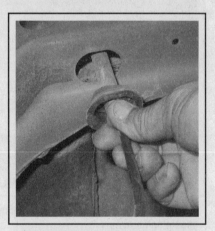

12.6c Lower the antenna from the fender

13 Rear window defogger - check and repair

1   The rear window defogger consists of a number of horizontal elements baked onto the glass surface.

2   Small breaks in the element can be repaired without removing the rear window.

CHECK

▶ **Refer to illustrations 13.4, 13.5 and 13.7**

3   Turn the ignition switch and defogger system switches to the ON position. Using a voltmeter, place the positive probe against the defog-

13.4 When measuring the voltage at the rear window defogger grid, wrap a piece of aluminum foil around the positive probe of the voltmeter and press the foil against the wire with your finger

ger grid positive terminal and the negative probe against the ground terminal. If battery voltage is not indicated, check the fuse, defogger switch and related wiring. If voltage is indicated, but all or part of the defogger doesn't heat, proceed with the following tests.

4   When measuring voltage during the next two tests, wrap a piece of aluminum foil around the tip of the voltmeter positive probe and press the foil against the heating element with your finger (see illustration). Place the negative probe on the defogger grid ground terminal.

5   Check the voltage at the center of each heating element (see illustration). If the voltage is 5 or 6-volts, the element is okay (there is no break). If the voltage is 0-volts, the element is broken between the center of the element and the positive end. If the voltage is 10 to 12-volts the element is broken between the center of the element and ground. Check each heating element.

6   Connect the negative lead to a good body ground. The reading should stay the same. If it doesn't, the ground connection is bad.

7   To find the break, place the voltmeter negative probe against the defogger ground terminal. Place the voltmeter positive probe with the foil strip against the heating element at the positive terminal end and

13.7 To find the break, place the voltmeter negative lead against the defogger ground terminal, place the voltmeter positive lead with the foil strip against the heating element at the positive terminal end and slide it toward the negative terminal end - the point at which the voltmeter reading changes abruptly is the point at which the element is broken

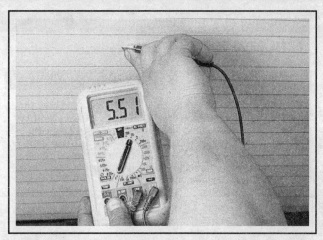

13.5 To determine if a heating element has broken, check the voltage at the center of each element - if the voltage is 5 or 6-volts, the element is unbroken - if the voltage is 10 or 12-volts, the element is broken between the center and the ground side - if there is no voltage, the element is broken between the center and the positive side

slide it toward the negative terminal end. The point at which the voltmeter deflects from several volts to zero is the point at which the heating element is broken (see illustration).

REPAIR

▶ **Refer to illustration 13.13**

8   Repair the break in the element using a repair kit specifically recommended for this purpose, available at most auto parts stores. Included in this kit is plastic conductive epoxy.

9   Prior to repairing a break, turn off the system and allow it to cool off for a few minutes.

10 Lightly buff the element area with fine steel wool, then clean it thoroughly with rubbing alcohol.

11 Use masking tape to mask off the area being repaired.

12 Thoroughly mix the epoxy, following the instructions provided with the repair kit.

13 Apply the epoxy material to the slit in the masking tape, overlapping the undamaged area about 3/4-inch on either end (see illustration).

14 Allow the repair to cure for 24 hours before removing the tape and using the system.

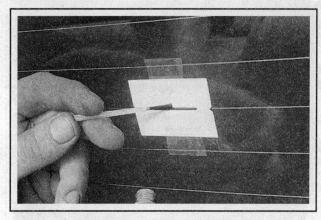

13.13 To use a defogger repair kit, apply masking tape to the inside of the window at the damaged area, then brush on the special conductive coating

14 Headlight bulb - replacement

▶ Refer to illustrations 14.3, 14.4 and 14.5

✳✳ WARNING:

Halogen bulbs are gas-filled and under pressure and may shatter if the surface is scratched or the bulb is dropped. Wear eye protection and handle the bulbs carefully, grasping only the base whenever possible. Don't touch the surface of the bulb with your fingers because the oil from your skin could cause it to overheat and fail prematurely. If you do touch the bulb surface, clean it with rubbing alcohol.

1 Open the hood.
2 If you're replacing the left headlight bulb, remove the headlight assembly (see Section 15).
3 Disconnect the electrical connector (see illustration).
4 Remove the retaining collar (see illustration).
5 Remove the bulb assembly (see illustration).
6 Install the new bulb in the headlight assembly.
7 Install the retaining collar and plug in the electrical connector.
8 Install the headlight housing assembly, if removed.

14.3 Disconnect the bulb electrical connector

14.4 Remove the bulb holder retaining collar by rotating it counterclockwise

14.5 Remove the bulb and holder unit

15 Headlight housing - removal and installation

▶ Refer to illustrations 15.2 and 15.4

✳✳ WARNING:

Halogen bulbs are gas-filled and under pressure and may shatter if the surface is scratched or the bulb is dropped. Wear eye protection and handle the bulbs carefully, grasping only the base whenever possible. Don't touch the surface of the bulb with your fingers because the oil from your skin could cause it to overheat and fail prematurely. If you do touch the bulb surface, clean it with rubbing alcohol.

1 Open the hood. Remove the grille (see Chapter 11).
2 Remove the two retaining bolts and pull the housing straight out (see illustration).
3 Disconnect the electrical connectors from the headlight assembly.
4 When installing the headlight housing, guide the post on the outboard end of the housing into its corresponding hole in the fender (see illustration). Installation is otherwise the reverse of removal. After you're done, check the headlight adjustment (see Section 16).

15.2 Remove the two headlight housing bolts

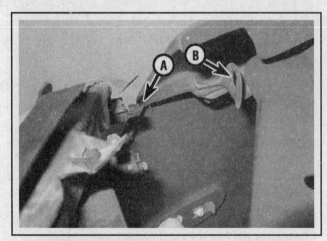

15.4 When installing the headlight housing, guide this post (A) into its corresponding hole (B)

16 Headlights - adjustment

▶ Refer to illustrations 16.1 and 16.4

➡Note: The headlights must be aimed correctly. If adjusted incorrectly they could blind the driver of an oncoming vehicle and cause a serious accident or seriously reduce your ability to see the road. The headlights should be checked for proper aim every 12 months and any time a new headlight is installed or front end body work is performed. It should be emphasized that the following procedure is only an interim step, which will provide temporary adjustment until a properly equipped shop can adjust the headlights.

1   These models have built-in horizontal adjustment screws located at each end of the headlight housing for adjusting the low and high beam simultaneously (see illustration).

2   Adjustment should be made with the vehicle on a level surface, with a full gas tank and a normal load in the vehicle.

3   There are several methods of adjusting the headlights. The simplest method requires masking tape, a blank wall and a level floor.

4   Position masking tape vertically on the wall in reference to the vehicle centerline and the centerlines of both headlight bulbs (see illustration).

5   Position a horizontal tapeline in reference to the centerline of all the headlights.

➡Note: It may be easier to position the tape on the wall with the vehicle parked only a few inches away.

6   Adjustment should be made with the vehicle parked 25 feet from the wall, sitting level, the gas tank half-full and no unusually heavy load in the vehicle.

7   Position the high intensity zone so it is two inches below the horizontal line and two inches to the side of the headlight vertical line, away from oncoming traffic. Adjustment is made by turning the horizontal adjusting screw to move the beam left or right. The high beams on these aero type headlights are automatically adjusted when the low beams are adjusted.

8   Have the headlights adjusted by a dealer service department or service station at the earliest opportunity.

16.1 Headlight adjusting screws are located at the ends of the housing (housing removed for clarity; the inner screws [indicated by the right arrow] are accessed through holes in the radiator support)

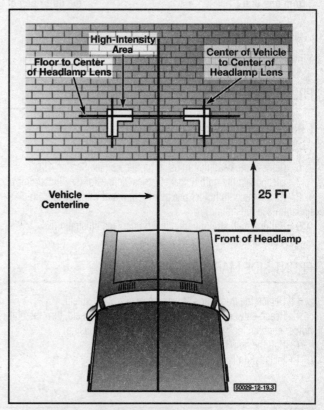

16.4 Headlight adjustment details

17 Bulb replacement

☀ WARNING:

Halogen bulbs are gas-filled and under pressure and may shatter if the surface is scratched or the bulb is dropped. Wear eye protection and handle the bulbs carefully, grasping only the base whenever possible. Don't touch the surface of the bulb with your fingers because the oil from your skin could cause it to overheat and fail prematurely. If you do touch the bulb surface, clean it with rubbing alcohol.

FOG LIGHT

▶ Refer to illustrations 17.2a, 17.2b, 17.3a and 17.3b

1   Remove the apron from between the bottom of the bumper cover and the lower radiator support.

2   Remove the retaining nuts and remove then housing from the bumper cover for access to the bulb holder (see illustrations).

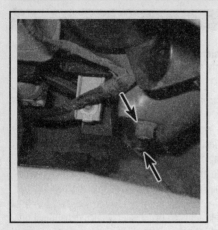

17.2a Remove the two nuts from one end . . .

17.2b . . . and the single nut from the other end of the fog light

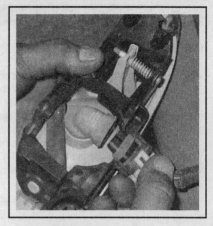

17.3a Disconnect the fog light connector

3   Disconnect the electrical connector and pull the bulb and holder straight out from the housing (see illustrations).

4   Installation is the reverse of removal.

FRONT PARK AND TURN SIGNAL LIGHT

▶ **Refer to illustrations 17.7, 17.8 and 17.9**

5   Open the hood.

6   Remove the headlight housing for access (see Section 15).

7   Disconnect the bulb holder electrical connector (see illustration).

8   Rotate the bulb holder one quarter turn and withdraw it from the housing (see illustration).

9   Pull the bulb straight out of the holder (see illustration).

10   Installation is the reverse of removal.

FRONT SIDE MARKER LIGHT

11   Detach the front edge of the wheelwell splash shield.

12   Reach between the bumper cover and splash shield, turn the bulb holder counterclockwise and withdraw it from the housing.

13   Pull the bulb out from the socket.

14   Installation is the reverse of removal.

CENTER HIGH-MOUNTED BRAKE LIGHT

▶ **Refer to illustrations 17.15, 17.16a and 17.16b**

15   Open the trunk and detach the trim panel for access (see illustration).

16   Rotate the bulb holder a quarter turn counterclockwise pull it out of the housing. Pull the bulb straight out of the holder (see illustrations).

17   Installation is the reverse of removal.

TAIL, STOP, BACK-UP AND TURN SIGNAL LIGHT

▶ **Refer to illustrations 17.20a and 17.20b**

18   Open the trunk.

19   Separate the trunk lining from the rear closure panel.

20   Rotate the bulb holder counterclockwise a quarter turn, pull it straight out from the housing, then pull the bulb straight out of the holder (see illustrations).

21   Installation is the reverse of removal.

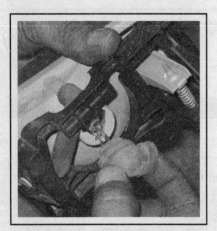

17.3b Remove the bulb from the housing

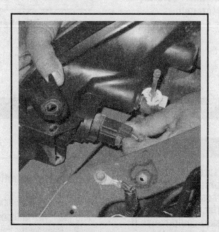

17.7 Disconnect the turn signal bulb electrical connector

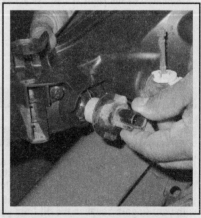

17.8 Withdraw the park/turn signal bulb holder from the headlight housing

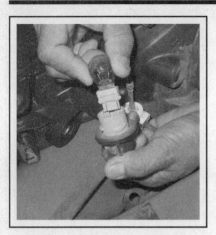

17.9 Remove the bulb from the holder

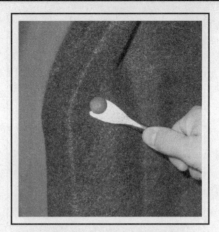

17.15 Pry the trim panel away for access to the brake light housing

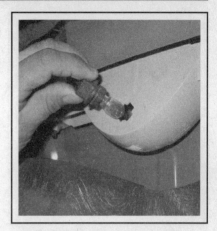

17.16a Turn the bulb holder and withdraw it from the lamp housing

LICENSE PLATE LIGHT

▶ **Refer to illustrations 17.22, 17.23a and 17.23b**

22 Remove the retaining screws and remove the housing from the

rear bumper for access to the bulb holder (see illustration).

23 Turn the bulb holder 1/4-turn and remove it from the housing, then pull the bulb straight out of the holder (see illustrations).

24 Installation is the reverse of removal.

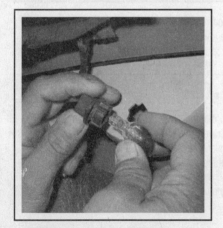

17.16b Remove the bulb from the holder

17.20a Remove the bulb holder from the tail light housing by rotating it counterclockwise

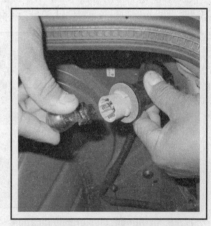

17.20b Pull the bulb straight out of the holder

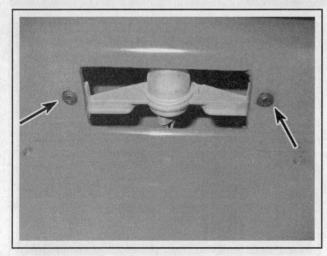

17.22 Remove the retaining screws and remove the license plate light housing from the bumper

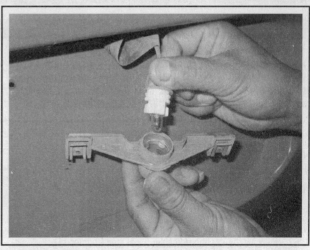

17.23a Turn the bulb holder 90-degrees and remove it from the housing

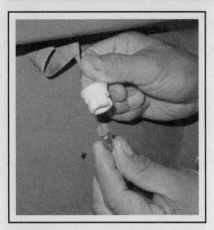

17.23b Pull the bulb from the holder

17.25 Detach the dome lamp lens by carefully prying the left side downward

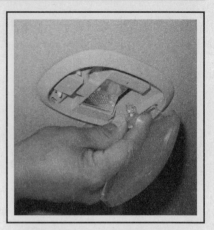

17.26 Pull the bulb straight out of its socket

DOME LAMP

♦ **Refer to illustrations 17.25 and 17.26**

25 Carefully insert a small screwdriver between the left side of the dome lamp lens and the housing (see illustration).

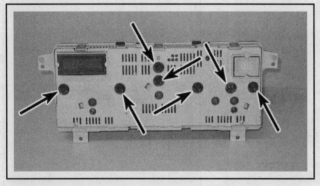

17.29 Instrument cluster bulb locations - 2000 model shown, other models similar

> ✳✳ **CAUTION:**
>
> Don't attempt to pry the right side of the lens out - you'll break the lens!

26 Grasp the bulb and remove it from the socket (see illustration).
27 Installation is the reverse of removal.

INSTRUMENT CLUSTER BULBS

♦ **Refer to illustrations 17.29**

28 To gain access to the instrument cluster illumination bulbs, the instrument cluster will have to be removed (see Section 10). The bulbs can be removed and replaced from the rear of the cluster.
29 Rotate the bulb holder counterclockwise to remove it (see illustration).
30 Installation is the reverse of removal.

---

18 Radio and speakers - removal and installation

> ✳✳ **WARNING:**
>
> These models have airbags. Always disconnect the negative battery cable and wait two minutes before working in the vicinity of the impact sensors, steering column or instrument panel to avoid the possibility of accidental deployment of the airbag, which could cause personal injury (see Section 25).

RADIO

♦ **Refer to illustrations 18.3, 18.4 and 18.5**

1 Disconnect and isolate the negative cable so that it can't accidentally come in contact with the battery negative terminal.
2 Remove the dashboard center bezel (see Chapter 11).
3 Remove the radio mounting screws (see illustration).

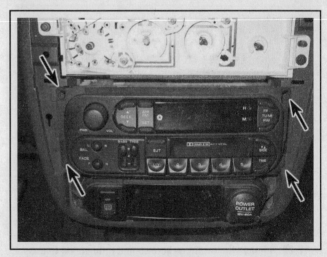

18.3 Radio mounting screw locations

4   Pull the radio out of the instrument panel. Release the clips and disconnect the electrical connectors and the ground wire (see illustration).

5   Carefully disconnect the antenna lead (see illustration), then remove the radio.

6   Installation is the reverse of removal.

SPEAKERS

Door

▶ **Refer to illustrations 18.8a and 18.8b**

7   Remove the door trim panel (see Chapter 11).

8   Remove the screws and detach the speaker (see illustration). Pull the speaker out of the door, unplug the electrical connector and remove the speaker from the vehicle (see illustration). Installation is the reverse of removal.

Instrument panel

▶ **Refer to illustration 18.10**

9   Remove the instrument panel top cover (see Chapter 11).

10 Remove the screws, pull the speaker up, then unplug the electrical connector and remove the speaker from the vehicle (see illustration). Installation is the reverse of removal.

18.4 Pull the radio out and disconnect the electrical connectors

Rear shelf

▶ **Refer to illustrations 18.12, 18.13 and 18.14**

11 Remove the entire rear seat assembly (see Chapter 11).

18.5 Disconnect the antenna cable from the radio by pulling it straight out

18.8a Remove the speaker screws

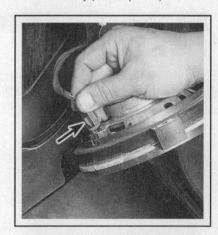

18.8b Pull the speaker out and disconnect the electrical connector

18.10 Disconnect the electrical connector and remove the dash speaker screws

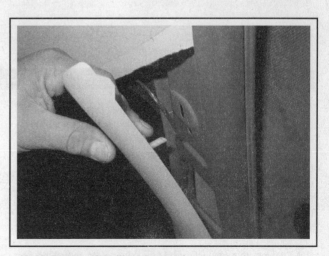

18.12 Remove the pillar trim panel

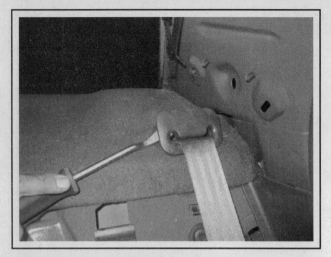

18.13 Pry out the seatbelt bezel

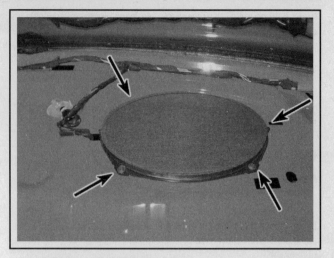

18.14 Rear speaker screws

12 Remove the pillar trim panel (see illustration).
13 Pry out the seat belt trim bezel (see illustration), then remove the rear parcel shelf trim panel

14 Remove the screws and detach the speaker (see illustration).
15 Installation is the reverse of removal, making sure that the wire connectors face toward the center of the vehicle.

19 Horn - replacement

♦ Refer to illustration 19.3

1 Remove the windshield wiper/washer fluid reservoir.
2 To replace the horn, disconnect the electrical connector and remove the bracket bolt (see illustration).
3 Installation is the reverse of removal.

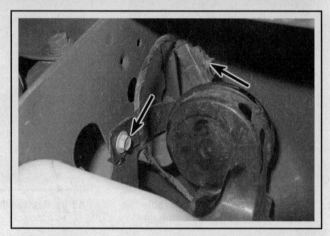

19.2 Disconnect the horn electrical connector and remove the bracket bolt

20 Electric side view mirrors - description

1 Most electric rear view mirrors use two motors to move the glass; one for up-and-down adjustments and one for left-right adjustments.
2 The control switch has a selector portion which sends voltage to the left or right side mirror. With the ignition ON but the engine OFF, roll down the windows and operate the mirror control switch through all functions (left-right and up-down) for both the left and right side mirrors.
3 Listen carefully for the sound of the electric motors running in the mirrors.
4 If the motors can be heard but the mirror glass doesn't move, there's a problem with the drive mechanism inside the mirror.

5 If the mirrors do not operate and no sound comes from the mirrors, check the fuse (see Section 3).
6 If the fuse is OK, remove the mirror control switch. Have the switch continuity checked by a dealership service department or other qualified automobile repair facility.
7 Check the ground connections.
8 Check the wires at the mirror for voltage when the switch is operated.
9 If there is no voltage in each switch position, check the circuit between the mirror and control switch for opens and shorts.
10 If there is voltage, remove the mirror and test it off the vehicle with jumper wires. Replace the mirror if it fails this test.

21 Cruise control system - description

♦ **Refer to illustration 21.5**

1   The cruise control system maintains vehicle speed with an electronically controlled, vacuum operated servo located in the engine compartment, which is connected to the throttle body by a cable. The system consists of the PCM, vacuum servo, brake switch, control switches and vehicle speed sensor. Some features of the system require special testers and diagnostic procedures, which are beyond the scope of this manual. Listed below are some general procedures that may be used to locate common problems.

2   Check the fuses (see Section 3).

3   Have an assistant operate the brake pedal while you check the brake lights (voltage from the brake light switch deactivates the cruise control).

4   If the brake lights don't come on or stay on all the time, correct the problem and retest the cruise control.

5   Visually inspect the control cable between cruise control servo and the throttle linkage for free movement (see illustration). Replace it if necessary.

6   The cruise control system uses inputs from the Vehicle Speed Sensor (VSS). Refer to Chapter 6 for more information on the VSS.

7   Test drive the vehicle to determine if the cruise control is now

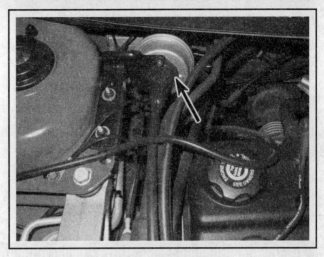

21.5 Location of the the cruise control servo

working. If it isn't, take it to a dealer service department or an automotive electrical specialist for further diagnosis.

22 Power window system - description

1   The power window system operates electric motors, mounted in the doors, which lower and raise the windows. The system consists of the control switches, relays, the motors, regulators, glass mechanisms and associated wiring.

2   The power windows can be lowered and raised from the master control switches by the driver or by remote switches located at the individual windows. Each window has a separate motor that is reversible. The position of the control switch determines the polarity and therefore the direction of operation.

3   The circuit is protected by a fuse and a circuit breaker. Each motor is also equipped with an internal circuit breaker; this prevents one stuck window from disabling the whole system.

4   The power window system will only operate when the ignition switch is ON. In addition, many models have a window lockout switch at the master control switch that, when activated, disables the switches at the rear windows and, sometimes, the switch at the passenger's window also. Always check these items before troubleshooting a window problem.

5   These procedures are general in nature, so if you can't find the problem using them, take the vehicle to a dealer service department or other properly equipped repair facility.

6   If the power windows won't operate, always check the fuse and circuit breaker first.

7   If only the rear windows are inoperative, or if the windows only operate from the master control switch, check the rear window lockout switch for continuity in the unlocked position. Replace it if it doesn't

have continuity.

8   Check the wiring between the switches and fuse panel for continuity. Repair the wiring, if necessary.

9   If only one window is inoperative from the master control switch, try the other control switch at the window.

Note: This doesn't apply to the driver's door window.

10 If the same window works from one switch, but not the other, check the switch for continuity. Have the switch checked at a dealer service department or other qualified automobile repair facility.

11 If the switch tests OK, check for a short or open in the circuit between the affected switch and the window motor.

12 If one window is inoperative from both switches, remove the trim panel from the affected door and check for voltage at the switch and at the motor while the switch is operated.

13 If voltage is reaching the motor, disconnect the glass from the regulator (see Chapter 11). Move the window up and down by hand while checking for binding and damage. Also check for binding and damage to the regulator. If the regulator is not damaged and the window moves up and down smoothly, replace the motor. If there's binding or damage, lubricate, repair or replace parts, as necessary.

14 If voltage isn't reaching the motor, check the wiring in the circuit for continuity between the switches and motors. You'll need to consult the wiring diagram for the vehicle. If the circuit is equipped with a relay, check that the relay is grounded properly and receiving voltage.

15 Test the windows after you are done to confirm proper repairs.

23 Power door lock system - description

1   A power door lock system operates the door lock actuators mounted in each door. The system consists of the switches, actuators, a control unit and associated wiring. Diagnosis can usually be limited to simple checks of the wiring connections and actuators for minor

faults that can be easily repaired.

2   Power door lock systems are operated by bi-directional solenoids located in the doors. The lock switches have two operating positions: Lock and Unlock. When activated, the switch sends a ground

signal to the door lock control unit to lock or unlock the doors. Depending on which way the switch is activated, the control unit reverses polarity to the solenoids, allowing the two sides of the circuit to be used alternately as the feed (positive) and ground side.

3　Some vehicles may have an anti-theft system incorporated into the power locks. If you are unable to locate the trouble using the following general Steps, consult a dealer service department or other qualified repair shop.

4　Always check the circuit protection first. Some vehicles use a combination of circuit breakers and fuses.

5　Operate the door lock switches in both directions (Lock and Unlock) with the engine off. Listen for the click of the solenoids operating.

6　Test the switches for continuity. Remove the switches and have them checked by a dealer service department or other qualified automobile repair facility.

7　Check the wiring between the switches, control unit and solenoids for continuity. Repair the wiring if there's no continuity.

8　Check for a bad ground at the switches or the control unit.

9　If all but one lock solenoids operate, remove the trim panel from the affected door (see Chapter 11) and check for voltage at the solenoid while the lock switch is operated. One of the wires should have voltage in the Lock position; the other should have voltage in the Unlock position.

10　If the inoperative solenoid is receiving voltage, replace the solenoid.

11　If the inoperative solenoid isn't receiving voltage, check the relay for an open or short in the wire between the lock solenoid and the control unit.

➡**Note: It's common for wires to break in the portion of the harness between the body and door (opening and closing the door fatigues and eventually breaks the wires).**

24 Daytime Running Lights (DRL) - general information

The Daytime Running Lights (DRL) system used on Canadian models illuminates the headlights whenever the engine is running. The only exception is with the engine running and the parking brake engaged. Once the parking brake is released, the lights will remain on as long as the ignition switch is on, even if the parking brake is later applied.

The DRL system supplies reduced power to the headlights so they won't be too bright for daytime use, while prolonging headlight life.

25 Airbag system - general information and precautions

GENERAL INFORMATION

▶ **Refer to illustrations 25.1a, 25.1b, 25.1c and 25.1d**

1　All models are equipped with a Supplemental Restraint System (SRS), more commonly known as an airbag. This system is designed to protect the driver, and the front seat passenger, from serious injury in the event of a head-on or frontal collision. It consists of a control module mounted in the center of the vehicle, under the floor console, an airbag module mounted on the steering wheel (see illustration), the top surface of the passenger's side dash and, on some models, in the sides of the front seats (see illustrations). The airbag control module is mounted under the center console (see illustration).

AIRBAG MODULE

Driver's side

2　The airbag inflator module contains a housing incorporating the cushion (airbag) and inflator unit, mounted in the center of the steering wheel. The inflator assembly is mounted on the back of the housing over a hole through which gas is expelled, inflating the bag almost instantaneously when an electrical signal is sent from the system.

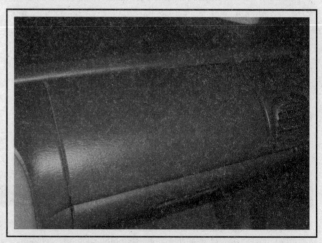

25.1a The driver's side airbag is located in the steering column horn pad

25.1b The passenger's airbag is located on the dashboard above the glove box

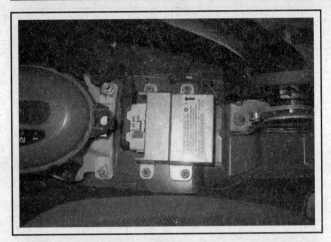

25.1c The airbag control module is located under the center console and mounted to the floorpan

3   A clockspring assembly on the steering column under the steering wheel carries this signal to the module.

4   This clockspring assembly can transmit an electrical signal regardless of steering wheel position. The igniter in the airbag converts the electrical signal to heat and ignites the powder, which inflates the bag.

Passenger's side

5   The airbag is mounted above the glove compartment and designated by the letters SRS (Supplemental Restraint System). It consists of an inflator containing an igniter, a bag assembly, a reaction housing and a trim cover.

6   The airbag is considerably larger than the steering wheel-mounted unit and is supported by the steel reaction housing. The trim cover is textured and painted to match the instrument panel and has a molded seam which splits when the bag inflates.

Side impact airbags

7   These airbags are mounted on the outboard side of the front seat back frames on some models. In the event of severe side impact they inflate, ripping open the front seat back trim covers and deploy to protect the occupants. They consist of a bag assembly, a canister-type inflator containing highly compressed gas and a mounting bracket.

AIRBAG CONTROL MODULE

8   The airbag control module supplies the current to the airbag system in the event of a collision, even if battery power is cut off. It checks this system every time the vehicle is started, causing the "SRS" light to go on then off, if the system is operating properly. If there is a fault in the system, the light will go on and stay on, flash, or the dash will make a beeping sound. If this happens, the vehicle should be taken to your dealer immediately for service.

DISARMING THE SYSTEM AND OTHER PRECAUTIONS

✳✳ WARNING:

Failure to follow these precautions could result in accidental deployment of the airbag and personal injury.

9   Whenever working in the vicinity of the steering wheel, steering column or any of the other SRS system components, the system must be disarmed. To disarm the system:
   a) *Point the wheels straight ahead and turn the key to the Lock position.*
   b) *Disconnect the cable from the negative terminal of the battery. Isolate the cable terminal so it won't accidentally contact the battery post.*
   c) *Wait at least two minutes for the back-up power supply to be depleted.*

10 Whenever handling an airbag module, always keep the airbag opening (the trim side) pointed away from your body. Never place the airbag module on a bench or other surface with the airbag opening facing the surface. Always place the airbag module in a safe location with the airbag opening facing up.

11 Never measure the resistance of any SRS component. An ohmmeter has a built-in battery supply that could accidentally deploy the airbag.

12 Never use electrical welding equipment on a vehicle equipped with an airbag without first disconnecting the yellow airbag connector, located under the steering column near the combination switch connector (driver's airbag) and behind the glove box (passenger's airbag).

13 Never dispose of a live airbag module. Return it to a dealer service department or other qualified repair shop for safe deployment and disposal.

COMPONENT REMOVAL AND INSTALLATION

Driver's side airbag module and clockspring

14 Refer to Chapter 10, Steering wheel - removal and installation, for the driver's side airbag module and clockspring removal and installation procedures.

Passenger's side airbag module

15 Refer to Chapter 11 for the passenger's side airbag removal and installation procedures.

Side impact airbag

16 This procedure will require removal and disassembly of the seat(s). Under normal circumstances there would never be a reason to remove a seat airbag. However, if it has been determined that there is a problem with the seat airbag module, the work should be left to a dealer service department or other qualified repair shop.

26 Wiring diagrams - general information

Since it isn't possible to include all wiring diagrams for every year covered by this manual, the following diagrams are those that are typical and most commonly needed.

Prior to troubleshooting any circuits, check the fuse and circuit breakers (if equipped) to make sure they're in good condition. Make sure the battery is properly charged and check the cable connections (see Chapter 1).

When checking a circuit, make sure that all connectors are clean, with no broken or loose terminals. When unplugging a connector, do not pull on the wires. Pull only on the connector housings.

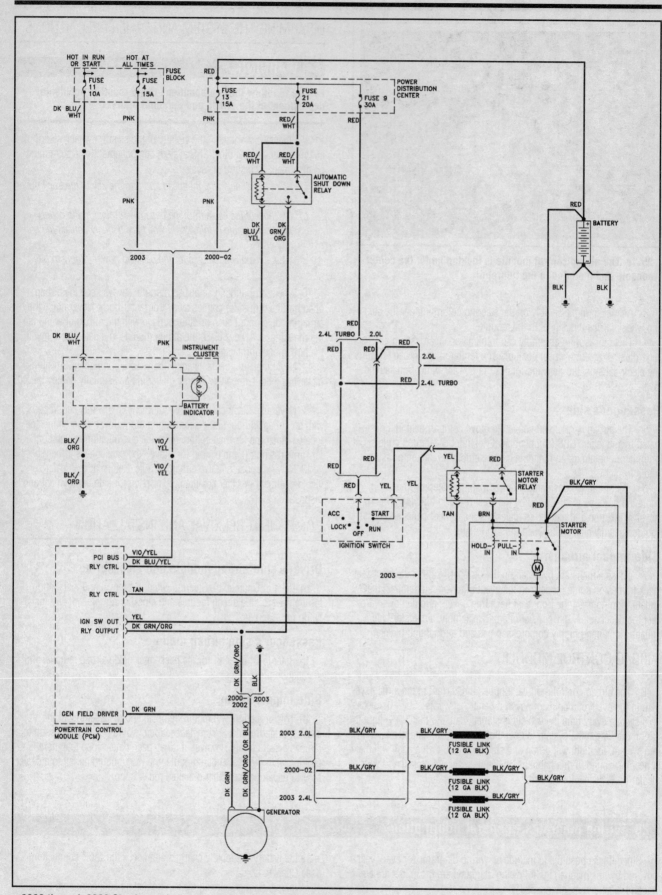

2000 through 2003 Starting and charging systems - later models similar

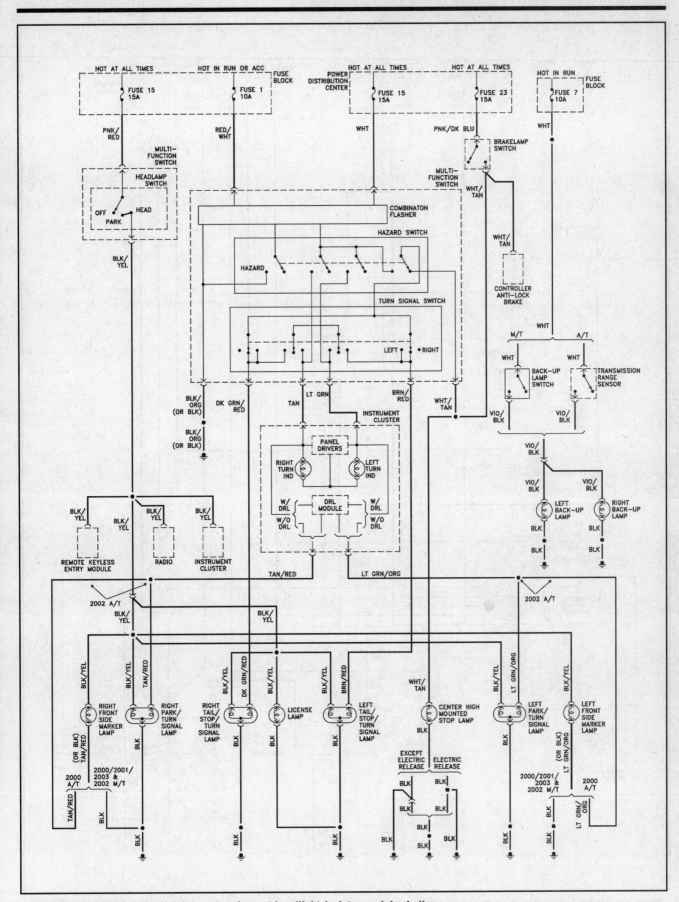

2000 through 2003 Exterior lighting system (except headlights) - later models similar

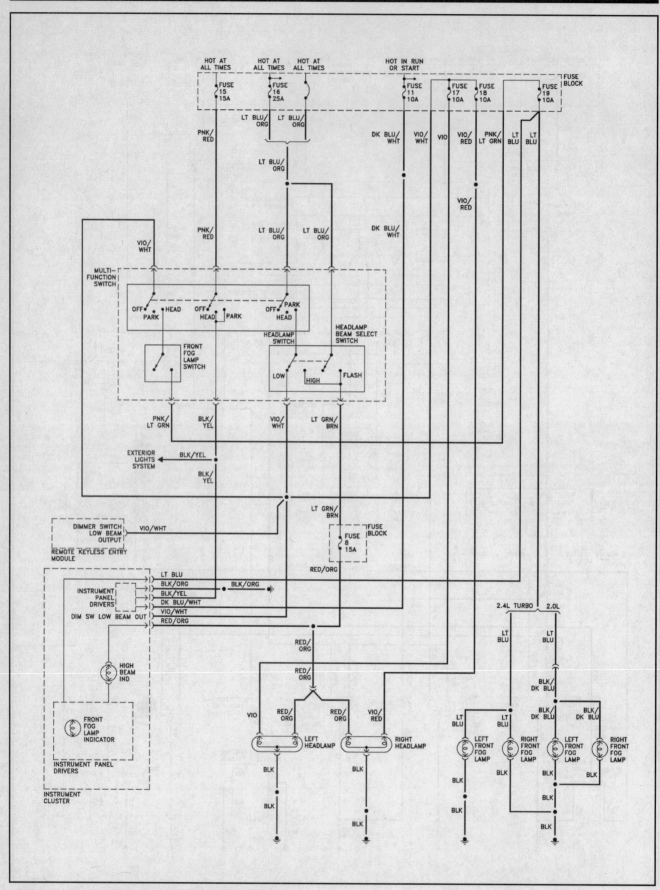

Headlight system - without Daytime Running Light system

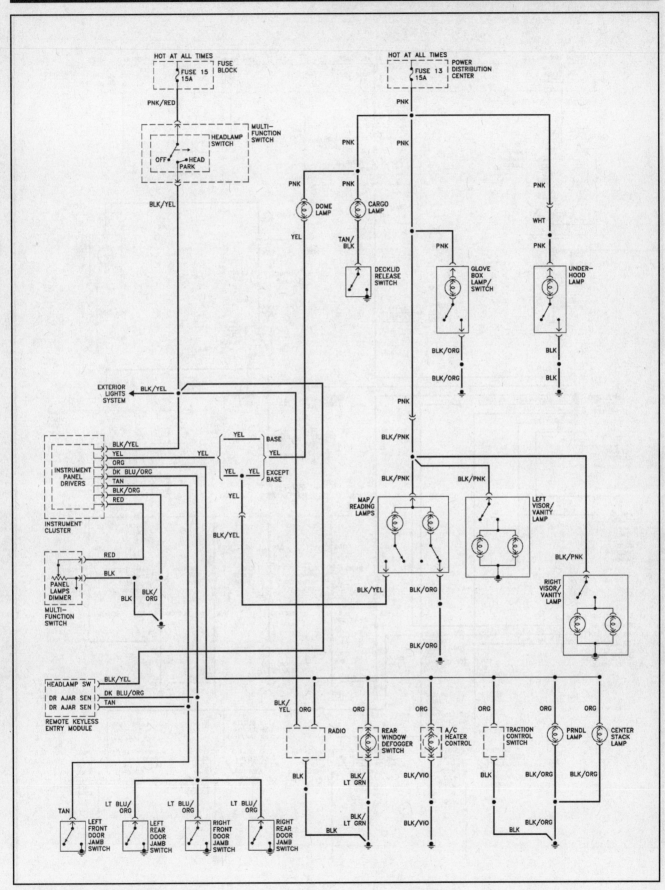

Interior lighting system (2000 models)

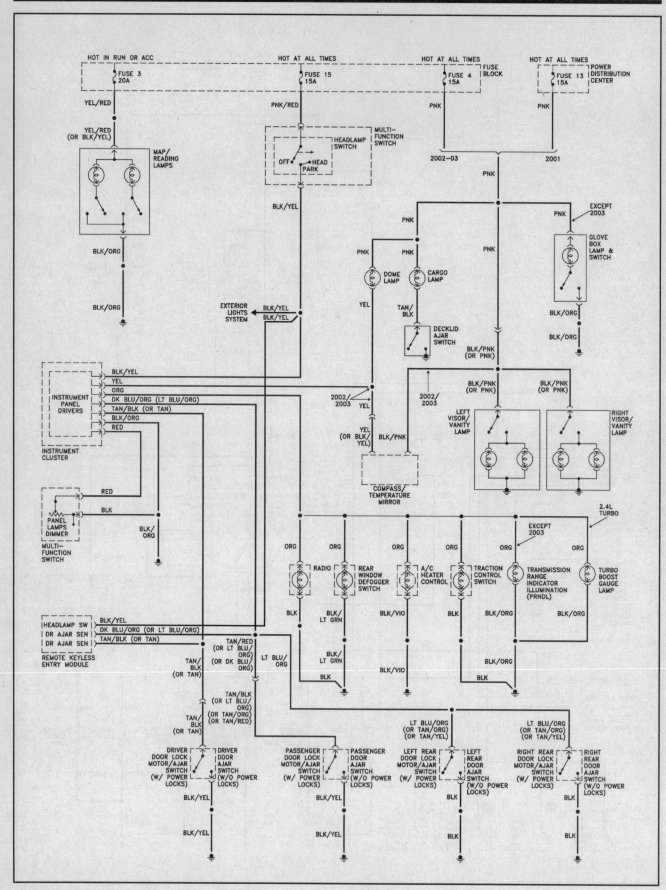

Interior lighting system (2001 and later models)

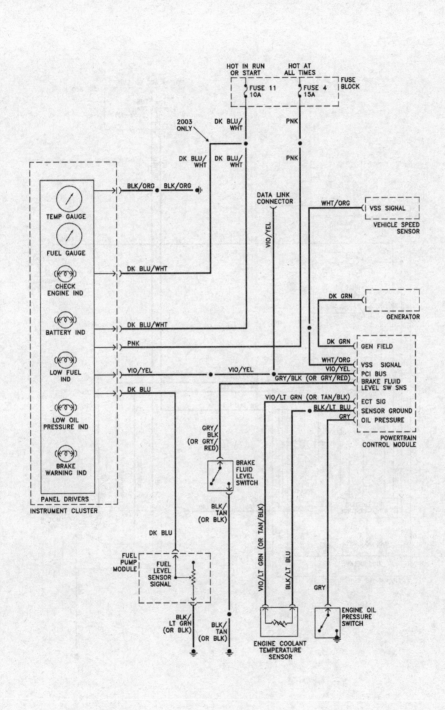

2000 through 2003 Warning systems - later models similar

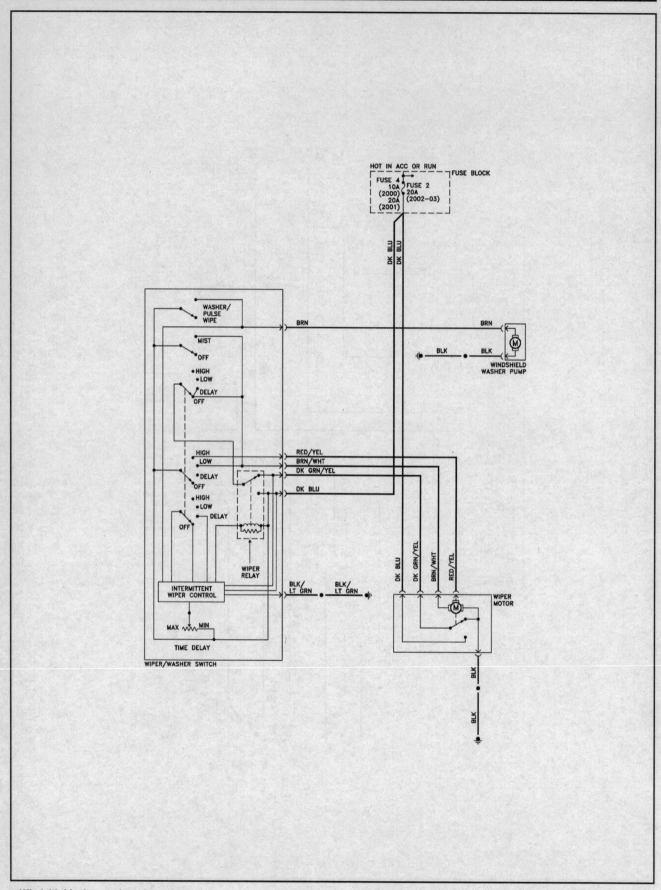

Windshield wiper and washer system

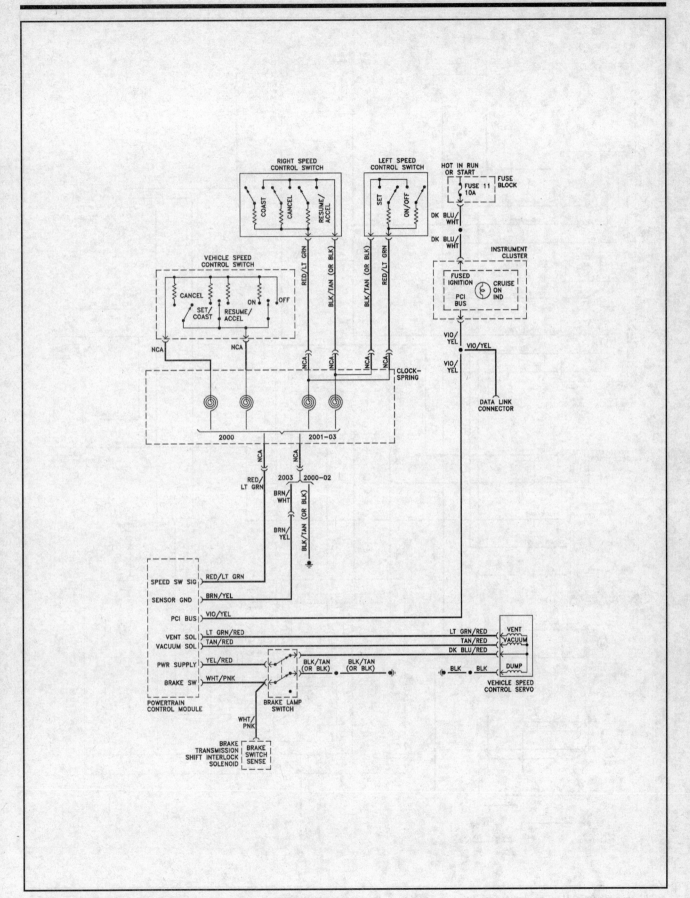

2000 through 2003 Cruise control system - later models similar

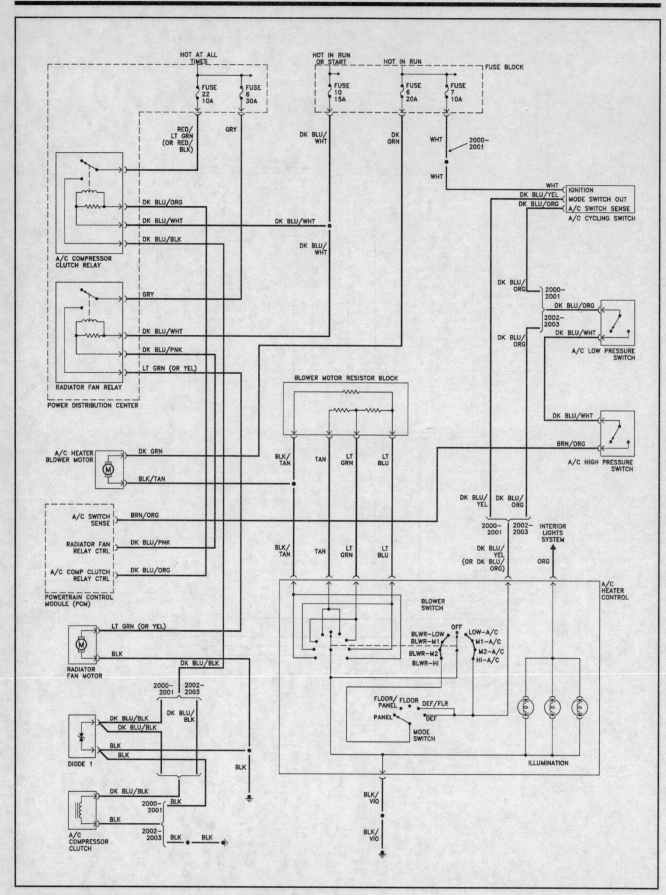

2000 through 2003 Heating and air conditioning system - later models similar

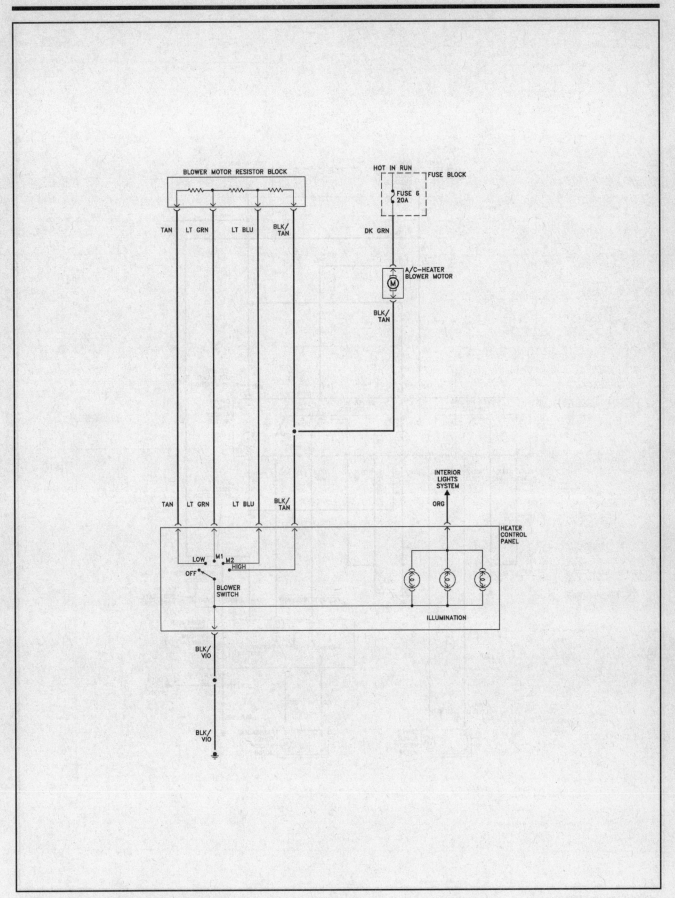

Heating system (without air conditioning)

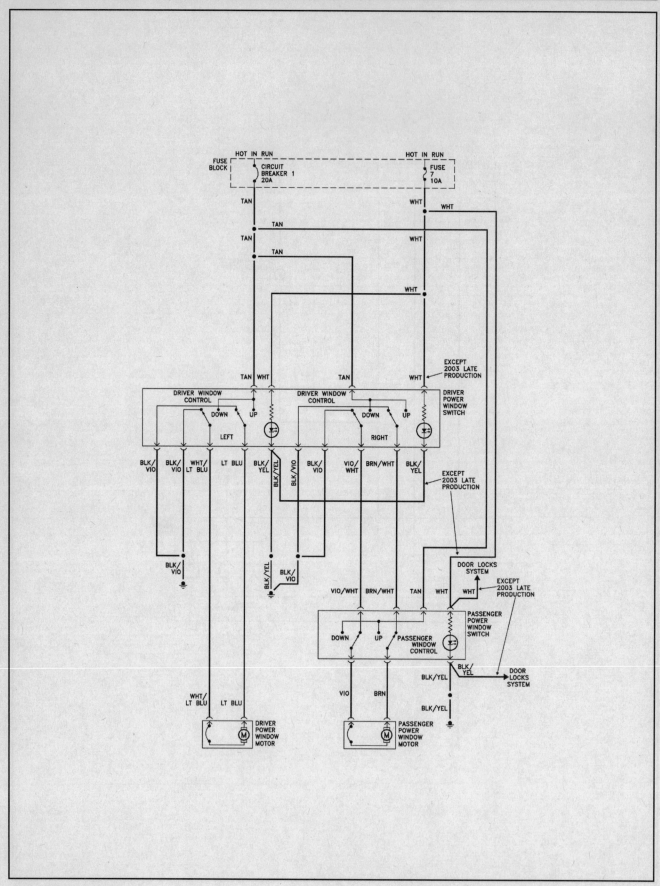

Power window system

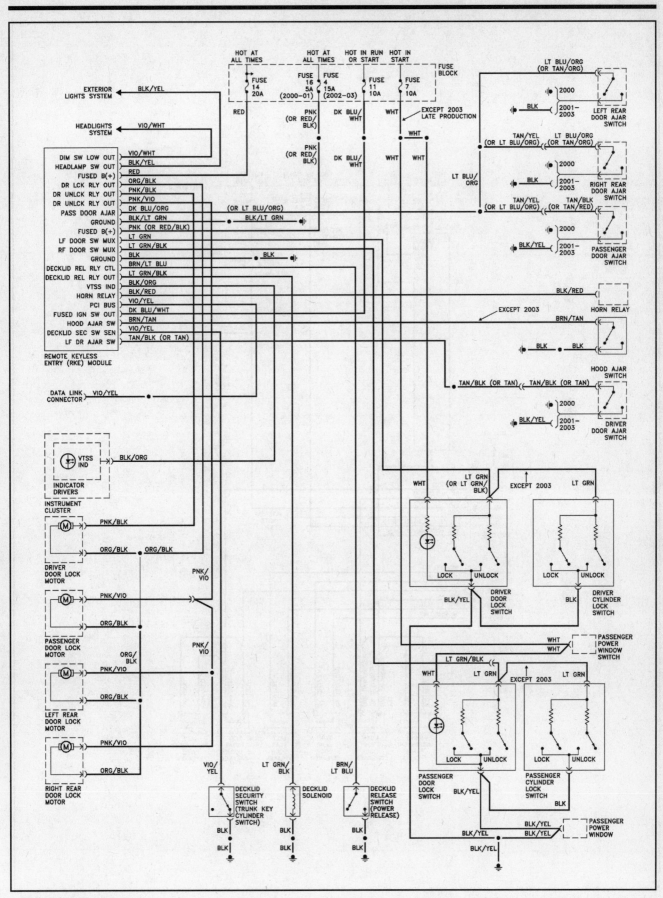

Power door lock system

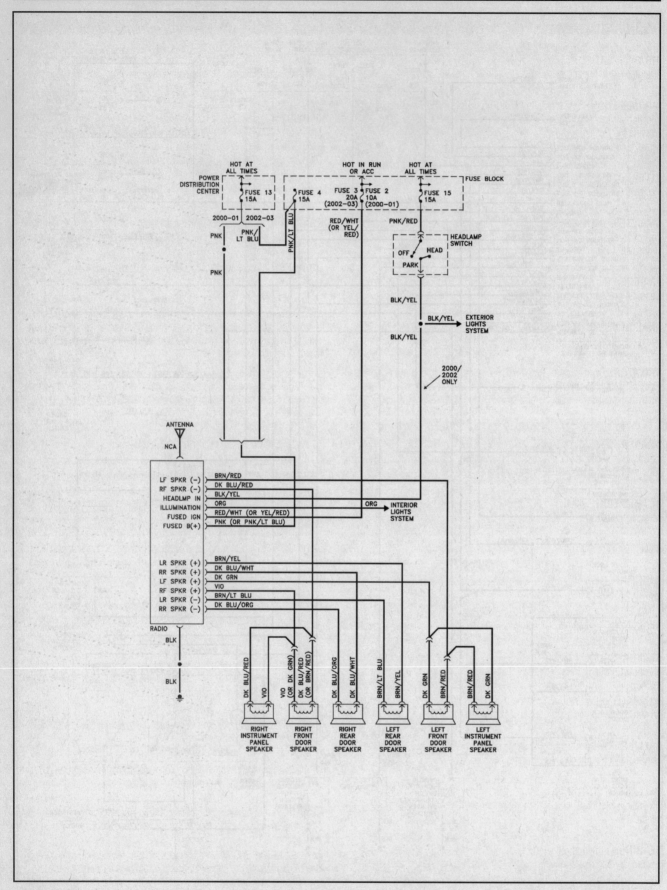

Audio system

GLOSSARY

AIR/FUEL RATIO: The ratio of air-to-gasoline by weight in the fuel mixture drawn into the engine.

AIR INJECTION: One method of reducing harmful exhaust emissions by injecting air into each of the exhaust ports of an engine. The fresh air entering the hot exhaust manifold causes any remaining fuel to be burned before it can exit the tailpipe.

ALTERNATOR: A device used for converting mechanical energy into electrical energy.

AMMETER: An instrument, calibrated in amperes, used to measure the flow of an electrical current in a circuit. Ammeters are always connected in series with the circuit being tested.

AMPERE: The rate of flow of electrical current present when one volt of electrical pressure is applied against one ohm of electrical resistance.

ANALOG COMPUTER: Any microprocessor that uses similar (analogous) electrical signals to make its calculations.

ARMATURE: A laminated, soft iron core wrapped by a wire that converts electrical energy to mechanical energy as in a motor or relay. When rotated in a magnetic field, it changes mechanical energy into electrical energy as in a generator.

ATMOSPHERIC PRESSURE: The pressure on the Earth's surface caused by the weight of the air in the atmosphere. At sea level, this pressure is 14.7 psi at 32°F (101 kPa at 0°C).

ATOMIZATION: The breaking down of a liquid into a fine mist that can be suspended in air.

AXIAL PLAY: Movement parallel to a shaft or bearing bore.

BACKFIRE: The sudden combustion of gases in the intake or exhaust system that results in a loud explosion.

BACKLASH: The clearance or play between two parts, such as meshed gears.

BACKPRESSURE: Restrictions in the exhaust system that slow the exit of exhaust gases from the combustion chamber.

BAKELITE: A heat resistant, plastic insulator material commonly used in printed circuit boards and transistorized components.

BALL BEARING: A bearing made up of hardened inner and outer races between which hardened steel balls roll.

BALLAST RESISTOR: A resistor in the primary ignition circuit that lowers voltage after the engine is started to reduce wear on ignition components.

BEARING: A friction reducing, supportive device usually located between a stationary part and a moving part.

BIMETAL TEMPERATURE SENSOR: Any sensor or switch made of two dissimilar types of metal that bend when heated or cooled due to the different expansion rates of the alloys. These types of sensors usually function as an on/off switch.

BLOWBY: Combustion gases, composed of water vapor and unburned fuel, that leak past the piston rings into the crankcase during normal engine operation. These gases are removed by the PCV system to prevent the buildup of harmful acids in the crankcase.

BRAKE PAD: A brake shoe and lining assembly used with disc brakes.

BRAKE SHOE: The backing for the brake lining. The term is, however, usually applied to the assembly of the brake backing and lining.

BUSHING: A liner, usually removable, for a bearing; an anti-friction liner used in place of a bearing.

CALIPER: A hydraulically activated device in a disc brake system, which is mounted straddling the brake rotor (disc). The caliper contains at least one piston and two brake pads. Hydraulic pressure on the piston(s) forces the pads against the rotor.

CAMSHAFT: A shaft in the engine on which are the lobes (cams) which operate the valves. The camshaft is driven by the crankshaft, via a belt, chain or gears, at one half the crankshaft speed.

CAPACITOR: A device which stores an electrical charge.

CARBON MONOXIDE (CO): A colorless, odorless gas given off as a normal byproduct of combustion. It is poisonous and extremely dangerous in confined areas, building up slowly to toxic levels without warning if adequate ventilation is not available.

CARBURETOR: A device, usually mounted on the intake manifold of an engine, which mixes the air and fuel in the proper proportion to allow even combustion.

CATALYTIC CONVERTER: A device installed in the exhaust system, like a muffler, that converts harmful byproducts of combustion into carbon dioxide and water vapor by means of a heat-producing chemical reaction.

CENTRIFUGAL ADVANCE: A mechanical method of advancing the spark timing by using flyweights in the distributor that react to centrifugal force generated by the distributor shaft rotation.

CHECK VALVE: Any one-way valve installed to permit the flow of air, fuel or vacuum in one direction only.

CHOKE: A device, usually a moveable valve, placed in the intake path of a carburetor to restrict the flow of air.

CIRCUIT: Any unbroken path through which an electrical current can flow. Also used to describe fuel flow in some instances.

CIRCUIT BREAKER: A switch which protects an electrical circuit from overload by opening the circuit when the current flow exceeds a predetermined level. Some circuit breakers must be reset manually, while most reset automatically.

COIL (IGNITION): A transformer in the ignition circuit which steps up the voltage provided to the spark plugs.

COMBINATION MANIFOLD: An assembly which includes both the intake and exhaust manifolds in one casting.

COMBINATION VALVE: A device used in some fuel systems that routes fuel vapors to a charcoal storage canister instead of venting them into the atmosphere. The valve relieves fuel tank pressure and allows fresh air into the tank as the fuel level drops to prevent a vapor lock situation.

COMPRESSION RATIO: The comparison of the total volume of the cylinder and combustion chamber with the piston at BDC and the piston at TDC.

CONDENSER: 1. An electrical device which acts to store an electrical charge, preventing voltage surges. 2. A radiator-like device in the air conditioning system in which refrigerant gas condenses into a liquid, giving off heat.

CONDUCTOR: Any material through which an electrical current can be transmitted easily.

CONTINUITY: Continuous or complete circuit. Can be checked with an ohmmeter.

COUNTERSHAFT: An intermediate shaft which is rotated by a mainshaft and transmits, in turn, that rotation to a working part.

CRANKCASE: The lower part of an engine in which the crankshaft and related parts operate.

CRANKSHAFT: The main driving shaft of an engine which receives reciprocating motion from the pistons and converts it to rotary motion.

CYLINDER: In an engine, the round hole in the engine block in which the piston(s) ride.

CYLINDER BLOCK: The main structural member of an engine in which is found the cylinders, crankshaft and other principal parts.

CYLINDER HEAD: The detachable portion of the engine, usually fastened to the top of the cylinder block and containing all or most of the combustion chambers. On overhead valve engines, it contains the valves and their operating parts. On overhead cam engines, it contains the camshaft as well.

DEAD CENTER: The extreme top or bottom of the piston stroke.

DETONATION: An unwanted explosion of the air/fuel mixture in the combustion chamber caused by excess heat and compression, advanced timing, or an overly lean mixture. Also referred to as "ping".

DIAPHRAGM: A thin, flexible wall separating two cavities, such as in a vacuum advance unit.

DIESELING: A condition in which hot spots in the combustion chamber cause the engine to run on after the key is turned off.

DIFFERENTIAL: A geared assembly which allows the transmission of motion between drive axles, giving one axle the ability to turn faster than the other.

DIODE: An electrical device that will allow current to flow in one direction only.

DISC BRAKE: A hydraulic braking assembly consisting of a brake disc, or rotor, mounted on an axle, and a caliper assembly containing, usually two brake pads which are activated by hydraulic pressure. The pads are forced against the sides of the disc, creating friction which slows the vehicle.

DISTRIBUTOR: A mechanically driven device on an engine which is responsible for electrically firing the spark plug at a predetermined point of the piston stroke.

DOWEL PIN: A pin, inserted in mating holes in two different parts allowing those parts to maintain a fixed relationship.

DRUM BRAKE: A braking system which consists of two brake shoes and one or two wheel cylinders, mounted on a fixed backing plate, and a brake drum, mounted on an axle, which revolves around the assembly.

DWELL: The rate, measured in degrees of shaft rotation, at which an electrical circuit cycles on and off.

ELECTRONIC CONTROL UNIT (ECU): Ignition module, module, amplifier or igniter. See Module for definition.

ELECTRONIC IGNITION: A system in which the timing and firing of the spark plugs is controlled by an electronic control unit, usually called a module. These systems have no points or condenser.

END-PLAY: The measured amount of axial movement in a shaft.

ENGINE: A device that converts heat into mechanical energy.

EXHAUST MANIFOLD: A set of cast passages or pipes which conduct exhaust gases from the engine.

FEELER GAUGE: A blade, usually metal, of precisely predetermined thickness, used to measure the clearance between two parts.

FIRING ORDER: The order in which combustion occurs in the cylinders of an engine. Also the order in which spark is distributed to the plugs by the distributor.

FLOODING: The presence of too much fuel in the intake manifold and combustion chamber which prevents the air/fuel mixture from firing, thereby causing a no-start situation.

FLYWHEEL: A disc shaped part bolted to the rear end of the crankshaft. Around the outer perimeter is affixed the ring gear. The starter drive engages the ring gear, turning the flywheel, which rotates the crankshaft, imparting the initial starting motion to the engine.

FOOT POUND (ft. lbs. or sometimes, ft.lb.): The amount of energy or work needed to raise an item weighing one pound, a distance of one foot.

FUSE: A protective device in a circuit which prevents circuit overload by breaking the circuit when a specific amperage is present. The device is constructed around a strip or wire of a lower amperage rating than the circuit it is designed to protect. When an amperage higher than that stamped on the fuse is present in the circuit, the strip or wire melts, opening the circuit.

GEAR RATIO: The ratio between the number of teeth on meshing gears.

GENERATOR: A device which converts mechanical energy into electrical energy.

HEAT RANGE: The measure of a spark plug's ability to dissipate heat from its firing end. The higher the heat range, the hotter the plug fires.

HUB: The center part of a wheel or gear.

HYDROCARBON (HC): Any chemical compound made up of hydrogen and carbon. A major pollutant formed by the engine as a byproduct of combustion.

HYDROMETER: An instrument used to measure the specific gravity of a solution.

INCH POUND (inch lbs.; sometimes in.lb. or in. lbs.): One twelfth of a foot pound.

INDUCTION: A means of transferring electrical energy in the form of a magnetic field. Principle used in the ignition coil to increase voltage.

INJECTOR: A device which receives metered fuel under relatively low pressure and is activated to inject the fuel into the engine under relatively high pressure at a predetermined time.

INPUT SHAFT: The shaft to which torque is applied, usually carrying the driving gear or gears.

INTAKE MANIFOLD: A casting of passages or pipes used to conduct air or a fuel/air mixture to the cylinders.

JOURNAL: The bearing surface within which a shaft operates.

KEY: A small block usually fitted in a notch between a shaft and a hub to prevent slippage of the two parts.

MANIFOLD: A casting of passages or set of pipes which connect the cylinders to an inlet or outlet source.

MANIFOLD VACUUM: Low pressure in an engine intake manifold formed just below the throttle plates. Manifold vacuum is highest at idle and drops under acceleration.

MASTER CYLINDER: The primary fluid pressurizing device in a hydraulic system. In automotive use, it is found in brake and hydraulic clutch systems and is pedal activated, either directly or, in a power brake system, through the power booster.

MODULE: Electronic control unit, amplifier or igniter of solid state or integrated design which controls the current flow in the ignition primary circuit based on input from the pick-up coil. When the module opens the primary circuit, high secondary voltage is induced in the coil.

NEEDLE BEARING: A bearing which consists of a number (usually a large number) of long, thin rollers.

OHM: (Ω) The unit used to measure the resistance of conductor-to-electrical flow. One ohm is the amount of resistance that limits current flow to one ampere in a circuit with one volt of pressure.

OHMMETER: An instrument used for measuring the resistance, in ohms, in an electrical circuit.

OUTPUT SHAFT: The shaft which transmits torque from a device, such as a transmission.

OVERDRIVE: A gear assembly which produces more shaft revolutions than that transmitted to it.

OVERHEAD CAMSHAFT (OHC): An engine configuration in which the camshaft is mounted on top of the cylinder head and operates the valve either directly or by means of rocker arms.

OVERHEAD VALVE (OHV): An engine configuration in which all of the valves are located in the cylinder head and the camshaft is located in the cylinder block. The camshaft operates the valves via lifters and pushrods.

OXIDES OF NITROGEN (NOx): Chemical compounds of nitrogen produced as a byproduct of combustion. They combine with hydrocarbons to produce smog.

OXYGEN SENSOR: Use with the feedback system to sense the presence of oxygen in the exhaust gas and signal the computer which can reference the voltage signal to an air/fuel ratio.

PINION: The smaller of two meshing gears.

PISTON RING: An open-ended ring with fits into a groove on the outer diameter of the piston. Its chief function is to form a seal between the piston and cylinder wall. Most automotive pistons have three rings: two for compression sealing; one for oil sealing.

PRELOAD: A predetermined load placed on a bearing during assembly or by adjustment.

PRIMARY CIRCUIT: the low voltage side of the ignition system which consists of the ignition switch, ballast resistor or resistance wire, bypass, coil, electronic control unit and pick-up coil as well as the connecting wires and harnesses.

PRESS FIT: The mating of two parts under pressure, due to the inner diameter of one being smaller than the outer diameter of the other, or vice versa; an interference fit.

RACE: The surface on the inner or outer ring of a bearing on which the balls, needles or rollers move.

REGULATOR: A device which maintains the amperage and/or voltage levels of a circuit at predetermined values.

RELAY: A switch which automatically opens and/or closes a circuit.

RESISTANCE: The opposition to the flow of current through a circuit or electrical device, and is measured in ohms. Resistance is equal to the voltage divided by the amperage.

RESISTOR: A device, usually made of wire, which offers a preset amount of resistance in an electrical circuit.

RING GEAR: The name given to a ring-shaped gear attached to a differential case, or affixed to a flywheel or as part of a planetary gear set.

ROLLER BEARING: A bearing made up of hardened inner and outer races between which hardened steel rollers move.

ROTOR: 1. The disc-shaped part of a disc brake assembly, upon which the brake pads bear; also called, brake disc. 2. The device mounted atop the distributor shaft, which passes current to the distributor cap tower contacts.

SECONDARY CIRCUIT: The high voltage side of the ignition system, usually above 20,000 volts. The secondary includes the ignition coil, coil wire, distributor cap and rotor, spark plug wires and spark plugs.

SENDING UNIT: A mechanical, electrical, hydraulic or electro-magnetic device which transmits information to a gauge.

SENSOR: Any device designed to measure engine operating conditions or ambient pressures and temperatures. Usually electronic in nature and designed to send a voltage signal to an on-board computer, some sensors may operate as a simple on/off switch or they may provide a variable voltage signal (like a potentiometer) as conditions or measured parameters change.

SHIM: Spacers of precise, predetermined thickness used between parts to establish a proper working relationship.

SLAVE CYLINDER: In automotive use, a device in the hydraulic clutch system which is activated by hydraulic force, disengaging the clutch.

SOLENOID: A coil used to produce a magnetic field, the effect of which is to produce work.

SPARK PLUG: A device screwed into the combustion chamber of a spark ignition engine. The basic construction is a conductive core inside of a ceramic insulator, mounted in an outer conductive base. An electrical charge from the spark plug wire travels along the conductive core and jumps a preset air gap to a grounding point or points at the end of the conductive base. The resultant spark ignites the fuel/air mixture in the combustion chamber.

SPLINES: Ridges machined or cast onto the outer diameter of a shaft or inner diameter of a bore to enable parts to mate without rotation.

TACHOMETER: A device used to measure the rotary speed of an engine, shaft, gear, etc., usually in rotations per minute.

THERMOSTAT: A valve, located in the cooling system of an engine, which is closed when cold and opens gradually in response to engine heating, controlling the temperature of the coolant and rate of coolant flow.

TOP DEAD CENTER (TDC): The point at which the piston reaches the top of its travel on the compression stroke.

TORQUE: The twisting force applied to an object.

TORQUE CONVERTER: A turbine used to transmit power from a driving member to a driven member via hydraulic action, providing changes in drive ratio and torque. In automotive use, it links the driveplate at the rear of the engine to the automatic transmission.

TRANSDUCER: A device used to change a force into an electrical signal.

TRANSISTOR: A semi-conductor component which can be actuated by a small voltage to perform an electrical switching function.

TUNE-UP: A regular maintenance function, usually associated with the replacement and adjustment of parts and components in the electrical and fuel systems of a vehicle for the purpose of attaining optimum performance.

TURBOCHARGER: An exhaust driven pump which compresses intake air and forces it into the combustion chambers at higher than atmospheric pressures. The increased air pressure allows more fuel to be burned and results in increased horsepower being produced.

VACUUM ADVANCE: A device which advances the ignition timing in response to increased engine vacuum.

VACUUM GAUGE: An instrument used to measure the presence of vacuum in a chamber.

VALVE: A device which control the pressure, direction of flow or rate of flow of a liquid or gas.

VALVE CLEARANCE: The measured gap between the end of the valve stem and the rocker arm, cam lobe or follower that activates the valve.

VISCOSITY: The rating of a liquid's internal resistance to flow.

VOLTMETER: An instrument used for measuring electrical force in units called volts. Voltmeters are always connected parallel with the circuit being tested.

WHEEL CYLINDER: Found in the automotive drum brake assembly, it is a device, actuated by hydraulic pressure, which, through internal pistons, pushes the brake shoes outward against the drums.

A

B

MASTER INDEX

P

R

S

T